BST

ACPL ITEM
DISCARDED

3 1833 05658 0886

P9-DGV-337

THE GREEN BUILDING BOTTOM LINE

McGRAW-HILL'S GREENSOURCE SERIES

Gevorkian
Solar Power in Building Design: The Engineer's Complete Design Resource

GreenSource: The Magazine of Sustainable Design
Emerald Architecture: Case Studies in Green Building

Haselbach
The Engineering Guide to LEED–New Construction: Sustainable Construction for Engineers

Melaver and Mueller (eds.)
The Green Building Bottom Line: The Real Cost of Sustainable Building

Yudelson
Green Building Through Integrated Design

About *GreenSource*

A mainstay in the green building market since 2006, *GreenSource* magazine and Green SourceMag.com are produced by the editors of McGraw-Hill Construction, in partnership with editors at BuildingGreen, Inc., with support from the United States Green Building Council. *GreenSource* has received numerous awards, including American Business Media's 2008 Neal Award for Best Website and 2007 Neal Award for Best Start-up Publication, and FOLIO magazine's 2007 Ozzie Awards for "Best Design, New Magazine" and "Best Overall Design." Recognized for responding to the needs and demands of the profession, *GreenSource* is a leader in covering noteworthy trends in sustainable design and best practice case studies. Its award-winning content will continue to benefit key specifiers and buyers in the green design and construction industry through the books in the *GreenSource* Series.

About McGraw-Hill Construction

McGraw-Hill Construction, part of The McGraw-Hill Companies (NYSE: MHP), connects people, projects, and products across the design and construction industry. Backed by the power of Dodge, Sweets, *Engineering News-Record (ENR)*, *Architectural Record*, *GreenSource*, *Constructor*, and regional publications, the company provides information, intelligence, tools, applications, and resources to help customers grow their businesses. McGraw-Hill Construction serves more than 1,000,000 customers within the $4.6 trillion global construction community. For more information, visit www.construction.com.

THE GREEN BUILDING BOTTOM LINE

THE REAL COST OF SUSTAINABLE BUILDING

EDITED BY

MARTIN MELAVER

PHYLLIS MUELLER

New York   Chicago   San Francisco   Lisbon   London   Madrid
Mexico City   Milan   New Delhi   San Juan   Seoul
Singapore   Sydney   Toronto

The McGraw·Hill Companies

Library of Congress Cataloging-in-Publication Data

The green building bottom line : the real cost of sustainable building / Martin
    Melaver, Phyllis Mueller, editors.
            p.   cm.
        Includes bibliographical references and index.
        ISBN 978-0-07-159921-4 (alk. paper)
        1. Sustainable buildings—Design and construction—Costs.   2. Sustainable
    architecture.   I. Melaver, Martin.   II. Mueller, Phyllis.
    TH880.G743   2009
    690′.8047—dc22   2008034435

Copyright © 2009 by The McGraw-Hill Companies, Inc. All rights reserved. Printed in the
United States of America. Except as permitted under the United States Copyright Act of
1976, no part of this publication may be reproduced or distributed in any form or by any
means, or stored in a data base or retrieval system, without the prior written permission
of the publisher.

1 2 3 4 5 6 7 8 9 0   DOC/DOC   0 1 4 3 2 1 0 9 8

ISBN 978-0-07-159921-4
MHID 0-07-159921-5

Sponsoring Editor: Joy Bramble Oehlkers
Production Supervisor: Pamela A. Pelton
Editing Supervisor: Stephen M. Smith
Project Managers: Nancy Dimitry and Joanna V. Pomeranz, D&P Editorial Services
Copy Editor: Gerard Farrell
Proofreaders: Donald Dimitry and Donald Pomeranz
Indexer: Seth Maislin
Art Director, Cover: Jeff Weeks
Cover Photo: John Clemmer Photography
Composition: D&P Editorial Services

Printed and bound by RR Donnelley.

McGraw-Hill books are available at special quantity discounts to use as premiums and
sales promotions, or for use in corporate training programs. To contact a special sales
representative, please visit the Contact Us page at www.mhprofessional.com.

 The pages within this book were printed on acid-free paper containing 100%
postconsumer fiber.

Information contained in this work has been obtained by The McGraw-Hill Companies, Inc. ("McGraw-Hill") from sources believed to be reliable. However, neither
McGraw-Hill nor its authors guarantee the accuracy or completeness of any information published herein, and neither McGraw-Hill nor its authors shall be responsible
for any errors, omissions, or damages arising out of use of this information. This work
is published with the understanding that McGraw-Hill and its authors are supply-
ing information but are not attempting to render engineering or other professional
services. If such services are required, the assistance of an appropriate professional
should be sought.

About the Editors

Martin Melaver is CEO and Vice President of Melaver, Inc., a position he has held since 1992. He is a graduate of Amherst College, earned M.A. and Ph.D. degrees at Harvard University and an M.B.A. from the J. L. Kellogg School of Management, Northwestern University, and is a LEED Accredited Professional. He lives in Savannah, Georgia and Tel Aviv, Israel with his wife and two children.

Phyllis Mueller is an editor and writer whose areas of interest and professional experience include environmental issues, construction, business, food, and health. A graduate of Emory University, she lives in Decatur, Georgia with her husband, George Eckard.

CONTENTS

Chapter 9   The Fine Print: Legal Issues in Green Building Projects

Chapter 10   Marketing Sustainable Development: A Million Shades of Green

Conclusion: Inscribing Key Lessons Learned into the Fabric of a Green Business

ABOUT THE CONTRIBUTORS

Denis Blackburne is the Chief Financial Officer of Melaver, Inc. Prior to joining Melaver, he held several senior financial positions with both private and public companies, including Tetra Pak and Du Pont de Nemours. Blackburne has an M.B.A. from the J. L. Kellogg School of Management at Northwestern University and is a LEED Accredited Professional. He lives with his family in Savannah, Georgia.

Colin M. Coyne, Chief Operating Officer of Melaver, Inc., is a LEED Accredited Professional and the author of numerous published articles. He received a B.S. in economics and business administration from Vanderbilt University and an M.B.A. with Dean's List honors from the J. L. Kellogg School of Management at Northwestern University. Coyne is a resident of Birmingham, Alabama, where he lives with his wife, Leslie, and their three children, Ariana, Dallas, and Zoe.

Scott Doksansky is Director of Portfolio Management for Melaver, Inc. and is charged with oversight of property management functions as well as the firm's acquisitions and dispositions. He has seventeen years of real estate background on behalf of some of the nation's largest firms and has worked for the chambers of commerce in Nebraska and California. A graduate of Dana College in Nebraska and the Institute for Organization Management, Doksansky is a CPM® (Certified Property Manager) and is LEED 2.0 accredited. He lives in Duluth, Georgia.

Clara Fishel is the sustainable brokerage associate at Melaver | Mouchet. After cooking professionally on three continents, she was a research associate at the University of California, Davis, contributing to articles on collaborative policy, the role of trust in environmental management, and sustainable cities. Fishel holds a B.S. in community and regional development from U.C. Davis and lives in Savannah, Georgia with her husband and two children.

Lisa Lilienthal is a public relations consultant with more than twenty years of experience in mission-based marketing. An award-winning communications professional with over twenty honors to her credit, she has a history of helping organizations elevate thoughts and ideas into brands. Lilienthal lives in Orange County, California with her husband and two children.

Tommy Linstroth is head of Sustainable Initiatives for Melaver, Inc., where he has overseen sustainable design solutions for fifteen LEED certified projects with a market value of a quarter-billion dollars. He is the founder and past president of the Savannah chapter of the U.S. Green Building Council and co-author of the book *Local Action: A New Paradigm in Climate Change Policy*. Linstroth received his M.A. in environmental studies from the University of Charleston. He lives in Savannah, Georgia.

Dan Monroe is one of the founding partners of Cayenne Creative Group, a strategic branding firm located in Birmingham, Alabama. Prior to forming Cayenne, he worked as a copywriter for Leo Burnett in Chicago and Slaughter Hanson Advertising in Birmingham. A writer by trade, he co-authored *Identifying Courage* and *The Good Books*. His poetry has been published in numerous literary reviews.

Rhett Mouchet is Broker-in-Charge at Melaver | Mouchet and has worked in commercial real estate in the Savannah, Georgia area for three decades. He holds a bachelor's degree from Georgia Southern University and did post-graduate work at the University of South Carolina and Armstrong Atlantic State University. Mouchet holds the Society of Industrial and Office Realtors (SIOR) designation, a professional symbol of the highest level of knowledge, production, and ethics in the real estate industry. He lives in Savannah, Georgia with his wife and three children.

Randy Peacock is head of Development and Construction at Melaver, Inc. and has worked in the construction industry for more than twenty years. He is a graduate of Georgia Southern and holds a B.S. degree in science, building construction, and contracting. He lives in Savannah, Georgia with his wife Stacey and their children, Abby and Reed.

Justin Shoemake, Esq. is an attorney with the law firm of Stanley, Esrey & Buckley, LLP in Atlanta, Georgia. He is a LEED Accredited Professional and holds a J.D. from Duke University School of Law. He lives in Atlanta, Georgia.

Robert E. Stanley, Esq. is a founding partner in the law firm of Stanley, Esrey & Buckley, LLP and a LEED Accredited Professional. His law practice is primarily focused on commercial real estate development, with an increasing emphasis on sustainable real estate development practices, including green leasing. He is a graduate of the University of Virginia School of Law and lives in Atlanta, Georgia.

Zelda Tenenbaum is owner and president of Tenenbaum Consulting. Prior to opening her business, she was an instructor in psychology at Armstrong Atlantic State University. Tenenbaum holds a master's degree in psychology and has thirty years of experience as a facilitator, coach, consultant, and educator. She has authored several training manuals and a chapter in the workbook *Leading by Stepping Back: A Guide for City Officials in Building Neighborhood Capacity*. She and her husband live in Savannah, Georgia and Telluride, Colorado.

ACKNOWLEDGMENTS

It takes a village. That's the essence of this book, the essence of the sustainability movement in general. It literally will take all hands on deck, from all sectors of society, to restore our planet to a type of balance viable for generations to come.

Like a natural order, in which everything is interconnected, it is hard—if not impossible—to designate all the influences and mentoring of the village that have gone into the crafting of this book. So from the outset, we'd like to acknowledge the support and guidance of friends, family members, colleagues, others writing in the field of sustainability, and our random contacts from unexpected quarters, all of whom have shaped *The Green Building Bottom Line*.

More specifically, we'd like to acknowledge the efforts of the various authors of the ensuing chapters, staff members at Melaver, Inc. and close outside partners who took significant time away from their day jobs not only to shape this book but to shape the journey upon which this book is based. Other colleagues weighed in with additional help. Cathy Rodgers, Communications Coordinator for Melaver, Inc., provided significant assistance with much of the imagery in this work. Karen Hudspeth, Director of Operations for Melaver, Inc., helped manage the business aspects of the project. Denis Blackburne coordinated the discounted cash flow analyses throughout each of the chapters to ensure that the various financial pieces of the story we are telling cohere. Marjorie Young of Carriage Trade PR provided us with various forums to present earlier drafts of the material. We thank Elizabeth Teel for her painstaking help with permissions and Ellen Glass for her invaluable advice and her assistance with manuscript preparation.

We also thank the team at McGraw-Hill—Joy Bramble Oehlkers, David Zielonka, Pamela Pelton, and Stephen Smith—for its help and guidance and Nancy Dimitry and her colleagues at D&P Editorial for their patience and attention to detail.

And we thank the authors whose works are quoted in this book, including:

Last Child in the Woods by Richard Louv; © 2005 by Richard Louv. Reprinted by permission of Algonquin Books of Chapel Hill.

When Corporations Rule the World by David C. Korten. Reprinted with permission of the publisher, Berrett-Koehler Publishers, Inc., San Francisco, Calif. All rights reserved; www.bkconnection.com.

What Is a Whole Community? A Letter to Those Who Care For and Restore the Land by Peter Forbes

Sincerity and Authenticity by Lionel Trilling; © 1971, 1972 by the President and Fellows of Harvard College. Reprinted by permission of the publisher, Harvard University Press, Cambridge, Mass.

The Triple Bottom Line: How Today's Best-Run Companies Are Achieving Economic, Social and Environmental Success—and How You Can Too by Andrew Savitz and Karl Weber. Reprinted with permission of John Wiley & Sons, Inc.

The Corrosion of Character: The Personal Consequences of Work in the New Capitalism by Richard Sennett; © 1998 by Richard Sennett. Used by permission of W. W. Norton & Company, Inc.

Green Marketing: Opportunity for Innovation, 2d ed., by Jacqueline A. Ottman, president, J. Ottman Consulting, Inc., New York.

Nature of Design: Ecology, Culture and Human Intention by David W. Orr; © 2002. By permission of Oxford University Press, Inc.

Regenerative Design for Sustainable Development by John Tillman Lyle. Reprinted with permission of John Wiley & Sons, Inc.

Greed to Green: The Transformation of an Industry and a Life by David Gottfried.

INTRODUCTION

MARTIN MELAVER

The genesis of *The Green Building Bottom Line* was an early Friday morning in Atlanta, Georgia—November 11, 2005—the closing day of the U.S. Green Building Council's fourth annual Greenbuild conference. A number of my colleagues at Melaver, Inc. and several close outside team members were appearing on a panel about the challenges of developing Abercorn Common, the first LEED shopping center in the country, in our company's hometown, Savannah, Georgia. The LEED program (LEED is an acronym for Leadership in Energy and Environmental Design) had been created a few years before by the U.S. Green Building Council as an independently verified, points-driven set of development criteria assessing five major areas: site management, water consumption, energy efficiency, materials used, and indoor air quality. Abercorn Common was our company's second LEED project, one that we had worked on for the previous four years (and it's the subject of Chapter 5 of this book).

Instead of a typical three-person Greenbuild panel presenting on three discrete projects, a team of eight—developer, architect, builder, general contractor, legal counsel, leasing agent, marketing representative, and property manager—gathered to discuss one single project. There wasn't enough space at the podium for the entire team to sit comfortably side by side, so the members were jammed into several rows of chairs, changing places as the time came for each to speak. Rather than a slick PowerPoint presentation about success, the whole orientation of the presentation was less about putting a pretty veneer on so-called best practices and more about sharing many of the mistakes the team had made along the way.

When we had decided as a company several years back to develop all of our projects to LEED standards, Abercorn Common was already under development. At the time, less than two dozen retailers in the country had gone on record indicating a desire to build their stores to green specifications, and none of these had modified their store prototypes to accommodate LEED guidelines. Nevertheless, we charged ahead, going green midstream to develop the project to LEED criteria. And here we were, in front of a national audience of designers, developers, retailers, and other real estate professionals admitting to a long litany of mistakes.

It should have been a recipe for disaster as far as a buttoned-up professional presentation was concerned. It wasn't. There seemed to be a pent-up demand to pull back the curtains on a green project to reveal the pitfalls and challenges of going green— not for the purpose of throwing cold water on this explosive new movement for green building, but rather to facilitate this movement by being transparent about the many challenges facing it so problems could be fixed, quickly.

Those engaged in environmental issues today, even on the most cursory of levels, understand the urgency. This environmental urgency is not confined to the familiar issues of global warming and greenhouse gas emissions but is linked to issues of consumption and waste of natural capital on a global scale. In the United States, we lose more than a million acres annually to urban sprawl, parking lots, and roads, while worldwide, we destroy 80,000 square miles of tropical forest each year.[1] Twelve million acres of once-productive land is transformed each year into desert.[2] Global warming, desertification, and loss of productive arable lands are in turn linked to the diminution of the quantity and quality of water supplies worldwide, with one-third of the world's irrigated lands suffering from saltwater intrusion. It is projected that by 2025, 40 percent of the world's population could be living with chronic water shortages.[3]

The building trade, which has played a significant role in this degradation, needs to find an alternative paradigm for shaping our lands and communities in the years to come. Part of that new paradigm-shaping needs to come not just from jettisoning an older, more consumptive, more wasteful, less ethical way of doing business, but also from being direct and forthcoming about the challenges and opportunities of a new approach to doing things. That was the impetus behind our presentation on Abercorn Common in 2005. It is also the impetus behind this book.

As I sat in the back of the room that day in November 2005, I felt proud of my company and the long, slow, patient years of work behind the presentation. Melaver, Inc. began with a corner grocery store founded by my grandmother Annie Melaver in 1940 in Savannah, Georgia. My father, Norton Melaver, grew the business over the following forty-five years into a supermarket business in and around the outlying environs of my hometown, eventually selling M&M Supermarkets to Kroger in 1985. Since then, the business has been devoted to real estate: development, acquisitions and sales, property management, and leasing. It's a family-owned business, one that since its inception has had a strong sense of core values having to do with community and the land of which that community is a part.

Early on in our entrance into the real estate business, family members felt quite strongly that we needed to practice real estate in ways that were more respectful of locale. Over time, that commitment evolved into a set of principles and practices encapsulated by the term sustainability. We develop to a minimum of LEED standards in everything we do. We avoid greenfield development, focusing our attention on urban core, largely in-fill projects. Our projects are guided by a set of triple bottom line metrics that not only call for viable economic returns, but also set threshold expectations for maximizing our positive impact on the communities in which we work (social bottom line) while minimizing our impact on the environment (environmental bottom line). Roughly 80 percent of our small staff of some thirty folks is LEED accredited, and everyone takes the exam at least once. And, despite our small size, we have participated in three of the U.S. Green Building Council's pilot programs for specific LEED programs (Core and Shell, Home, and Neighborhood Development) and account for roughly 1 percent of all LEED certifications in the country.

But even as my colleagues were making that presentation in Atlanta in 2005, I felt that there was so much more we needed to say and share with others—lessons learned, mistakes to be avoided, and opportunities to be realized for businesses that not only build greener buildings but also embrace a values-centric orientation focused on social and environmental consequences of what and how and where we build. Despite the growing awareness in the United States of all things green, the sustainability movement is still very much in its infancy, with tremendous work and a short time frame within which to do that work facing us all.[4] Moreover, there is still a prevalent belief out there in the business world today that a values-centric, sustainable orientation is not financially viable.

Political policy planner, one-time U.S. Secretary of Labor, and author Robert B. Reich laments in *Supercapitalism*:

> For many years I have preached that social responsibility and profitability converge over the long term. That's because a firm that respects and values employees, the community, and the environment eventually earns the respect and gratitude of employees, the community, and the larger society—which eventually helps the bottom line. But I've never been able to prove this proposition nor find a study that confirms it.[5]

Marc Gunther, senior writer at *Fortune* magazine and author of *Faith and Fortune,* echoes Reich's lament, noting:

> The truth is, no study has proven convincingly that social responsibility is good for business, and it may be that none ever will. The definitions are fluid and fuzzy, and the interplay between values and profitability may be too complex to be reduced to numbers.[6]

The Green Building Bottom Line is an effort to redress this lack: To provide a financially based business case study for how doing the right thing for land and community also means doing well.

The Green Building Bottom Line is composed of interwoven parts. It is partially a deep dive into the greening of Melaver, Inc., looking at the time, effort, and resources expended in shaping a company culture built from the ground up around an ethos of sustainability (Chapters 1 through 4). It is partially devoted to the examination of specific green projects we have been engaged in developing, looking particularly at the costs and benefits derived from each one (Chapters 5 through 7). And it is partially focused on the time, effort, and resources we have expended outside the company's specific projects in an effort to help the green movement become more widespread (Chapters 8 through 10). Each chapter is written either by a colleague at Melaver, Inc. or by a close associate (legal counsel, human resources counselor, marketing team member, public relations advisor) who is aligned with our values and passion and has worked with us for an extended period of time. My co-editor Phyllis Mueller and I worked with the authors to shape the stories told in each chapter—me from the inside looking out, she from the outside looking in. Each chapter has its own flavor, reflective of the particular roles the authors play in our company and of their slightly different philosophical orientations toward our collective endeavor.

The chapters build upon one another, much like our various contributions to our collective work, as author after author adds layers to the story we are trying to tell about the value of going green. The individual authors range across the political spectrum. Each has his or her own particular sense of the challenges we face and the role business can and should play in that overall effort. Many of us differ philosophically about the pace of growth for business generally, as well as the specific slow-growth strategies of Melaver, Inc. Each of us in our own lives walks the talk of sustainable practices differently. As a company, we try to give voice to these differences, even as we try to shape a synthesis from them. That is our particular strategy for how a sustainable business is shaped.

Throughout the book, there is a loose use of the terms "green" and "sustainable," often with the notion that the two terms are clearly defined and synonymous. They aren't. There's a whole body of literature devoted to the problems associated with each term and the fact that they have become so widely and uncritically used so as to mean most anything these days.[7] For the purposes of this book, "green," when applied to specific projects, typically refers to the LEED criteria upon which a green development is based. Used more broadly and generically, "green" is used in this book to convey a more systemic approach to sustainable practices having to do with minimizing our environmental footprint (the environmental bottom line) and optimizing our positive impacts upon our communities (the social bottom line). These usages reflect more or less everyday, colloquial understandings of the terms. We felt it was more useful to dig down into the practices underscoring this terminology rather than debate the fine points of the nomenclature.

Taken as a whole, *The Green Building Bottom Line* may be viewed as one extended financial analysis: a discounted cash flow statement that cumulatively tallies all of the costs our company has expended in the course of becoming a green company and views those costs in the context of the total return on our investment. The financial bottom line of this analysis is likely to surprise (and hopefully delight) most readers, as we show that it makes economic sense not only to develop a specific green project, but also to develop an entire business around sustainable values.

Because Melaver, Inc. prides itself on a culture of transparency, all of the costs itemized in the chapters are real dollars that we have expended over the past decade in our efforts to become a more sustainably oriented company. Having said that, as a privately held company, there are some sensitivities to providing complete financial disclosure. As such, we have created a fictionalized company called Green, Inc. that mirrors—but is not an exact replica of—Melaver, Inc. Green, Inc. is a smaller version of our own company. Its revenues and profitability understate the performance of our own company while using our actual cost structure. Such an approach provides a conservative and understated picture of the value of a green bottom line, though it does not provide a precise audited analysis of the value of going green.

Although virtually all of the authors who appear in this book are part of Melaver, Inc.'s speaker's bureau—a group of colleagues who devote a portion of each month to educating, advocating, and in general doing outreach into the community on sustainable practices—we rarely (if ever) get together to share what we know. In that sense,

The Green Building Bottom Line has served as a virtual meeting place for me and my colleagues, with serendipitous results.

Frankly, the results of the financial analysis undertaken in this book surprised me. My colleagues and I have long felt that we were doing well financially by doing the right things. But until we all pooled our knowledge, we had never verified our assumptions. This book has, among other things, provided us with the opportunity to examine critically every aspect of what we do, integrate those findings, and objectively assess our intuitive sense of creating value through a values-centric orientation. I had assumed that someday, perhaps far in the distant future, our early investment in green practices would make financial sense. I'm pleased to discover that this value realization has occurred much sooner than I would have believed possible.

How is it that the CEO of a company was not specifically aware of the financial benefits derived from the sustainable orientation of his company? Isn't that just a little irresponsible (not to say unacceptable)? Maybe. My group of family shareholders by and large is less focused on quarter-to-quarter performance and more attuned to long-term value creation. As such, we have been able to invest time and resources in various ways that, from time to time, have made our quarterly performance look rather miserable. Nevertheless, our year-to-year returns have been in line with standard benchmarks for our industry, enough so that we were comfortable with the feeling that our investment in green was making sense even if we didn't take the time to quantify it.

This book has enabled us to see that shorter-term and longer-term value creation are not as distinct in time as we had always thought them to be. That is good news for us, for the real estate development profession, and for a whole new paradigm for how this profession conducts its business. The story of our green bottom line is partially about the community we have created within our business, a community we think other businesses should at least consider as part of their overall strategy. Our story is partially about creating time and space for imagination and creativity to take hold.

It is also about the powerful hold of place and the need to help nurture communities that are beloved. Our story, told in the multiple voices of colleagues and associates, is about nurturing specific places through the sustainable projects we develop. It is a story of a business trying to restore its sense of wonder for the world around it. And it is a story of a community of individuals trying to serve the larger community.

NOTES

[1] David W. Orr, *Earth in Mind: On Education, Environment, and the Human Prospect* (Washington, D.C.: Island Press, 2004), p. 203.

[2] David C. Korten, *When Corporations Rule the World* (Bloomfield, Ct. and San Francisco, Calif.: Kumerian Press and Berrett-Koehler Publishers, 1995), p. 36.

[3] Edward O. Wilson, *Consilience: The Unity of Knowledge* (New York: Random House, 1998), p. 311.

[4] Edward O. Wilson, *The Future of Life* (New York: Random House, 2002), p. 151.

[5] Robert B. Reich, *Supercapitalism: The Transformation of Business, Democracy, and Everyday Life* (New York: Alfred A. Knopf, 2007), p. 171.

[6] Marc Gunther, *Faith and Fortune: The Quiet Revolution to Reform American Business* (New York: Crown Publishing, 2004), p. 261.

[7] See, for example, Eric T. Freyfogle, *Why Conservation Is Failing and How It Can Regain Ground* (New Haven: Yale University Press, 2002), Chapter 4, especially p. 114.

THE GREEN BUILDING BOTTOM LINE

NARRATING VALUES, SHAPING VALUES, CREATING VALUE

MARTIN MELAVER

SUMMARY

This chapter is divided into three parts.

Part one looks at the values embedded in two overarching master narratives about capitalism—a narrative of indifferent capitalism and a narrative of capitalism with a difference. Both consider the role of business in fostering financial prosperity as well as its responsibilities toward the social and environmental consequences of its practices. The argument is made that embracing one of these narratives—the narrative of capitalism with a difference—is a critical precursor to the shaping and practice of values within a business.

Part two considers the organizational substructure of a business focused on shaping values. Values in a company are shaped by three systemic layers: basic guiding principles that serve as a bill of rights for the values-driven company, structural concepts that form the basis for more concrete values, and specific values themselves. With these foundational layers in place, the company focused on a green bottom line has the capacity to create actual financial value from the values it practices.

Part three focuses on the specific value creation a company realizes by undergoing the foundational work described previously.

My father Norton Melaver, who for four decades ran M&M Supermarkets, the predecessor to our real estate company, Melaver, Inc., has a favorite story that will be familiar to many readers. As he tells it, a businessman is strolling along a beach one day when he encounters a fisherman. A nice if modest number of trout he's caught rest in a peach basket beside him.

"You ought to hire yourself some people to help you fish," said the businessman.

"What would that bring me?" came the laconic reply from the fisherman.

"Well, you could sell the fish and make a nice profit," was the answer.

"What would I do with that?"

"Well, you could hire a few more people to help with the fishing."

"And with that?"

"Well, you'd make a good bit of money over time."

"What would I do with that?"

"Well," said the businessman, warming to the task of creating a business plan for this obvious rube, "you could buy yourself a fishing boat."

"And what would I do with that?"

"You could catch a whole lot more fish and make a lot more money."

"And what would I do with that?"

"Well, you could sit back and let others do the fishing for you."

"And what would I do then?"

"Well, then you would have time for leisurely pursuits instead of just working all the time."

"What would I do with all that leisure time?"

"Well, you could sit back and relax and maybe just enjoy the day and do a little fishing," said the businessman, feeling a bit exasperated by the man's obtuseness.

"But I'm doing that now," the fisherman replied.

I swear the way my father tells this story, it seems to go on indefinitely. You can feel the punch line coming ten minutes before his drawling delivery gets you there. He never seems to tire of telling it, perhaps even relishing the fact that most of us have heard the story a thousand times and are champing at the bit at his slow windup and delivery.

But I have to admit the story is a good one. For one thing, there's the strong synergy between the story and the storyteller, my father. As someone who grew up during the tail end of the Depression, finished college, did a stint in the Navy, and then returned home to run the family grocery business without a pause for forty years, he probably feels personally connected to this story. *What are you waiting for to do what matters most,* the story seems to be saying to him as much as my father's audience, a self-reflective question echoed in his own drawling raconteurship. And certainly, the story broaches the idea about doing those things that are most important to you *now,* rather than deferring them to the future.

The story is also germane to the workplace more generally, begging the question not only about the meaning of our toil but of the overall purpose of business. Do we, as workers or owners, envision long careers creating wealth after which we can *then* invest ourselves in finding meaning? Or do we look for ways to integrate value creation and the practice of values into the present moment? Are the two inseparable? How might they coalesce? What is the *raison d'etre* of business? My father's story, as it turns out, lies at the crux of business practices today, squaring a short-term focus on growth and profits with a longer-term focus on having that profit mean something. The story invites us to think more deeply about integrating value creation with practicing values.

Narrating Values:
A Tale of Two Capitalisms

All of us tell stories. Stories such as my father's help us make sense of the complex world we live in, provide order and meaning. The historian Hayden White, in his seminal study *Metahistory,* examines five famous versions of the French Revolution. Depending on the particular facts each of the five historians selected (and the facts each historian ignored) and the way in which each historian ordered those facts, a very different version of the French Revolution emerges: a tragedy, a romance, a comedy.[1]

Writing almost a decade later, the psychologists Kahneman and Tversky more or less underscored White's work by pointing out that the stories we shape to provide meaning are informed by personal biases.[2] Biases are the basis of our decision-making process. One bias Kahneman and Tversky discuss is something they termed "confirmation bias," which entails the strong tendency among all of us to scan a field of data for those pieces of information that confirm what we already believe to be true and ignore whatever data do not fit our belief system. White's argument about selection and ordering as the key components of storytelling and Kahneman and Tversky's notion of confirmation bias are two distinct academic versions of the same thing: All of us are actively engaged in creating stories that confirm our various views of the world. As the historian Jeremy Rifkin notes:

> I would suggest that throughout history, people's experience of reality begins with creating a story about themselves and the world and that story acts as a kind of cultural baseline DNA for the all the evolutionary permutations that follow.[3]

What does all of this have to do with sustainable business practices? A lot, as it turns out.

All of us are bombarded with facts and information and stories every day of our lives. A colleague or friend is laid off from work after many years of loyal service to a company and now, in her forties, has the experience but lacks certain skills such that finding a new job seems hopeless. We get an e-mail from an acquaintance trying to raise money for someone who lacks the necessary medical insurance for an expensive operation. We share with neighbors and friends stories about ourselves and our kids as they spend seemingly endless hours on homework and recreation and community service and as we try to balance our own work lives with spending time with them. We hear news about job creations here and job cuts there, rising mortgage rates and lower cost of consumer goods, the competition to get into good universities and hence the competition to get into good secondary schools and even good pre-schools. We read stories about the increasing tendency among younger and younger persons to undergo plastic surgery and the surge in obesity. On and on it goes.

At work, during casual conversation over a meal, piped into our headphones as we listen to the news during a morning or evening workout, in the general atmosphere in the city or town where we live and play, most of us—consciously or unconsciously—are trying to make personal sense of all this information: What does this stuff mean to me? How am I doing physically, financially, emotionally? Where do I fit in this society I see and feel around me? Are things getting better or worse? Am I better off this year as opposed to last year? What will happen next year? What will life be like for my kids or grandkids, and how does that compare with how I grew up?

The challenge is organizing all of this data into a meaningful and coherent personal story, selecting and ordering the plethora of information we receive so that it makes sense. This challenge is aided particularly by an overarching sense of our political, economic, and social order. Is that order fair? Just? Secure? Does it provide happiness, hope, fulfillment? Does it offer opportunity? What should our culture look like to provide these things? We all, I think, have different responses to these questions.

There are, however, two current narratives about our culture that inform the ways we answer these questions. Both narratives focus on the role of business in the shaping of our lives as citizens today. We can characterize these two narratives, perhaps a bit fancifully, as a narrative of *indifferent capitalism* and a narrative of *capitalism with a difference*. The narrative of indifferent capitalism is a narrative of danger. The narrative of capitalism with a difference is a narrative of opportunity.

THE NARRATIVE OF INDIFFERENT CAPITALISM

The narrative of indifferent capitalism goes something like this.[4] In less than fifteen generations, we have seen a profound shift from medieval sensibilities to a modern ethos in which theology, faith, and salvation gave way to reason and a belief in progress. Cyclical time, marked by the change of seasons, was supplanted by linear time focused on the clock and efficient productivity. Personal, covenantal relationships, bound by fealty, became redefined by contracts, and individual rights emanated from notions of property and ownership. Emphasis on good works came to be supplanted by a work ethic, particularly a Protestant work ethic that linked material rewards in this life with spiritual redemption in the world to come. Belief in a free market economy, with free circulation of money, goods, and labor, served as the basis for this nascent capitalism, with Adam Smith's *Wealth of Nations* (1776) its seminal text. The Industrial Revolution, often thought to begin with the invention of the first practical steam engine by Thomas Newcomen around 1710, set the stage not only for a new way of conducting business, but for a new world order.

America, the inheritor of the European enlightenment, served as the fullest realization of such changes, with its seemingly endless supply of abundant land offering the potential to make good on John Locke's linkage between private property and the natural right of individuals. A capitalist economy based on small yeoman farmers and a democratic ideology based on the individual were joined. It would take over two centuries for this bond between capitalism and democracy to unwind. In the early going of the new republic, individual property rights including the right to quiet enjoyment were balanced against the government's right to regulate land use in the public interest. The legal doctrine of *sic utero tuo ut alienum non laedas* ("use your own so as to cause no harm") held sway. Over time, though, two critical changes tipped the balance, with individual rights superseding those of the polis and corporations slowly developing rights of their own synonymous with those of the individual. The tipping point was probably the 1886 Supreme Court case *Santa Clara County v. Southern Pacific Railroad*, which held that a private corporation is a natural person under the U. S. Constitution. The decision, which would assist greatly in the growth and maturation of capitalism in the United States, would in many ways lead to the pendulum swinging strongly in favor of business to the detriment of larger social interests.

The final years of the nineteenth century were marked by the growth of huge corporations such as J. P. Morgan and Sons, Carnegie Steel, the Standard Oil Company, and Chase Manhattan, ushering in unprecedented economic growth but also fostering problematic social conditions (sweatshops, child labor, unsafe working conditions). Of the Fortune 500 largest corporations today, more than half were founded between 1880 and 1930. The growth of a largely unaccountable business sector briefly raised discussions (and some legislation) as to how capitalism might better serve the populace. By the mid-20th century, America settled in to what might be termed a *modus vivendi* between business and government. Most business sectors were dominated by oligopolies, government played a role in regulating jobs and wages, a third of the workforce was represented by unions that tended to work closely with management to avoid strikes, and top corporate executives played a role as "corporate statesmen," responsible for balancing the claims of various stakeholders (stockholders, employees, the public). During the post-World War II period until the mid-1970's, economies of scale were large, with a high degree of productivity and profit. Tens of millions of steady jobs were created. There was a wide distribution of profits to blue-collar workers and others, and a growing middle class helped fuel this economic growth through increasing consumption of goods and services. It was, perhaps, the high point of democratic capitalism.

From the 1970s on, however, the picture has changed dramatically. Old oligopolies have given way to intense competition on a global scale. Manufacturing in the developed world has given way to a service economy, with manufacturing relocated to over seventy developing countries, primarily in the 1,000-plus free-trade zones abroad, where environmental and labor regulations are lax, where workers' wages of eighty-seven cents per hour are a fraction of labor costs in the U.S., and where production costs can be cut to a minimum. Domestically, such changes have resulted in the growth of a temporary work force by 400 percent since 1982, as 83 percent of America's fastest growing companies are now outsourcing part of their work. Only 8 percent of the American work force is now represented by unions. The average corporation replaces its entire workforce every four years. Abroad, the creation of manufacturing jobs that can quickly move in and out of export development zones in search of the cheapest available labor has had the effect of creating uncertain, unstable employment.

It draws rural, agricultural work forces into large urban areas, putting added pressure on these cities as every week 1.4 million people are drawn into the world's slums and *favelas,* while moving developing economies away from local subsistence food production and toward greater use of farm land for cash crop exports.

The intense competition to reduce the cost of goods has gone hand-in-hand with similarly intense competition for capital. With the breakdown of the Bretton Woods talks on currency regulation in the mid-1970s, capital began to move fluidly across national borders in search of the best returns. Institutional investors, particularly pension funds, became the predominant holders of stock, with high expectations of return from quarter to quarter. Those CEOs who delivered on such high expectations were compensated richly. Those who did not were fired. Among other things, such competitive pressures have spilled over into government, as businesses invest more and more dollars in an effort to shape legislation and regulation that will give them a competitive advantage in the marketplace.

While consumers and investors have gained by this global competition in the form of lower prices of goods and strong economic returns, most citizens have lost ground both financially and in terms of their overall welfare. The robust growth of capitalism has benefited only a few at the very top of the food chain, while the median income of American households has remained largely static for the past three decades. Moreover, most citizens have come to see that government, once careful to balance economic power with other considerations, is no longer looking after their interests. Such indifference, as public policy expert Robert Reich notes, comes at a particularly inopportune juncture in history:

> As companies drop health insurance and pension coverage for example, public provision of them becomes more important. As jobs and incomes grow less secure, public safety nets become more essential. As companies are pressured to show profits, tougher measures are needed to guard public health, safety, the environment, and human rights against the possibility that executives may feel compelled to cut corners.[5]

The larger environmental and social consequences of this growth of global capitalism can be seen in myriad ways. While our per capita GDP has increased three-fold since the 1950s, that growth has occurred alongside unprecedented waste and consumption. Americans comprise 5 percent of the global population but consume 25 percent of the world's oil and coal, 80 to 90 percent of which is wasted through inefficiencies in delivery systems and poor design. We spend ten kilocalories of hydrocarbons to produce one kilocalorie of food. We have 30 percent of the world's supply of automobiles that contribute over 50 percent of the world's carbon emissions from transportation. The average American disposes of one million pounds of waste each year. We have 172 deep well injection sites strewn across the country storing our hazardous waste. We have lost more than 50 percent of the country's wetlands since European settlement, lose more than one million acres each year to urban sprawl, roads and parking lots, and now have less than 2 percent of land mass in the continental United States designated as wilderness. It takes twenty-four acres to service the needs of one American, five times the global average.

The social consequences of our economic growth are similarly troubling. Health care? Approximately fifty million Americans are without health insurance. Sense of community? Forty-seven million Americans live in private or gated communities, with some speculation that such housing arrangements will become the dominant form of community by 2050. Commuting time from our suburban enclaves to work and back are up, the time we spend on civic involvement is down, and the time we now spend outdoors as opposed to thirty years ago has been reduced by 75 percent. Our sense of being unmoored, of being adrift has never been greater. Those now leaving college can expect to change jobs eleven times and change basic skill sets three times in the ensuing forty years. The very flexible, dynamic nature of business today leads the vast majority of Americans feeling that work has become emotionally illegible. Mobility is validated, "success" is oftentimes associated with frequent job changes, and yet most of us find these are lateral changes with very little substantive difference. Work provides little in the way of developing complex skills over an extended period of time, less a sense of accomplishment and achievement for one's craft and more about shared superficiality among teammates at work, jumping from one diverse task to the next. Our feeling of overall happiness is down, our overall sense of failure high:

> Failure to make one's life cohere, failure to realize something precious in oneself, failure to live rather than merely exist.... The short-term, flexible time of the new capitalism seems to preclude making a sustained narrative out of one's labors, and so a career.[6]

Such is the narrative of indifferent capitalism. It is a narrative of economic efficiency, drawing upon rules and regulations business itself has had an active hand in shaping to provide subsidies for and minimize costs of its operations, to open new markets and maintain its profitability in existing ones. It depends on a constant furthering of technological innovation, on free and fast exchange of information, and unfettered flow of capital. Its sole focus is the financial bottom line. To the extent that it is attuned to the social and environmental consequences of its performance, it looks to externalize them as costs that business, in the interest of market efficiency, should not have to bear.

THE NARRATIVE OF CAPITALISM WITH A DIFFERENCE

The narrative of capitalism with a difference, like the narrative of indifferent capitalism, also begins with the Protestant reformation and the liberating opportunities afforded by a new theology centered on the primacy of the individual. It too highlights the transformation in commerce played by the Industrial Revolution and the critical role played by coal in fueling a new economy and liberating humankind in so many disparate ways.[7] And finally, the narrative of capitalism with a difference also grounds its story in the writings of Adam Smith as the intellectual forebear of the modern economic era. But at this point, the two narratives part company.

For one thing, the narrative of capitalism with a difference points out that Adam Smith's theories of a free market, espoused in his *The Wealth of Nations* (1776) are misinterpreted without reading this work in the context of Smith's earlier study, *The Theory of Moral Sentiments* (1759). Read together, a different picture of capitalism emerges, one in which commerce is "composed of small, owner-managed enterprises located in the communities where the owners reside,"[8] where market efficiency complements moral sympathy for all members of a community. Capitalism, so the argument goes, has missed the boat by historically focusing on free markets to the exclusion of collective moral sympathy. The various freedoms that capitalism has provided us have not come without costs. The opportunity to "make something of ourselves" has resulted in losing the sense that our individual identities are wrapped in others—family, community, connection to society. Instead of creating simply the greatest number of goods, or even creating the greatest number of goods for the greatest number of people, the narrative of capitalism with a difference holds out the vision, as Gandhi put it, of simply the greatest good for all.

The narrative of capitalism with a difference holds that businesses have not only created great value but have also destroyed value, in terms of environmental degradation and social disintegration. Rather than focus on short-term horizons and financial returns measured in ever-diminishing increments, capitalism with a difference takes a longer view of value creation. Instead of a financial bottom line focus, the narrative of capitalism with a difference emphasizes a multiple bottom line orientation, taking into account stewardship of the environment, nurturing the social fabric of community, providing a living livelihood, job security, diverse work opportunities, craftsmanship, and purpose. Rather than focus on the sheer ongoing survival of a company—on its longevity—the narrative of capitalism with a difference wonders to what uses such longevity is put. It asks, are the practices of capitalism *sustainable* for our land and for our community? And (to paraphrase architect Carol Venolia's definition of sustainability) it asks simply this: If you keep doing what you're doing, will you be able to keep doing what you're doing?[9]

Thoughts about a different type of capitalism began to emerge in the early 20th century as people such as Herbert Croly, Walter Lippmann, Adolf Berle, and Gardiner Means variously suggested notions of stakeholder capitalism, whereby business would either voluntarily or through government mandate consider the interests of a broader spectrum of society than simply its shareholders. At about the same time, the notion of socially responsible investing first took hold, with Gary Moore's publication in the 1920s of *The Thoughtful Christian's Guide to Investing* and the launching of the morally focused Pioneer Fund in 1928.

But it was not until indifferent capitalism took off in the mid-1970s that an alternative story began to take shape and become fully formed. At that time, the first United Nations Conference on the Human Environment took place in Stockholm (1972), Pax—the first of what would evolve into a $2 billion socially responsible investing (SRI) mutual fund industry—was formed (1971), and small though now iconic alternative businesses such as Tom's of Maine (1970), Patagonia (1972), and Ben & Jerry's (1978) all got their start. Environmental author Paul Hawken, who got his start running the natural foods store Erewhon Trading Company in Boston in the late 1960s, noted that such new, small businesses were and are particularly adept at reexamining how business and culture are intertwined and how a business can serve as an agent of social change.[10]

This different approach to business, which took into account the social and environmental aspects of how a business created value, seemed to move in lockstep with a growing international awareness of such issues as CFCs and their impact on the ozone layer (Montreal Protocol, 1987), sustainable development of the urban environment ("Agenda 21" as part of the Earth Summit in Rio, 1992 and the *Aalborg Charter of European Cities and Towns Toward Sustainability*, 1994), and the overall need to increase efficiencies geometrically to reduce humankind's ecological footprint (*Carnoules Declaration*, 1994). During this period, a different approach to capitalism seemed to become more widespread and mainstream. In 1988, 3M voluntarily committed to reducing its hazardous emissions by 90 percent. In the early 1990s Monsanto and DuPont followed suit with similar commitments. Before the end of the last decade, the notion of a triple bottom line became more of a familiar concept in the corporate world, numerous transnational companies began to draft statements of social responsibility, and some even began to commission third-party audits evaluating their social and environmental impacts. In just three short decades, the narrative of capitalism with a difference had come of age.

While the narrative of indifferent capitalism reads very much like a grand epic—think Homer's *Iliad*—the narrative of capitalism with a difference is more akin to a loose collection of stories—like Chaucer's *Canterbury Tales,* for example—that collectively convey a sense that business can and should be conducted with an eye toward more than narrow economic efficiencies. Some of these stories evolve around companies driven by a form of spiritual calling.[11] Others focus on companies that have evolved almost reluctantly into business endeavors.[12] Some of these stories focus on small businesses that have deliberately limited their growth so as to remain true to critical core values.[13] Other stories have focused on large, global companies and their change-management strategies for doing business differently.[14] Each company's story has a different rationale for the changes the company has made and is still going through. Each has different objectives. But they all comprise an overarching narrative, one that sets itself in direct contrast to the narrative of indifferent capitalism. What all these stories are saying, in their disparate ways, is that business and ethics, profit and social/environmental concerns, doing well and doing good, are not necessarily at odds with one another.

The narrative of indifferent capitalism is a fully fleshed-out story that has unfolded over centuries. It's a mature narrative, with a beginning, a middle and, if not an end, at least a well-evolved sense of an ending. The narrative of capitalism with a difference is a relatively new kid on the block, with barely a beginning to its credit. The narrative of indifferent capitalism is largely descriptive, characterizing the way our economic system functions today. By contrast, the narrative of capitalism with a difference is largely optative, expressing a hope or desire for our economic system to function differently than it does today. Much of contemporary writing on the environment (and on social justice issues as well) points to the need for a positive vision for the future. What is yearned for is a narrative that doesn't merely critique the inadequacies of a system to address such problems as global warming, energy inefficiencies, water quality and quantities, fast disappearing biological and cultural diversity, deforestation and desertification, and global poverty, but provides vistas of how to redress such critical concerns. The narrative of capitalism with a difference is an effort in that direction. These two different versions and visions of capitalism are distinct from one another, as can be seen in a side-by-side comparison in Table 1.1.

TABLE 1.1   A TALE OF TWO CAPITALISMS

CONCEPT	INDIFFERENT CAPITALISM	CAPITALISM WITH A DIFFERENCE
Basis in economic theory	Adam Smith, *The Wealth of Nations* (1776): Free circulation of money, goods, labor, specialization providing for market efficiency	Adam Smith, *The Wealth of Nations* (1776) plus *The Theory of Moral Sentiments* (1759): Balance commercial liberty with moral sympathy, holding that the efficient market is composed of small enterprises located in community
Bottom line	Financial performance only	Includes social and environmental metrics with financial performance
Governance	Shareholder theory: Business is answerable to its shareholders	Stakeholder theory: Business is answerable to a broader range of people beyond shareholders to employees, community
Scope	Globalization	Internationalization (each country sets its own rules) or localization (self-sustaining local economies)
Notion of growth	Maximize growth: The bigger the pie, the better	Optimize growth: Smaller pie, more equitably distributed
Integration	Efficient integration of global supply chain	Effective integration of economic, social, and environmental issues
Mobility	Highly mobile in terms of investment, production, capital; will move quickly in and out of markets based on cheap costs and high returns	Place-centric, tending to put down long-term roots in a particular locale. Strong investment in community
Value & investment threshold	Focus on quarter-to-quarter short-term returns; value based on stock price	Greater focus on long-term value, total return
Company profile	Large transnational companies, typically publicly traded, with short-term investment horizon	Often (though not necessarily) smaller companies or larger publicly traded companies with stable, long-term shareholders

3 1833 05658 0886

Having said that, however, there are multiple roads by which a capitalist system can respond positively to the many environmental and social ills we are facing. Both narratives can get us there. They just do so differently (and at different paces).

Adherents to the narrative of indifferent capitalism might look to government and its use of regulations, subsidies, and suasion to change the way the game of business is played, so as to address environmental and social justice issues more effectively. Of course, for government to effectuate such changes, it would need to substantively reduce the powerful lobbying influence currently exercised by the business sector on legislation and regulation. In addition to government, there is also the factor of the marketplace itself, which has in certain situations pushed companies toward multiple bottom line expectations. Market forces, such as rules of commercial engagement in a geographical region, competition within an industry, demands placed on business from financial institutions evaluating risk exposure, and shareholder expectations, can and do make business more responsive to and more responsible for the social and environmental consequences of its actions.

Adherents to the narrative of capitalism with a difference tend to be more skeptical of governmental good intent as well as of market forces to catalyze change. Instead, they tend to place more faith (or hope) in the efforts of business either to take a higher ethical stand in its commercial dealings or to recognize that it is just plain good business sense to move in that direction. This direction—referred to variously as a sustainable orientation or a focus on the triple bottom line of economic, environmental, and social viability—is one we will be referring to throughout this book as a company's green bottom line. The multiple roads to a green bottom line can be seen in Figure 1-1.

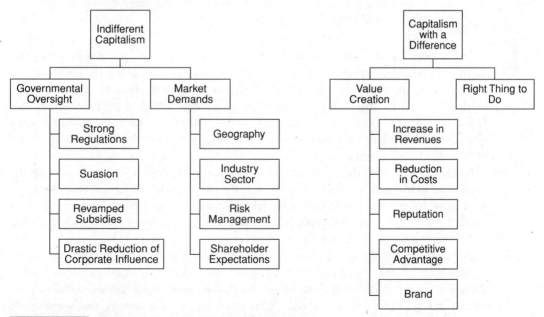

Figure 1.1   **Multiple roads to a green bottom line.**

So, if a variety of paths could get us to a green bottom line, does it matter all that much which particular narrative speaks to each of us? It matters greatly. The question hinges on issues of pragmatism and timing and on two distinct s-shaped diffusion curves projecting how change occurs.

While there is no question that government has the capacity to move the practice of business toward a more sustainable orientation, it has shown a general resistance to doing so. For example, proposed innovative legislation, such as the Apollo Project on clean energy (calling for a $300 billion investment in clean tech alternatives over ten years) and the Health Care for Hybrids Act (linking fuel economy to providing health care for autoworkers) were apparently killed by political maneuverings.[15] For government to play a timely leadership role in shaping the way business is conducted calls for a significant transformation in government itself, a transformation that is unlikely to occur from within the system.[16]

Looking to market forces to move business toward more sustainable practices is more promising, though only slightly so. One the one hand, market forces, such as the 2003 Equator Principles, in which ten global banks have joined together to require large-scale development projects to account for their environmental and social impacts, have influenced businesses to move toward a green bottom line. On the other hand, such influences are hardly broad or rapid enough to address such urgent issues we all face. As environmental educator David Orr notes:

> The things that need to happen rapidly such as the preservation of biological diversity, the transition to a solar society, the widespread application of sustainable agriculture and forestry, populations limits, the protection of basic human rights, and democratic reform occur slowly, if at all, while ecological ruin and economic dislocation race ahead. What can be done?[17]

Orr has his finger at the crux of the problem: Sole reliance upon the current paradigm of indifferent capitalism to respond to exigent environmental issues is a strategy that is simply too slow. The strategy's s-shaped curve—a tool used by economists and market analysts to project the time it takes for an innovation to become widely diffused, hit a tipping point, and become adopted—is too flat, taking too long to reach a tipping point. Lacking is a critical mass of innovators and early adopters that would accelerate change (see Figure 1-2).

Instead, what is needed is a strategy—or a set of strategies—that ratchets up the number of innovators and early adopters and thus accelerates the time it takes for business to be more responsible for its social and environmental consequences. What is needed is the s-shaped curve depicted in Figure 1-3. That is where the narrative of capitalism with a difference offers so much promise.

Accelerating change in the way business is conducted, creating a shift in an s-shaped diffusion curve, calls for a recognition that a crisis is at hand, a true paradigm shift, a change in systemic thinking.[18] As one executive in charge of corporate social responsibility for a large company noted, "Can you think of a company that's embraced Corporate Social Responsibility (CSR) *without* feeling pressure from outsiders first?"[19]

OF ADOPTEES

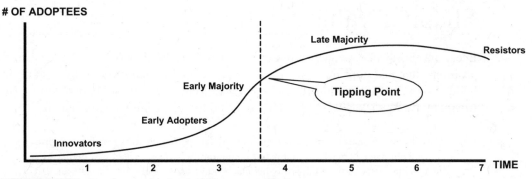

Figure 1.2   Tipping point for indifferent capitalism.

Some of that pressure is surely being exerted successfully by the more than 1.5 million nonprofit or civil service organizations working at the grassroots level for social and environmental justice issues.[20] And some of that pressure and influence comes from companies large and small that have *chosen* to adopt a narrative of capitalism with a difference—companies that believe sustainable practices provide opportunities for significant value creation and are the right thing to do. These two constituencies—grassroots organizations and companies conducting business differently—comprise the early adopters in an overall sustainability movement. Increasing the number and influence of early adopters affects the overall pace at which the diffusion of sustainable business practices occurs.

Reliance on the narrative of indifferent capitalism to restore our lands and communities is in many ways a dangerous proposition. Reliance on an alternative narrative, the narrative of capitalism with a difference, offers up much in the way of an opportunity to avert the significant crises facing us. But what does it take for a business to embrace such an alternative narrative? Does it make business sense? These two questions will shape the remainder of this chapter.

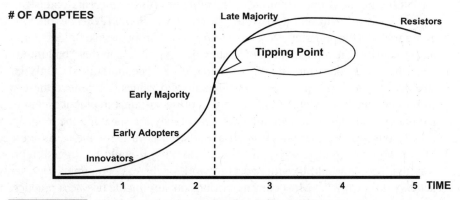

Figure 1.3   Tipping point for capitalism with a difference.

Shaping Values

Every once in a while, I come across a piece of art—a painting, a story, a musical composition—that just stops me in my tracks. A comment by John Abrams, founder of the build/design firm South Mountain Company on Martha's Vineyard and author of *The Company We Keep,* has this effect on me. I find myself reading and re-reading it:

> The question at hand is this: can small business, supported by strong underlying principles, help make better lives and better communities? To go farther, and perhaps too far, can business conducted this way help us be kinder to ourselves and to one another, to the planet, and especially to our children? Is it a stretch to say that the more fully we are fulfilled in our work, the more fully we can love both our children and our community? And that the more fulfilled we are, the more we can help build a future that's sane and just? If I overreach, it is only my enthusiasm for the possibility that is at fault.[21]

If I had to choose one passage that for me summed up the essence of managing a sustainable business, it would be this. For me, any discussion of capitalism with a difference boils down to the set of values embraced by a company and the uses to which those values are put. A green bottom line is rooted in people and the values they fully express. Abrams understands that about as well as any business person today writing about a company's responsibility to land and community.

THREE SYSTEMIC LAYERS OF THE VALUES-DRIVEN COMPANY

It has become trendy among various writers on triple bottom line practices to address first the financial benefits that accrue to a sustainably oriented company and then, only as an afterthought, to examine the deeper, underlying motivations for a business doing the right thing. Daniel Esty and Andrew Winston, in their book *Green to Gold*, argue that the case for thinking and acting environmentally is not based primarily on values but on the business opportunities such thinking affords to create competitive advantage.[22] Andrew Savitz and Karl Weber, in *The Triple Bottom Line,* devote virtually their entire focus to the pragmatics of managing a broad array of stakeholders, drawing upon a case study of the almost-sale of Hershey as a moral lesson that "bad things can happen when companies fail to take a broad view of accountability." Only late in their study do the authors get around to leadership and values.[23] I believe these writers are well intentioned. They are largely trying to bring sustainable practices and beliefs into the mainstream business community. As such, they are speaking the language of business to business people, which is all about what is in it for them. As the economist Steven D. Levitt trenchantly points out, "Morality, it could be argued, represents the way that people would like the world to work—whereas economics represents how it actually does work."[24] Most current writings on sustainable business practices are simply a pragmatic recognition of how best to make the business case stick.

Despite the good intentions, such an approach is "bass ackwards," as my kids would say. While the remainder of this book is devoted to making the business case for sustainability, we begin here with values. That's the way our own company charted its evolution. It's how it continues to evolve. To do otherwise, to begin with a business plan for a green bottom line and then perhaps later deepen that approach with ethical considerations, is a bit like binge dieting before an important occasion: It is a short-term strategy built around a specific moment in time and is likely to last only until the next trendy occasion comes along. A business built around a truly substantive green bottom line is about taking steps, behavior-altering steps that endure. It is a regimen of diet and exercise for the long haul.

But how does this happen? How are a company's values determined? Do they pass from one generation of leadership to the next, unchanged? Should those values remain unchanged over time? Or should each generation of company leaders be free to shape the company's values anew? What's the role of employees in the process of determining values, if any? Do other stakeholders have a say in this process? Why or why not? And how?

Layer One: Six Guiding Principles

Our company, Melaver, Inc., has gone through a formal values-determination process three times in the fifteen years I have run the business, twice with staff and once with shareholders. Each process has been somewhat different in terms of the length of time involved (from one month to almost a year) and the methodologies utilized. All have involved the use of an outside facilitator to guide the process along. And all have these guiding principles in common.

1. Values Belong to People, Not to the Company: This concept runs contrary to numerous writings on business culture[25] and even contradicts language our own company uses from time to time. We tend to speak about our company's core values as if the company itself were some anthropomorphic character having a unique set of beliefs that are removed from the people who shape them. That's unfortunate and potentially harmful, removing human agency from our sense of what a business stands for. But a company is nothing more and nothing less than the people who comprise it. Despite the legal fiction that enables a company to abrogate to itself the rights of an individual, the values a company articulates do not materialize out of thin air but arise out of actions that individuals within the company put in motion. This principle is a critical foundation stone. Without it, individuals within a company lack the sense that they are the ones responsible for true leadership and vision. With it, individuals within a company feel empowered to make something meaningful, purposive, and lasting. Peter Senge, author of *The Fifth Discipline*, captures this sensibility as well as anyone when he notes:

> The problem lies in our reactive orientation toward current reality. Vision becomes a living force only when people truly believe they can shape their future. The simple fact is that most managers do not experience that they are contributing to creating their current reality. So they don't see how they can contribute toward changing that reality.[26]

Our own company is small enough (fewer than three dozen people) that we can and do engage everyone in discussions about what is of utmost importance. Responses vary. Some focus on the intimate value of belonging, having to do with self-worth, connectedness to family, and the community at work. Others emphasize the value of self-actualization connected to a sense of liberation and the capacity to integrate the various aspects of one's life into a complete whole. Still others value the knowledge and insight connected to wisdom and the capacity to shape a new world order. It takes time to elicit expressions of value from every person at a company. It takes even more time to synthesize those disparate values into the daily practices of a business. Many companies are simply too large or lack the inclination to listen to their workforce. But in the absence of deliberate effort to query what its people value, a company drifts without definition. As the old saying goes, you cannot manage what you do not measure.

2. Knowledge Is Critical: Individual and collective knowledge about what we want is critical to the values-centric business. One notable aspect of my father's story of the fisherman and the businessman is that the fisherman already knows what he wants and he's doing it. In the somewhat lofty language of philosophy, his sense of being and his sense of becoming are integrated.[27] Most of us, I think, view work as something of an unfortunate necessity at worst and a profession at best. Few see work as a vocation or calling, providing what we need and desire. Among those who do, however, is Catholic priest and business ethicist Oliver Williams, who notes that the key purposes of a business are to "enable human flourishing" and to "provide opportunities for us to develop character and virtues, as well as talents and skills."[28]

This enabling of human capacity, like the surveying of values of people within an organization, takes time and effort to bring forth from one's workforce. First and foremost, most of us have been conditioned not to think about what we truly expect from work except in the rather narrow context of receiving a wage and perhaps developing professional skills. Secondly, it is rare for others, particularly management, to ask their co-workers what it is they truly want and need. The immediate tasks at hand almost always take precedent. However, the acts of asking and listening have the corollary effect of empowering a work force to reflect upon the nature of work. And that is a good thing. Almost always during our interviewing process for a new hire, one of my colleagues will ask a candidate what it is he or she really wants to do in life. Sometimes, the answer suggests a trajectory in life dramatically different from work at our company. And so we respond with a question that evokes the story of the fisherman: *If that is what you truly want, why aren't you pursuing it instead of talking with us?* It is in the interest of our company that the interests of each potential employee take center stage.

3. Values Are Not to Be Deferred: Another noteworthy element to the story of the businessman and the fisherman is that the fisherman is clearly not deferring his values to a latter stage of life. He knows what he wants and he's acting on it in the present time. Most Americans, by way of contrast, know everything there is to know about pursuing material wealth in the here and now and deferring the stuff that really matters (call it meaning or spiritual contentment or salvation) until much later in the game. Most of us simply refer to it nonchalantly, even proudly, as our Protestant work ethic. It's not that we are unaware that we are deferring what we really value. We are. The problem, as Paulo Coelho points out in the introductory essay to his allegorical novel *The Alchemist,* is fear.[29]

When our company first began in the early 1990s to discuss changing the way we practiced real estate to become more in tune with our vision of a sustainable business, we found all sorts of reasons to drag our feet: It seemed to be impossible, other companies weren't doing those sorts of things, we were likely to fail, what would happen if others viewed us as tree-huggers, etc. Later on, when we debated whether or not to commit all of our development to adhere to LEED criteria, a similar resistance was found. More recently, as we considered the pros and cons to signing on to the 2030 Challenge (committing to progressive reductions in carbon emissions over a twenty-plus year period), these same fears and concerns again surfaced.

These moments of resistance within our own company remind me that the fear of change will always be with us, and that we might as well get accustomed to it as a part of the furniture. It helps to have this longer term historical or institutional memory so we can tell ourselves, *Oh, we always go through this period of initial fear whenever we think about trying something new. It'll be OK*. Typically, our fears are overstated and things indeed are OK. Finally, it helps enormously that my colleagues and I are able to mutually support one another, so the weight of fear and potential failure do not inhibit us from deferring the valuable work that needs to be done today. We know that we are all in this together. No sense putting off till tomorrow what needs doing now.

4. Authenticity Is Key: When I read John Abrams' words, quoted above, I am struck above all by the authenticity of his voice. Nothing seems to be held back. He uses language few business leaders would dare give voice to, speaking about kindness to ourselves and to others, building a world that is more sane and just, creating a community where love prevails. Here is a businessman talking about love, for god's sake. Abrams not only has found his voice, he has the courage to put it out there irrespective of the consequences. It's powerful stuff. Abrams' work would have been warmly welcomed by the literary scholar Lionel Trilling, one of whose seminal studies was about the search for authenticity in the modern world. In *Sincerity and Authenticity,* Trilling writes about the process by which the effort to be true to one's self—sincerity—evolves into the much more challenging effort to be morally centered in an increasingly complex world. "Now and then," he notes, "it is possible to observe the moral life in process of revising itself."[30] One can see and feel that process at work at South Mountain Company, the company John Abrams co-founded. Businesses similarly aspiring to be values-centric need to find their own processes for enabling their moral sensibility to evolve.

It probably takes most people at our company two to three years to find their voice. Some take a good bit longer than that. I have no way of knowing for sure, no method for evaluating this process objectively. It's what I notice around me—a gradual process of individuals, as Trilling would say, revising themselves. Newly hired colleagues seem both intrigued by the openness of our dialogue and highly skeptical, as if to say *I'm not going to be brainwashed by this corporate mumbo-jumbo about values.* After a while (and it can take quite a while for this transition to take place) there is a growing awareness that others are true to who they are, and that this might be a safe environment to speak openly about what matters most. A still longer period of time elapses before this evocation of sincerity evolves into a much more liberating sense that this is a community of people whom I not only can trust but who will enable me to trust myself.

This continual process of revising oneself toward greater authenticity is a critical part of values-shaping. It has taken time for us, as a company, to get our hands around the idea that aspiring to be a sustainable real estate company means more than simply developing LEED certified buildings. A fly on the wall at one of our company retreats would get an earful of debate regarding why we should confine our development to urban core areas or why we should limit our growth or how we can be a real estate company and not degrade our natural environment to some extent. At heart, we are a learning company, learning primarily about ourselves and our relationships to our land and community. Such learning, as Peter Senge points out, that

> changes mental models is immensely challenging. It is disorienting. It can be frightening as we confront cherished beliefs and assumptions. It cannot be done alone. It can occur only within a community of learners.[31]

This community of learners can be most effective when mutual trust allows for authenticity of belief, expression, and action.

5. *Tension Can Be Good:* Tension among values can be good, creative, and productive. Unless there's a deliberate effort to hire an entire staff of like-minded souls all holding to the same precise values, there is bound to be tension among the various values espoused by staff members. Such tension, like cholesterol, can be healthy or problematic. Frans Johansson, whose book *The Medici Effect* considers how we create "an explosion of extraordinary ideas" with paradigm-shifting consequences, notes that viewing situations and problems from multiple perspectives and providing for the random combination of ideas are two critical components of positive, ground-breaking creativity. Both are facilitated by a diverse working team that knows how to depersonalize the conflicts that arise by virtue of their different disciplines, cultures, styles, attitudes, and values:

> ...if you wish to generate intersectional ideas, you should see environments where you will work with people who are different from you. Put another way: A sure path to inhibit your own creativity is to seek out environments where people are just like you.... Your team will get along great and it will get a lot of things done. But will it be innovative?... At the Intersection, we need as many opportunities for random combinations of ideas as possible. A team of diverse people who feel free to exchange and combine their ideas is exactly what can make that happen.[32]

Some of us find strong community within the fellowship of our work environment. Others prefer to find that community within the more intimate setting of family. Some of us thrive in an environment that is constantly trying to innovate beyond what our most recent project entailed. Others are much more content to repeat a winning formula over and over. There is an ongoing tension within our company about where we should and should not develop, about the pace of our growth and the merits of growth in general, about whether or not we would be willing to forego LEED certification on a building in order to devote more financial resources to creating a building with super high performance energy efficiency with minimal carbon emissions. The key is to manage this tension of values so that mutual respect is always present and yet does not inhibit the creative challenging of one another, so that each person's voice is given authentic expression and yet does not lead to a divisive, negative atmosphere. It's not as difficult as it sounds.

The curious thing about a values-oriented company is this: While it lends itself to a highly charged, passionate environment full of diverse perspectives, the very fact that these perspectives have a nurturing environment in which to be articulated means that the adherents of various perspectives tend to have great respect for one another. It's something akin to the ethos, *I might disagree strongly with your opinion but I will defend completely and absolutely your right to express it.*

6. *Values Are Revisited Regularly:* Doing a deep dive into the specific values of individuals at a company doesn't occur very often. The three values-exploration exercises we've undergone over the past fifteen years are probably sufficient based on our own growth, adding a few new people each year. However, it is probably fair to say that we conduct informal values-determination exercises almost daily, as we go about our regular decision-making. Reviews of policies—vacation, sick leave, maternity leave, time off to tend to an ill family member or close friend—almost automatically trigger larger discussions having to do with our values system. Retooling our compensation package—moving costs, severance pay, a short-term bonus plan that fairly recognizes the team element in most individual performances, a long-term incentive plan that adequately recognizes the company's value creation over an extended period of time, investments in continuing education for all staff members, how pension plans will be managed—also inevitably calls upon all of us to think of balancing immediate and long-term financial security for all. Debate about whether or not to pursue a specific real estate development inevitably begs the question of *why*, beyond the pure economics of a deal, we feel this project adds needed value to a community. Cobbling together complex deals entails bringing in outside parties—financial institutions, designers, architects, contractors—which raises immediate questions regarding these parties' congruence with our own values. From big decisions to the seemingly mundane, basic questions of our own values are always present, questions that have to be addressed for these decisions to be made.

Often, the decision on any particular issue is less important than the process we go through in making that decision and the stated or unstated rationale behind it. The extent of inclusion (or exclusion) of staff members, the ways in which we listen to one another, how various opinions are weighed (based on the merit of an argument or seniority or who speaks loudest): All of these details of daily business life speak volumes about the values a company truly embraces. The practice of our business, day in and day out, is fundamentally that of re-visiting and refining what we hold to be truly important.

While these six principles serve to guide us toward a deeper understanding of the roles individual values play in an organization, our discussion still lacks specificity. What are the specific concepts that help shape a green bottom line? Can a company realistically take into account the wide, disparate values of individual staff members without falling into dysfunctional chaos? How do these values, pragmatically speaking, function together? To address these questions, we need to consider the second layer of values building within an organization: structural concepts.

Layer Two: Structural Concepts That Shape a Green Bottom Line

There is one overarching organizational concept—the ring to rule all rings, as it were—that has guided the development of our company from its inception. The concept originates from the philosopher John Rawls, whose foundational work linking social justice to fairness (*A Theory of Justice,* 1971) is a landmark of 20th century political thought. Rawls' concept begins with a deceptively simple question: If we began with a blank slate, what type of system would we create without knowing in advance what our roles would be in that system? Translated into a business organizational context, the question becomes this: How would we design a business without knowing in advance what role we would play in that company? Most of us, I believe, would design a business that we would want to work for over the course of a lifetime. Simple, really. But too often neglected.

Companies that begin to embrace a sustainable ethos often start by looking at their operations differently through what is called life cycle assessment (LCA). Life cycle assessment looks at the total costs (and benefits) over the lifetime of an expenditure, not just first cost (and recovery). To me, however, a true life cycle assessment begins not with the materials and equipment a company purchases but with the investment a company makes from the get-go in its underlying culture. GE, under Jack Welch's leadership, was known for having a forced rating system for all its employees, with the top 10 percent rated as As, the next 70 percent as Bs, and the bottom 20 percent as Cs. It was Welch's and GE's practice to jettison the Cs on a regular basis and replace them with other workers. It's a system that may seem efficient, focused on having the very best work for the company. But what's the cost of such a system, with all of its built-in churn and internal competitiveness? Hard to know precisely. But the waste of human capital, of talent and training, of the potential for collaboration was, in all likelihood, huge. GE's approach to business culture is the very opposite of a life cycle approach to culture-creation. It is the antithesis of Rawls' theory of justice, which, translated into a business setting, comes down to the life cycle investment a company makes in its fundamental design. What is needed is a life cycle values assessment (LCVA). What would a life cycle values assessment look like? What would it measure?

What would a company look like, what would it need to offer us in order to satisfy those things that we feel are most critical? What would a company need to look like in order to be considered just for all, irrespective of the particular roles we play in it? To my mind, such a company would need to consider seven concepts as part of its underlying structure.

1. Sustenance and Security: In the wide-ranging literature devoted to sustainable practices, there is near universal agreement on one topic: An agenda for social and environmental justice issues cannot be adequately addressed until basic needs—food, clothing, shelter, security—are taken care of.[33] A business devoted to the stewardship of land and community must first take care of its own staff before it can hope to tend to larger needs. Every few years, for example, our company surveys total compensation packages nationwide and pegs our own compensation program to around the eighty-fifth percentile of what companies elsewhere are doing. The logic is simply this. We want to ensure that everyone on staff is sufficiently compensated to afford home ownership, to receive adequate healthcare, to be able to set aside funds for future needs such as kids' education, retirement, etc. By the same token, we do not want to have a compensation program that is at the very highest end of the spectrum. I personally feel that were I to take a job solely for its compensation package, I would also leave that job for a better offer elsewhere. We feel monetary compensation should not be a deterrent to working for us, but neither should it be the primary driver. For that, we rely on other, intangible aspects of working together. An overview of our compensation program can be seen in Table 1.2.

TABLE 1.2 MELAVER, INC. COMPENSATION PROGRAM	
COMPENSATION	**TERMS**
Base salary	Approximately 85th percentile, based on national survey
Annual bonus	Up to 10% of base salary, based on performance
401(k)	Employer match up to 5% of base salary
Pension plan	6% of base salary
Long-term incentive	Percentage of company growth, based on position
Health insurance	Full coverage for staff member
Purchase of hybrid car	$3,000–4,000 subsidy, based on emissions
Carbon reduction compensation	Under review
Maternity/paternity leave	Under review

Another aspect of sustenance and security worth dwelling on briefly is the notion of long-term employment, which is something of an anachronism in this day and age, given the high turnover rate among companies and the corollary notion that many of us are likely to hold on to any job no more than 3.6 years over the next forty years. This alarming aspect of globalized trade has even strong supporters of globalization such as Thomas Friedman suggesting in a clever verbal sleight of hand that American companies, while they cannot provide lifetime employment, should provide for life-time employability.[34]

We certainly support this notion of lifetime employability, investing money each year for the continuing education and training of everyone on staff. But here's the kicker: We hope that lifetime employability is with us. Such an approach to long-term employment provides security for our staff and also provides security and stability for the company. Our turnover rate is practically nil. There's considerable value to be gar-nered from such stability.

2. Belonging: Drift has become a household term these days, conveying the sense that most of us have of being unmoored from a sense of community. There are two pri-mary sociological interpretations of this fact of contemporary life. Richard Florida, who writes about people he terms "cultural creatives," numbering some thirty-eight million, tends to celebrate the loose ties these cultural creatives have with their social context:

> In place of the tightly knit urban neighborhoods of the past or alienated and generic suburbs, we prefer communities that have a distinctive character. These communities are defined by the impermanent relationships and loose ties that let us live the quasi-anonymous lives we want rather than those that are imposed on us.... The kinds of communities that we desire and that generate economic prosperity are very different from those of the past.... Where old social structures were once nurturing, now they are restricting.[35]

Where Florida celebrates the autonomy that comes from such loose-tie communi-ties, the sociologist Richard Sennett bemoans the lack of connective tissue:

> "Who needs me?" is a question of character which suffers a radical challenge in mod-ern capitalism. The system radiates indifference. It does so in terms of the outcomes of human striving, as in winner-take-all markets, where there is little connection between risk and reward. It radiates indifference in the organization of absence of trust, where there is no reason to be needed. And it does so through reengineering of institutions in which people are treated as disposable. Such practices obviously and brutally diminish the sense of mattering as a person, of being necessary to others.[36]

Interestingly, while Florida embraces our latest, bravest new world and Sennett criti-cizes it, both acknowledge the powerful yearning for a sense of belonging.

Constructed from this strong sensibility, our company might be best described as "intimate" or "familial." We try to be looser-knit than the claustrophobic small-town ethos of a generation ago but tighter-knit than the drift that seems so pervasive these days. We collectively celebrate birthdays, visit with each other on weekends. We counsel one another with regard to love gone right and love gone wrong and love perhaps heading in a positive direction. We hire each other's kids for summer internships and see them in a slightly different light than their parents (our colleagues) do. We ring a brass captain's bell to celebrate when good things happen and SMS (short message service) and e-mail and BlackBerry each other at ridiculous hours of the night and weekend when we are concerned. We try to stay out of each other's hair and probably do a lousy job of it. But the effort is appreciated nevertheless. Like the members of a family or kinship group, we verbally beat up on each other day in and day out, but get defensive and supportive when someone from outside the company even slightly criticizes one of our colleagues.

The sense of belonging within an organization encompasses a code of engagement that nurtures intimacy. Things like mutual respect, civility of behavior and discourse, empathy, trust, and transparency are all necessary components of the type of community we collectively shape and share. So, too, is the expectation that each of us is involved and needed.[37] Things easily stated, but much less easy to deliver. In an intimate setting such as ours, oftentimes it is more convenient to push contentious issues under a rug rather than address one another directly. No one likes confrontation. No one wants to unnecessarily make it awkward for oneself and other colleagues by practicing the type of straight talk we feel is necessary for our code of engagement to work. And so we struggle with being open and honest with one another all the time. It's a continual work in progress.

The concept of belonging we try to foster extends beyond our company into the community at large. Over half of our staff sits on nonprofit board or is otherwise involved in civic organizations. It's not mandated, but it's encouraged. We match staff members' charitable contributions and allow for a week's paid leave to anyone wishing to volunteer services to a civic organization. Wherever feasible, we provide pro-bono consulting on projects where our knowledge of sustainable practices can come in handy. And we try our damndest to ensure that everyone strikes a good work-family balance—not always an easy thing to manage. Our culture is definitely not for everyone. But it is for those for whom everyone else matters. We cannot be good stewards of our land and community if we are not, first and foremost, good stewards of one another.

3. Craftsmanship: Most of us are probably familiar with the scene in the movie *Modern Times* where Charlie Chaplin literally becomes a cog in an assembly-line operation. While comic on film, it's not a life circumstance any of us are likely to find appealing. If there were one thing I wish I might benefit from at work, it is the development of a particular craft I was really good at. Unfortunately, that's not the case. I'm an English major who runs a real estate company, a generalist who wears lots of different hats at a small company, pitching in wherever help is needed. If I had to look for a job today, I'd be in a peck of trouble. I don't want that to be the case for the rest of my colleagues, who by and large feel the need to have a craftsman-like expertise for some facet of our business. It is, I feel, our obligation, to meet that need.

The notion of craftsmanship I have in mind goes well beyond the capacity to do something well, something Richard Sennett describes beautifully when he notes:

> Craftsmanship has a cardinal virtue.... It is commitment. It's not simply that the obsessed, competitive craftsman may be committed to doing something well, but more that he or she believes in its objective value.... Getting something right, even though it may get you nothing, is the spirit of true craftsmanship.... Commitment entails closure, forgoing possibilities for the sake of concentrating on one thing.[38]

The craft we ply is that of creating and nurturing community. How we engage the various people who work with us—designers, architects, land planners, engineers, builders, financiers—is part of that craft. How we draw up our legal documents and analyze our financial returns is also part of that craft. As is the actual construction. Through our specific roles with the company, by means of our various vocations, we look at what we develop from a long-term, life cycle perspective. In that sense, our sense of values and our practice of values become inseparable.

4. Diversity of Work Opportunities: At first blush, the opportunity for diverse work experiences is directly at odds with the notion of craftsmanship. To some extent, that's true. While craftsmanship is about focusing on one thing over an extended period of time, diverse work opportunities convey the sense of jumping around from one type of task to another. But the type of opportunity I am talking about here is not jumping broadly around at any given point in time, but probing deeply into the various levels of what we do. It's about deep time, not broad time. Layering complex skills one upon the other over an extended period of time. Developing sophistication in one's capacities. In this sense, diversity of work opportunities and craftsmanship are powerfully connected.

As odd as it sounds, we have a succession plan in place for key management positions that goes out fifteen years from the present time. Strong young leadership in place today is positioned to be senior management in the distant future. For this young leadership cadre to develop fully, they need that diversity of work experiences. It's something that larger companies have a much easier time structuring, but it's not an easy thing to provide in small companies such as ours.

But the rationale for diverse work opportunities goes well beyond succession planning. Development work, done well, calls for integrating a wide knowledge of disciplines—history and geography, geology, biology, and hydrology, sociology and communications, finance, ecology and organizational design, psychology and land planning—just to name a few. Diversity of work opportunities provides exposure to these various disciplines, and it's a good fit with a business looking to harmonize its financial orientation with social and environmental concerns. More fundamentally, providing diverse work opportunities, with the objective of achieving craftsmanship, helps replace the more rote sense of having a job (or a profession, even) with having a vocation for life.

5. Meaning: Beyond the basics of work—sustenance and security, a spirit of belonging, and a sense of having a vocation—there is the need to find personal meaning in work. Personal meaning differs with the individual. For me, having my financial needs met, sharing a sense of community, and having a vocation collectively shape a sense of meaning, but the sense that what I'm doing is redemptive or restorative as just as essential. Max De Pree, former CEO of Herman Miller, captures this sensibility about as well as anyone I know when he notes that a company needs to be a place where "they will let me do my best," where I can realize my highest potential.[39]

6. Usefulness or Purpose: While meaning involves internal validation, a sense of usefulness or purpose has more to do with external validation. I would find it difficult to work for a company, for example, that manufactured landmines. The work might provide financial security, it might have a great sense of community, it might satisfy my sense of having a vocation, and it even might provide a sense that I am realizing my highest potential as an individual. But I would not find gratification in knowing the work I do serves no useful social purpose. Is my work a legitimate calling (in more than a mere legalistic sense)? Can I honestly say that what I am doing ennobles the human condition, perhaps by enabling others to have a livelihood or to provide for a healthy habitat or to have access to inexpensive and healthy food or to become literate or to have basic rights of governance? I think we shy away from thinking about work in these terms in part because of the seeming grandiosity of it all, in part because we have been conditioned early on to put aside these seemingly romantic or quixotic ideals in order to "get on with life." In so doing, we're not getting on with living, we're getting on with the process of dying. Because something essential dies when we allow someone else's sense of the pragmatic to quell our capacity to dream and enact.

7. Legacy: There is a deep paradox embedded in Western culture today. On the one hand, advances in medicine have lengthened life spans and enhanced the quality of life for so many of us. We have a material prosperity unprecedented in history. And yet, with all of our progress—in medicine and science and technology—our general sense of happiness has diminished. Max De Pree captures this conundrum elegantly:

> To be a part of a throwaway mentality that discards goods and ideas, that discards principles and law, that discards persons and families, is to be at the dying edge. To be at the leading edge of consumption, affluence, and instant gratification is to be at the dying edge. To ignore the dignity of work and the elegance of simplicity, and the essential responsibility of serving each other, is to be at the dying edge.[40]

The implication of De Pree's argument is clear: At the end of a professional career that has aided a throwaway culture, we too will be discarded.

Our work and our works, though, should stand for something positive and good. We should not indenture our children with our bad choices and practices. Thomas

Jefferson understood that well, noting that no generation has the right to encumber the future with debt,[41] something we clearly violate today as we deplete our natural capital to the detriment of those who will succeed us.

Part of the problem lies in our ever-shrinking sense of time. We change jobs with ever-increasing frequency, hold stocks for barely six months, fire CEOs who do not measure up to quarterly expectations. We need a more positive vision for work and business, one that takes the longer view and challenges us now to be positive stewards of that longer time frame. Marco Polo, in his travels in the Far East in the fourteenth century, encountered just such a long-term vision that is relevant for us today:

> In a city called Tinju, they make bowls of porcelain, large and small, of incomparable beauty. They are made nowhere else except in this city, and from here are exported all over the world.... These dishes are made of a crumbly earth or clay which is dug as though from a mine and then stacked in huge mounds and left for thirty or forty years exposed to wind, rain, and sun. By this time the earth is so refined that dishes made of it are of an azure tint with a very brilliant sheen. You must understand that when a man makes a mound of this earth he does so for his children; the time of maturing is so long that he cannot hope to draw any profit from it himself or put it to use, but the son who succeeds him will reap the fruit.

Work that takes such a long view is at the heart of the values-centric business.

Thus far, we have discussed two layers of the values-driven business: guiding principles that serve as a bill of rights for all and structural concepts that can shape a company's values. Together, they are the machine language of a business system, serving as a foundation for the more concrete "software" of a company, the more visible, articulated values that a company embraces. It now remains to address this third layer, albeit briefly since this layer is very company-specific.

Layer Three: Shared Core Values

If you ask members of Melaver, Inc. about our company values, they will not talk about guiding principles or structural concepts. However, everyone at our company will tell you that we have four core values: ethical behavior, learning, service, and profitability. They might also tell you that each core value has a belief statement that elucidates the concept and that everyone at the company had a hand in writing those belief statements. And they will also be able to tell you a story about how these core values and belief statements were shaped, a process that involved everyone at the company meeting regularly for almost a year until we all agreed these were the values we *all* felt strongly about.

While the process a business engages in to elucidate shared values will differ with every company, there are a few guidelines worth keeping in mind as the process unfolds.

1. Realize the Process Takes Time: Be prepared to devote considerable time, at least initially, to the process. We devoted a good bit of time to talking about what each of us felt was of particular importance. The overall process took the better part of a year. In the early going, we probably spent two to four hours a week, which amounts to

5 to 10 percent of staff time. In the middle stages of the process, we probably devoted a half-day each month or about 3 percent of staff time. Toward the end, we had a flurry of meetings during which we rolled out a trial version of our core values and belief statements, and then refined them. This probably took about three days over the course of a month. All told, we probably devoted about 4 percent of staff time over the course of a year in developing our core values.

2. Be Inclusive: Try to include as many staff members as possible in the process. How inclusive a company can be depends in part on the overall size of a company, as well as the geographical dispersion of staff members. That said, the shaping of values should be a bottom-up process involving as many individuals at the company as possible.

3. Encourage Storytelling: Not everyone thinks in the broad abstract language of values. Most, however, are adept at telling stories from personal experience—background, life-changing moments, the guidance provided by family members and/or mentors—that speak poignantly about the values we hold dear. Elucidating those stories is a critical part of the process.

4. Use a Framework: Work from guiding principles to structural concepts to core values. The systemic framework provided in this chapter can aid the process of shaping values. What should a bill of rights for staff members look like? What would a just organization look like if we built it from John Rawls' orientation, creating the company from a blank slate? These are some of the questions that help frame the shaping of values.

5. Engage Top Management: Leadership should engage but not dominate. It is critical to have the active engagement of top management throughout the process. To do otherwise, to have top management delegate this process, signals its lack of importance in the grand scheme of things. However, once top management is at the table, it should not dominate the discussion. To do so undermines the bottom-up essence of the process and belies the sense that the company's values are really the values of all staff members.

6. Don't Force the Process: There's a natural tendency to plan the overall course of this process, including a set deadline by which the process ends. Such planning can, in fact, be helpful. But the timetable should be a guideline. The process takes time. Many surprising and unexpected thoughts and feelings are voiced, it takes a while to learn to hear one another, and people need the opportunity to absorb the ideas being expressed. Try working at the speed of the slowest assimilators in the room. The point is to optimize the process, not maximize the efficiency of your time together.

7. Hire an Outside Facilitator: Outside facilitation helps transcend the trap of top management dominating the process. It also helps ensure that ample time is allotted and that the process does not come to an abrupt conclusion before it has run its natural course. Outside facilitation should ensure that all voices have been heard and that deep listening is taking place. It should also ensure that office politics are kept out of the discussion and facilitate the voicing of the "elephants in the room" that many might be fearful of articulating.

8. Have a Communication Plan: It's a good idea to have a communication plan at the end of the process. We rolled out our core values and belief statements at a company retreat with everyone in attendance. But that was only the beginning of a follow-up communication plan that involved getting everyone not just conversant with but knowledgeable about the values we chose for ourselves. What did those values mean for the various development projects we were vetting? How about our selection of outside partners? How would we articulate those values to the outside world? The process of shaping values sets in motion a whole series of questions involving how we use those values and how we communicate them to others.

Bringing the Layers Together

The way a company goes about the process of shaping values is likely to be unique and specific to that company. Having said that, it is helpful to consider the systemic layers that comprise this process: guiding principles that articulate staff members' basic expectations, structural concepts that serve as foundation stones for the specific values a company's workforce chooses to embrace, and finally the specific, concrete values themselves. An overview of these systemic layers can be seen in Figure 1-4.

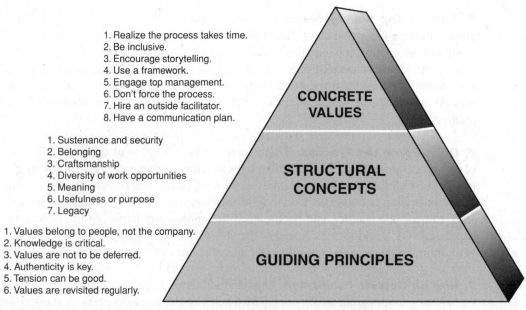

1. Realize the process takes time.
2. Be inclusive.
3. Encourage storytelling.
4. Use a framework.
5. Engage top management.
6. Don't force the process.
7. Hire an outside facilitator.
8. Have a communication plan.

1. Sustenance and security
2. Belonging
3. Craftsmanship
4. Diversity of work opportunities
5. Meaning
6. Usefulness or purpose
7. Legacy

1. Values belong to people, not the company.
2. Knowledge is critical.
3. Values are not to be deferred.
4. Authenticity is key.
5. Tension can be good.
6. Values are revisited regularly.

CONCRETE VALUES

STRUCTURAL CONCEPTS

GUIDING PRINCIPLES

Figure 1.4   Three systemic layers for shaping company values.

Having considered how a company goes about shaping values, big questions still remain: Can a company create added value by the values it creates? Does it pay? The answer to both questions is yes. We now turn our attention to the business case itself.

Creating Value

It's time to crunch some numbers. Since Melaver, Inc. is a privately held company and there are sensitivities among shareholders about using our company's data, we'll create a fictional company called Green, Inc., whose performance is not that dissimilar from our company's.

This is Green, Inc.'s profile:

- It has twenty employees, with a payroll of around $2.4 million annually.
- It has an asset base of $100 million, with 75 percent debt ($75 million) and 25 percent equity ($25 million).
- It has gross revenues of $12 million, which yields a 12 percent return on the total asset base ($100 million divided by $12 million).
- It has net income after debt service of $3.75 million, which yields roughly a 15 percent return on the equity ($3.75 million divided by equity of $25 million).
- The cost of capital (i.e., the interest paid on debt) is 7 percent.
- The annual growth of the company's assets tends to be around 7 percent after distributions to shareholders.
- The discount rate—the rate at which multiple year cash flow projections are discounted back to Year 0—is 10 percent.

In Year 0, Green, Inc. decides to undergo a values-shaping process similar to the one described in the previous section of this chapter. There is the cost of using a consultant for the values-shaping process. Management calculates that 80 hours of consulting time at $200 per hour is probably a good estimate, for a cost of $16,000. To have all of its employees involved in this process will require about 4 percent of total company time for the year. This means that 4 percent of projected net income will not materialize because of the lost opportunity of devoting company time to this non-productive endeavor. Lost opportunity of income amounts to $150,000 (which is 4 percent of the 15 percent of the equity of the portfolio, or $4\% \times 15\% \times \$25,000,000 = \$150,000$). All totaled, the cost of the values-shaping exercise in Year 0 is $166,000 ($150,000 + $16,000).

In thinking about the costs of this project a little more carefully, senior management also realizes that there is a carry-over effect in future years as a result of the original lost opportunity for revenue. The reasoning is as follows. In Year 0, the staff devoted some of its time to values shaping and, as such, did not devote this time to developing a project that would have brought in an additional $150,000 in revenue. But that's not all. This lost revenue in Year 0 could have been reinvested in the company in future years (Years 1 through 10), bringing with it additional growth of around 15 percent

Years	0	1	2	3
COSTS				
Lost Opportunity	(150,000)	(22,500)	(3,375)	(3,611)
Facilitator	(16,000)			
Sub-total Costs	(166,000)	(22,500)	(3,375)	(3,611)
Total Cashflow	(166,000)	(22,500)	(3,375)	(3,611)
Discount Factor	1.000	0.909	0.826	0.751
PV Cashflow	(166,000)	(20,455)	(2,789)	(2,713)
NPV	(208,987)			

Figure 1.5   Cash flow analysis of basic expenses.

annually (plus an annual compounding factor). In fact, over ten years, this lost revenue stream from the original lost opportunity amounts to about $213,000, consisting of $150,000 in Year 0 and additional lost opportunity of $63,000 in Years 1 through 10.

Total sticker price over ten years for this original values-sharing exercise amounts to about $229,000 ($16,000 + $150,000 + $63,000). When discounted back into Year 0 dollars (earlier dollars are more valuable than later dollars), the investment amounts to nearly $209,000 as shown in Figure 1-5 (see the net present value (NPV) line under Year 0).

But we aren't quite done yet. Green, Inc., as a result of its values-sharing process, has decided to add some bells and whistles to its overall program. It decides to invest in continuing education of all staff members, which results in about $1,000 per person plus travel and expenses, for a total of $2,000 per employee or $40,000 annually. The continuing education also results in time out of the office, about two days per person, so there's the cost of that plus the lost opportunity cost of that productivity. Green, Inc. has also decided to provide a subsidy of $4,000 for any staff member who buys a

Years	0	1	2	3
COSTS				
Lost Opportunity	(150,000)	(22,500)	(3,375)	(3,611)
Facilitator	(16,000)			
Continuing education		(40,000)	(40,000)	(40,000)
Lost opportunity from cont. ed.		(30,000)	(34,500)	(39,675)
Hybrids		(12,000)	(4,000)	(4,000)
Employer match contributions		(10,000)	(10,000)	(10,000)
Sub-total Costs	(166,000)	(114,500)	(91,875)	(97,286)
Total Cashflow	(166,000)	(114,500)	(91,875)	(97,286)
Discount Factor	1.000	0.909	0.826	0.751
PV Cashflow	(166,000)	(104,091)	(75,930)	(73,093)
NPV	(883,909)			

Figure 1.6   Cash flow analysis of total expenses (all bells and whistles).

4	5	6	7	8	9	10 TOTALS	
(3,864)	(4,135)	(4,424)	(4,734)	(5,065)	(5,420)	(5,799)	(212,926)
							(16,000)
(3,864)	(4,135)	(4,424)	(4,734)	(5,065)	(5,420)	(5,799)	(228,926)
(3,864)	(4,135)	(4,424)	(4,734)	(5,065)	(5,420)	(5,799)	
0.683	0.621	0.564	0.513	0.467	0.424	0.386	
(2,639)	(2,567)	(2,497)	(2,429)	(2,363)	(2,298)	(2,236)	

hybrid vehicle. Three staff members take advantage of the offer immediately, and it is projected that every year after that, one person will buy a hybrid. Cost of that program over ten years is $48,000. Green, Inc. has a company match of $500 per employee for anyone making a contribution to a non-profit organization, part of its values-oriented focus on community. The cost of this company match is $10,000 per year.

The cost of Green, Inc.'s values-oriented program, fully loaded, over ten years amounts to about $1.39 million. Translated back into Year 0 dollars, the cost is approximately $884,000. A breakdown of these various costs can be seen in Figure 1-6.

Green, Inc.'s investment in a values-oriented program is as follows:

- $166,000 initial investment in the actual values-sharing process
- An additional $63,000 in the ensuing ten years as a result of lost opportunity cost of this year-long valuing exercise.
- An additional $1,157,000 in follow-on programs ("bells and whistles") and lost opportunity costs.

The cost of the baseline program is relatively inexpensive (at $229,000). Granted, Green, Inc. did not have to make the add-on "bells and whistles" investment of

4	5	6	7	8	9	10 TOTALS	
(3,864)	(4,135)	(4,424)	(4,734)	(5,065)	(5,420)	(5,799)	(212,926)
							(16,000)
(40,000)	(40,000)	(40,000)	(40,000)	(40,000)	(40,000)	(40,000)	(400,000)
(45,626)	(52,470)	(60,341)	(69,392)	(79,801)	(91,771)	(105,536)	(609,112)
(4,000)	(4,000)	(4,000)	(4,000)	(4,000)	(4,000)	(4,000)	(48,000)
(10,000)	(10,000)	(10,000)	(10,000)	(10,000)	(10,000)	(10,000)	(100,000)
(103,490)	(110,605)	(118,765)	(128,125)	(138,866)	(151,190)	(165,335)	(1,386,037)
(103,490)	(110,605)	(118,765)	(128,125)	(138,866)	(151,190)	(165,335)	
0.683	0.621	0.564	0.513	0.467	0.424	0.386	
(70,685)	(68,677)	(67,040)	(65,749)	(64,782)	(64,119)	(63,744)	

$1,157,000, but it nevertheless decided to do so. Is the entire investment worth it? Why or why not?

In the same way that I have been conservative in my estimate of the costs of a values-oriented program by *overstating* expenses, I will be conservative in my estimate of value creation by *understating* the value addition. In doing so, I will adopt several assumptions.

First, we'll assume that no value creation occurs in the first five years after the values-shaping program occurs. Change management, as many businesses know, takes time. There's an initial "storming" period before practices become normalized, a period in which some people leave the organization because they're uncomfortable or don't fit with the new orientation. There's also an inevitable lag time before principles are put into practice. For instance, when Melaver, Inc. decided that it would construct all of its buildings to a minimum of LEED standards, it took several years before our first LEED building was certified. It took several years before 80 percent of our staff became LEED accredited. It simply takes time for a company to establish credibility both within the organization and externally, time before beliefs morph into actions that translate into value creation.

Value creation is typically parsed into four major categories, two that tend to be tangible and short term, and two that tend to be more intangible and long term. Those four categories are:

- Reduced costs (short-term and tangible)
- Reduced risks associated with various types of liability (long-term and intangible)
- Increased revenues (short-term and tangible)
- Increased value through brand build and reputation (long-term and intangible)[42]

I want to focus on the value creation that occurs specifically because of a company's ideology and not on the carryover value creation that occurs as a result of practices that are the result of this ideology. That's a tricky judgment call to make, since the two are closely interrelated. Let's look at some examples.

While Green, Inc., like Melaver, Inc., only builds to LEED specifications as a result of its sustainable orientation, the cost savings it will see as a result of greater energy and water efficiencies are not factored in to our calculations. Similarly, although Green, Inc., also because of its LEED orientation, is able to reduce its liability exposure by not building near floodplains, the savings it sees in lower insurance premiums and lower underwriting costs for construction loans is also not factored in. Both of these components result in clear savings for the company and will be addressed elsewhere in this book (see Chapter 4). But because they have to do with what the company *does* as opposed to what the company *values*, I excluded them from this economic analysis.

The third assumption I use in this analysis is that any value creation that cannot be determined concretely will not be considered. So intangible value created through brand build and reputation, while certainly important, is left out of our analysis. It's not that such intangibles are unimportant. Herb Kelleher, former CEO of Southwest

Airlines, is absolutely spot-on when he notes that "the intangibles are more important than the tangibles because you can always imitate the tangibles."[43] But these intangibles have simply not been measured with any analytical discipline.

With these three conservative assumptions in mind, let's consider the value that Green, Inc. can concretely realize as a result of its values-shaping orientation. Two categories are concrete and measurable: value creation through employee retention and value creation through business development.

The average large company today turns over its workforce every four years.[44] While statistics aren't available on the rate of turnover of much smaller companies, I would assume it to be much lower—let's say every ten years, to err on the side of caution. Melaver, Inc. has virtually no turnover, averaging over the course of the past fifteen years around 3 percent annually or complete turnover in thirty-three years. That low turnover rate has everything to do with our values orientation, as we have deliberately built and continue to build our company around the values each person espouses. But again, for the sake of being conservative, let's say Green, Inc. sees complete turnover in its workforce every fifteen years. With twenty people on staff, that means 1.3 persons leave the company each year, as opposed to a more conventional small business, which we are estimating sees two people leave annually.

Bob Willard, who was written about hiring and retention in *The Sustainability Advantage*, estimates that it takes $7,000 to recruit a new employee. It costs, according to the accounting firm KPMG, about $25,000 to replace any worker but according to Willard, there's a premium for replacing a good worker.[45] While I think there is, indeed, a premium to be extracted for retaining a good worker, we will consider this savings in the context of the following chapter, when we analyze more closely the value a green bottom line business can add by creating a culture of shared leadership. With these data points in mind, Green, Inc., because of its values orientation, will save conservatively about $21,000 annually as compared to a similar-sized conventional company. The calculations can be seen in Table 1.3.

TABLE 1.3   COMPARATIVE COST SAVINGS ON RETENTION

ITEM	STANDARD SMALL COMPANY	GREEN, INC.	ANNUAL SAVINGS (DIFFERENCE)
Complete company turnover (# years)	10	15	
Annual loss of personnel (# people)	2	1.333	
Cost of replacing a worker ($25,000)	$50,000	$33,333	$16,667
Cost of recruitment ($7,000)	$14,000	$9,333	$4,667
TOTALS	$64,000	$42,667	$21,333

In addition to realizing savings through greater employee retention, Green, Inc. also will see increased revenues by additional development work. At our own company today, we vet around ten potential third-party development deals each year where we have been directly sought out by an owner because of our values-oriented reputation and our sustainable practices. In 2007 alone, we selected four such projects to develop for others, each providing us with a development fee of 4 percent. In the interest of being conservative, let's consider only two of those projects, costing $15 million each and providing gross income of about $1.2 million ($30,000,000 × 4% = $1,200,000). Granted, there are costs associated with these projects, mostly the cost of labor in overseeing things from start to finish. Moreover, it's difficult to say to what extent we are being approached because of our values-orientation per se versus our experience in developing sustainably. Clearly, if we lacked sustainable development experience, we would not have been considered in the first place. However, while this experience has been a driver in being invited to make a proposal to potential clients, it is the way in which our staff members conduct themselves during the selection process—the way we engage with ourselves and others, the care we take to listen, the way in which we transmit our people-oriented values to a potential client—that closes the deal. Conversely, we won't accept a development deal if we feel there isn't a strong convergence of our values and those of our potential client.

Again, for the sake of being conservative, let's say that of the 4 percent development fee we receive on any one of these projects, a quarter (25 percent) of it is eaten up by the cost of labor and time, thus reducing it to 3 percent net income ($900,000). Let's also say that of the 3 percent net fee, fully half of it is attributable to our experience with sustainable development with the other half ($450,000) attributable to our values orientation. Finally, let's assume that 2007 was something of an anomaly in doing four such projects and that in future years, Green, Inc. will be successful in garnering only two $15 million development projects each year. In short, Green, Inc., as a direct result of its

Years	0	1	2	3
REVENUES/SAVINGS				
Additional Development		0	0	0
Reinvestment of Added Revenue from yr 1			0	0
Reinvestment of Added Revenue from yr 2			0	0
Reinvestment of Added Revenue from yr 3			0	0
Reinvestment of Added Revenue from yr 4			0	0
Reinvestment of Added Revenue from yr 5			0	0
Reinvestment of Added Revenue from yr 6			0	0
Reinvestment of Added Revenue from yr 7			0	0
Reinvestment of Added Revenue from yr 8			0	0
Reinvestment of Added Revenue from yr 9			0	0
Savings from Retention		21,300	21,300	21,300
Sub-total Revenues		21,300	21,300	21,300

Figure 1.7   Total revenue from values-centric orientation.

values orientation, will receive additional income of $450,000 in Years 6 through 10 of operations (remember, we are assuming no additional revenues for the first five years).

One additional revenue line needs to be factored in to our analysis. The additional revenue accruing to the company's development work ($450,000 annually) will be re-invested in the company, presumably providing a return consonant with the company's hurdle rate of 15 percent. Total revenue stream from the company's values-centric orientation of roughly $2.8 million can be seen in Figure 1-7.

A few final comments on the assumptions underlying the revenue for Green, Inc., since my colleagues and I debated these figures interminably and because these assumptions will wind their way throughout the remainder of this book. It could be argued that an alter ego to Green, Inc., one of similar size and nature but conventional in its approach, would garner the same amount of revenues from doing conventional third-party development work. It is our contention, however, that Green, Inc., because of its green orientation, has successfully differentiated itself in the market to garner work that simply would not have been available to it, were it practicing conventional real estate. This is certainly true in terms of Melaver, Inc.'s actual experience. We are afforded the opportunity to do work that simply would not have come our way as a conventional developer. Secondly, we could have reduced our assumption to just one green additional project each year rather than two. As a theoretical exercise, this more conservative approach appeals to us, simply because realizing one new third-party deal each year (rather than two) has the appeal of truly minimizing our upside assumptions. As a matter of interest, if we assume just one additional third-party development deal each year starting in Year 3 (as opposed to Year 6), the overall returns cited throughout the remainder of this book are equivalent to the ones we decided to report. However, once again, we decided to stick with our actual experience, prolonging the initial time it takes for revenue-generation to kick in but assuming two new third-party deals each year in Years 6 through 10, rather than just one. In short, our assumptions for Green, Inc., mimic the actual experience of Melaver, Inc.

4	5	6	7	8	9	10	TOTALS
0	0	450,000	450,000	450,000	450,000	450,000	2,250,000
0	0	0	0	0	0	0	0
0	0	0	0	0	0	0	0
0	0	0	0	0	0	0	0
0	0	0	0	0	0	0	0
0	0	0	0	0	0	0	0
0	0	0	67,500	10,125	1,519	228	79,372
0	0	0	0	67,500	10,125	1,519	79,144
0	0	0	0	0	67,500	10,125	77,625
0	0	0	0	0	0	67,500	67,500
21,300	21,300	21,300	21,300	21,300	21,300	21,300	213,000
21,300	21,300	471,300	538,800	548,925	550,444	550,672	2,766,640

ASSEMBLING THE DATA

Now let's assemble the various data. We started with a fictionalized company, Green, Inc., which closely resembles Melaver, Inc. We looked at the various first costs associated with shaping the company's values as well as the carryover costs associated with the development of various programs that support those values. We then turned our attention to value creation. We limited this value creation to the last five years of a ten-year analysis on the assumption that it would take a significant amount of time before beliefs translated into practices that would accrue to the bottom line. We further limited this value creation by including only those elements we could link directly to the values orientation of the company—employee retention and additional development work. All of the numbers and assumptions utilized throughout this analysis are based on actual experience at Melaver, Inc. With that summary in mind, we can see the performance of Green, Inc. in the ten years following its values-shaping process (Figure 1-8).

I realize that there are a lot of numbers to absorb in this analysis. First, the net present value (NPV) is a positive number ($441,189), which indicates that the process Green, Inc. has gone through makes financial sense. Indeed, its internal rate of return

Years		0	1	2	3
REVENUES/SAVINGS					
Additional Development			0	0	0
Reinvestment of Added Revenue from yr 1				0	0
Reinvestment of Added Revenue from yr 2				0	0
Reinvestment of Added Revenue from yr 3				0	0
Reinvestment of Added Revenue from yr 4				0	0
Reinvestment of Added Revenue from yr 5				0	0
Reinvestment of Added Revenue from yr 6				0	0
Reinvestment of Added Revenue from yr 7				0	0
Reinvestment of Added Revenue from yr 8				0	0
Reinvestment of Added Revenue from yr 9				0	0
Savings from Retention			21,300	21,300	21,300
Sub-total Revenues			21,300	21,300	21,300
Years		0	1	2	3
COSTS					
Lost Opportunity		(150,000)	(22,500)	(3,375)	(3,611)
Facilitator		(16,000)			
Continuing education			(40,000)	(40,000)	(40,000)
Lost opportunity from cont. ed.			(30,000)	(34,500)	(39,675)
Hybrids			(12,000)	(4,000)	(4,000)
Employer match contributions			(10,000)	(10,000)	(10,000)
Sub-total Costs		(166,000)	(114,500)	(91,875)	(97,286)
Total Cashflow		(166,000)	(93,200)	(70,575)	(75,986)
Discount Factor		1.000	0.909	0.826	0.751
PV Cashflow		(166,000)	(84,727)	(58,326)	(57,090)
NPV		441,189			
IRR		22.50%			

Figure 1.8   Projected discounted cash flow for Green, Inc.

(IRR) is 22.5 percent, which is above the company's threshold return (or hurdle rate) for investments of 15 percent. When a company weighs the pros and cons of a particular deal, it typically wants to know, among other things, if the project makes financial sense. In the case of Green, Inc., it wants to know this: If we engage in a process of values-shaping, is our investment of time and money and human capital worth it, or would we be better off investing our time and money and human capital elsewhere? The answer here is, yes, the investment is worth it. Had Green, Inc. decided to remain a conventional real estate company and not engage in a values-shaping process, it would have simply seen returns at or around its hurdle rate of 15 percent.

In actuality, the value creation to be realized by Green, Inc. is significantly higher than the analysis provided above. I've simply been extremely conservative in my use of assumptions, not taking into account a number of other value-added elements that do, in fact, elevate the value of a values-oriented company. We see savings from our energy- and water-management programs; our insurance premiums and cost of construction financing are lower than those a conventional development company would face; and our own pipeline of development business is greater than that projected by

4	5	6	7	8	9	10	TOTALS
0	0	450,000	450,000	450,000	450,000	450,000	2,250,000
0	0	0	0	0	0	0	0
0	0	0	0	0	0	0	0
0	0	0	0	0	0	0	0
0	0	0	0	0	0	0	0
0	0	0	0	0	0	0	0
0	0	0	67,500	10,125	1,519	228	79,372
0	0	0	0	67,500	10,125	1,519	79,144
0	0	0	0	0	67,500	10,125	77,625
0	0	0	0	0	0	67,500	67,500
21,300	21,300	21,300	21,300	21,300	21,300	21,300	213,000
21,300	21,300	471,300	538,800	548,925	550,444	550,672	2,766,640

4	5	6	7	8	9	10	TOTALS
(3,864)	(4,135)	(4,424)	(4,734)	(5,065)	(5,420)	(5,799)	(212,926)
							(16,000)
(40,000)	(40,000)	(40,000)	(40,000)	(40,000)	(40,000)	(40,000)	(400,000)
(45,626)	(52,470)	(60,341)	(69,392)	(79,801)	(91,771)	(105,536)	(609,112)
(4,000)	(4,000)	(4,000)	(4,000)	(4,000)	(4,000)	(4,000)	(48,000)
(10,000)	(10,000)	(10,000)	(10,000)	(10,000)	(10,000)	(10,000)	(100,000)
(103,490)	(110,605)	(118,765)	(128,125)	(138,866)	(151,190)	(165,335)	(1,386,037)
(82,190)	(89,305)	352,535	410,675	410,059	399,254	385,336	
0.683	0.621	0.564	0.513	0.467	0.424	0.386	
(56,137)	(55,451)	198,997	210,741	191,296	169,322	148,564	

Green, Inc. The question regarding a values-oriented company is not does it pay, but rather, does it pay to proceed any other way?

Concluding Remarks

I'd like to conclude this chapter where I began, with my father's story of the businessman and the fisherman. The story asks us to consider what is truly important to us and if we are doing the important things now or deferring them indefinitely. For me, it is important to recognize that we, as a company, have a hand in determining what type of story we wish to embrace about the role of capitalism and its consequences, not only in terms of financial well-being, but also of the well-being of our land and our community. We can blithely accept a narrative of indifferent capitalism, keep making money, and tell ourselves that someday we'll change if government regulations and market forces force us to. Or we can embrace the notion of capitalism with a difference and take it upon ourselves to change now. In so doing, we have the capacity to envision a more just business, linking work and meaning and purpose. And we have the chance to reflect upon and respond to the crises of the present time, sidestep the dangers of old narratives that have run their courses, and avail ourselves of the opportunities offered by a story of a different hue and stripe.

NOTES

[1] Hayden White, *Metahistory: The Historical Imagination in Nineteenth-Century Europe* (Baltimore: The Johns Hopkins University Press, 1975).

[2] Amos Tversky and Daniel Kahneman, "Judgment under uncertainty: Heuristics and biases," Daniel Kahneman, Paul Slovic, and Amos Tversky (eds.), *Judgment Under Uncertainty: Heuristics and Biases* (Cambridge: Cambridge University Press, 1982). Kahneman won the Nobel Prize in economics for his work, the only psychologist to do so. (Tversky had since died and so did not share in the award.)

[3] Jeremy Rifkin, *The European Dream: How Europe's Vision of the Future Is Quietly Eclipsing the American Dream* (New York: Jeremy P. Tarcher/Penguin, 2004), p. 236.

[4] The narrative of indifferent capitalism has been synthesized from the following sources: Jeremy Rifkin, *The European Dream: How Europe's Vision of the Future Is Quietly Eclipsing the American Dream* (New York: Jeremy P. Tarcher/Penguin, 2004); Robert B. Reich, *Supercapitalism: The Transformation of Business, Democracy, and Everyday Life* (New York: Alfred A. Knopf, 2007); Eric T. Freyfogle, *The Land We Share: Private Property and the Common Good* (Washington, D.C.: Island Press, 2003); Naomi Klein, *No Logo: Taking Aim At the Brand Bullies* (New York: Picador, 1999); David C. Korten, *When Corporations Rule the World* (Bloomfield, Conn. and San Francisco: Kumarian Press & Berrett-Koehler Publishers, 1995); Bill McKibben, *Deep Economy: The Wealth of Communities and the Durable Future* (New York: Henry Holt & Co., 2007); Thomas Friedman, *The Lexus and the Olive Tree: Understanding Globalization* (New York: First Anchor Books, 2002).

[5] Robert B. Reich, *Supercapitalism: The Transformation of Business, Democracy, and Everyday Life* (New York: Alfred A. Knopf, 2007), p. 126.

[6] Richard Sennett, *The Corrosion of Character: The Personal Consequences of Work in the New Capitalism* (New York: W. W. Norton, 1998), pp. 119, 122.

[7] Bill McKibben, *Deep Economy: The Wealth of Communities and the Durable Future* (New York: Henry Holt & Co., 2007) pp. 95–96.

[8] David C. Korten, *When Corporations Rule the World* (Bloomfield, Conn. and San Francisco: Kumarian Press & Berrett-Koehler Publishers, 1995), p. 84.

[9] See John Abrams, *The Company We Keep: Reinventing Small Business for People, Community, and Place* (White River Junction, Vt.: Chelsea Green, 2005), p. 158

[10] Paul Hawken, *Growing a Business* (New York: Fireside Books, 1987), pp. 52, 93.

[11] Marc Gunther, *Faith and Fortune: The Quiet Revolution to Reform American Business* (New York: Crown Publishing, 2004); Nikos Mourkogiannis, *Purpose: The Starting Point of Great Companies* (New York: Palgrave Macmillan, 2006).

[12] Yvon Chouinard, *Let My People Go Surfing: The Education of a Reluctant Businessman* (New York: The Penguin Press, 2005).

[13] Bo Burlingham, *Small Giants: Companies That Choose to Be Great Instead of Big* (New York: Portfolio, 2005); John Abrams, *The Company We Keep: Reinventing Small Business for People, Community, and Place* (White River Junction, Vt.: Chelsea Green, 2005); Michael Shuman, *The Small-Mart Revolution: How Local Businesses Are Beating the Global Competition* (San Francisco: Berrett-Koehler Publishers, Inc., 2006).

[14] Andrew Savitz and Karl Weber, *The Triple Bottom Line: How Today's Best-Run Companies Are Achieving Economic, Social, and Environmental Success—and How You Can Too* (San Francisco: Jossey-Bass, 2006); Daniel C. Esty and Andrew S. Winston, *Green to Gold: How Smart Companies Use Environmental Strategy to Innovate, Create Value, and Build Competitive Advantage* (New Haven, Conn.: Yale University Press, 2006); Ray Anderson, *Mid-Course Correction: Toward a Sustainable Enterprise: The Interface Model* (Atlanta: The Peregrinzilla Press, 1998); Brian Natrass and Mary Altomare, *The Natural Step for Business: Wealth, Ecology, and the Evolutionary Corporation* (Gabriola Island, B.C.: New Society Publishers, 1999); Brian Natrass and Mary Altomare, *Dancing With the Tiger: Learning Sustainability Step by Natural Step* (Gabriola Island, B.C.: New Society Publishers, 2002).

[15] Ted Nordhaus and Michael Shellenberger, *Break Through: From the Death of Environmentalism to the Politics of Possibility* (New York: Houghton Mifflin Co., 2007), pp. 8–9, 11, 257–60, 228–30. See also Amory Lovins, *Winning the Oil Endgame: Innovation for Profits, Jobs, and Security* (Snowmass, Colo.: Rocky Mountain Institute, 2005).

[16] Robert L. Reich, *Supercapitalism: The Transformation of Business, Democracy, and Everyday Life* (New York: Alfred A. Knopf, 2007), p. 211.

[17] David W. Orr, *The Nature of Design: Ecology, Culture, and Human Intention* (New York: Oxford University Press, 2002), pp. 50–51.

[18] Donella Meadows, "Places to Intervene in a System," *Whole Earth Magazine*, Winter 1997.

[19] Daniel C. Esty and Andrew S. Winston, *Green to Gold: How Smart Companies Use Environmental Strategy to Innovate, Create Value, and Build Competitive Advantage* (New Haven, Conn.: Yale University Press, 2006), pp. 67–68.

[20] Paul Hawken, *Blessed Unrest: How the Largest Movement in the World Came into Being and Why No One Saw It Coming* (New York: Viking, 2007).

[21] John Abrams, *The Company We Keep: Reinventing Small Business for People, Community, and Place* (White River Junction, Vt.: Chelsea Green, 2005), p. 22.

[22] Daniel C. Esty and Andrew S. Winston, *Green to Gold: How Smart Companies Use Environmental Strategy to Innovate, Create Value, and Build Competitive Advantage* (New Haven, Conn.: Yale University Press, 2006), p. 14.

[23] Andrew Savitz and Karl Weber, *The Triple Bottom Line: How Today's Best-Run Companies Are Achieving Economic, Social, and Environmental Success—and How You Can Too* (San Francisco: Jossey-Bass, 2006), p. 15.

[24] Steven D. Levitt and Stephen J. Dubner, *Freakonomics: A Rogue Economist Explores the Hidden Side of Everything* (New York: HarperCollins, 2005), p. 13.

[25] James C. Collins and Jerry I. Porras, *Built to Last: Successful Habits of Visionary Companies* (New York: HarperCollins, 1994); Rob Goffee and Gareth Jones, "What Holds the Modern Company Together?" *Harvard Business Review*, Nov.–Dec., 1996.

[26] Peter M. Senge, *The Fifth Discipline: The Art & Practice of The Learning Organization* (New York: Doubleday, 1990), p. 231.

[27] The synthesis of being and becoming is often attributed to Alfred North Whitehead, *Science and the Modern World* (New York: Free Press, 1967).

[28] Cited in Marc Gunther, *Faith and Fortune: The Quiet Revolution to Reform American Business* (New York: Crown Publishing, 2004), p. 124.

[29] Paulo Coehlo, *The Alchemist* (New York: HarperCollins, 1988).

[30] Lionel Trilling, *Sincerity and Authenticity* (Cambridge, Mass.: Harvard University Press, 1971), p. 1.

[31] Peter Senge, The Fifth Discipline: The Art & Practice of The Learning Organization (New York: Doubleday, 1990), p. xv.

[32] Frans Johansson, The Medici Effect: What Elephants and Epidemics Can Teach Us About Innovation (Boston: Harvard Business School Press, 2006), pp. 7, 47, 67–72, 82–3.

[33] See, for example, two otherwise very different pieces that nevertheless agree that baseline economic needs have to be met before environmental issues can be addressed: Bill McKibben, Deep Economy: The Wealth of Communities and the Durable Future (New York: Henry Holt & Co., 2007); Ted Nordhaus and Michael Shellenberger, Break Through: From the Death of Environmentalism to the Politics of Possibility (New York: Houghton Mifflin Co., 2007).

[34] Thomas L. Friedman, The World Is Flat: A Brief History of the Twenty-first Century (New York: Farrar, Straus, & Giroux, 2005).

[35] Richard Florida, The Rise of the Creative Class: And How It's Transforming Work, Leisure, Community, and Everyday Life (New York: Basic Books, 2002), pp. 15, 269.

[36] Richard Sennett, The Corrosion of Character: The Personal Consequences of Work in the New Capitalism (New York: W. W. Norton, 1998), p. 146.

[37] Max De Pree, *Leadership Is an Art* (New York: Doubleday, 1987).

[38] Richard Sennett, *The Culture of the New Capitalism* (New Haven, Conn.: Yale University Press, 2006), pp. 195–6.

[39] Max De Pree, *Leadership Is an Art* (New York: Doubleday, 1987), p. 42.

[40] *Ibid.,* p. 22

[41] David W. Orr, *The Nature of Design: Ecology, Culture, and Human Intention* (New York: Oxford University Press, 2002), pp. 104–117.

[42] Daniel C. Esty and Andrew S. Winston, *Green to Gold: How Smart Companies Use Environmental Strategy to Innovate, Create Value, and Build Competitive Advantage* (New Haven, Conn.: Yale University Press, 2006), pp. 101–4.

[43] Cited in Marc Gunther, *Faith and Fortune: The Quiet Revolution to Reform American Business* (New York: Crown Publishing, 2004), p. 72.

[44] Michael Shuman, *The Small-Mart Revolution: How Local Businesses Are Beating the Global Competition* (San Francisco: Berrett-Koehler Publishers, Inc., 2006), p. 87.

[45] Bob Willard, *The Sustainability Advantage: Seven Business Case Studies of a Triple Bottom Line* (Gabriola Island, B.C.: New Society Publishers, 2002), p. 34.

"GREEN GLUE": HR PRACTICES AND PROCESSES THAT MAKE SUSTAINABLE VALUES STICK

ZELDA TENENBAUM

SUMMARY

This chapter describes how a company builds a cohesive culture once its value system is in place. Told from the perspective of Zelda Tenenbaum, Melaver, Inc.'s outside human resources (HR) consultant for over a decade, the chapter focuses on the company's realization, early on, that in order to be a viable profitable business focused on the health and well-being of the larger community, it first had to be a self-sustaining community or culture in its own right. While a conventional business has a strategic blueprint that it executes linearly, the green bottom line company has a greenprint, an integrated approach to building culture that drives its strategic plan. The focus of this chapter is this greenprint: a disciplined approach to culture creation through myriad HR processes.

The HR processes addressed are organized around four areas critical to culture building. Section one considers governance issues such as the shaping of concrete shared values, organizational design, collective envisioning, empathic leadership, decision making, and shared leadership. Section two looks at the various rituals that shape a green bottom line culture. Section three looks at the special language of a green bottom line culture, which is as much a language of practicing values as it is a tangible language of rhetoric. Section four takes into account the history of a business' culture, focusing on the synthesis of change and continuity. In section five, the author looks at the costs and benefits of the glue binding a green culture, updating the financial analysis begun in Chapter 1. In the final section, she summarizes her study.

Building a Culture of Community

How does a fast-moving, innovative, values-driven company that is focused on the well being of the land and community around it take care of its own small community of people? Can it afford to spend time (and money) on so-called soft issues when there is so much work out there to be done? Can it afford not to? Can a company aspiring to sustainable real estate practices be sustainable if it is constantly turning over its workforce or otherwise engaging in practices that do not nurture and cherish its own human capital?

This chapter addresses such questions, telling the story of how Melaver, Inc. developed the culture of community now in place. The company journey required leadership's awareness of the company's future direction, the provision of internal resources needed to follow that direction, and the ability to recruit and develop the right people to get them there. My role as a consultant included guiding Melaver, Inc.'s management team toward an assessment of individual and corporate values, creation of a values-centric organizational design, articulation of a shared vision, methodologies for working collaboratively despite distinct and diverse management styles, and the development of HR processes that would move the company toward its vision of becoming a thought and product leader in sustainable practices, internally and externally. As the journey evolved, people went through personal transitions that required leaders to learn how to coach their colleagues through change as well as manage such change themselves.

My work started at the top with Martin Melaver, CEO of the company, and then broadened over time to include every staff member at Melaver, Inc. I engaged in one-on-one coaching with the management team, worked to develop a performance team, facilitated meetings and retreats, and collaborated with other Melaver partners. My focus has been to assist this innovative company to become the sustainable pioneer and learning company it aspired to be as well as to ensure that leaders focused on people and process as they decided *what* to become and *how* to become what they envisioned. This chapter speaks to years of dedicated, diligent work by the Melaver team. Their work was accomplished at day-long meetings, annual retreats, in one-on-one coaching and feedback sessions, and by internally driven team efforts.

Problems and Opportunities

Change. Such a funny thing. It's everywhere around us, part of our daily lives. A part of us feels it and resists. And a part of us isn't even aware of it, so gradually does it seem to occur. And then blink—we're part of a new way of thinking and acting and belonging. And almost as instantaneously, it feels as if this new world has been with us always, like a well-worn pair of jeans that feels good on us. Did you know that it took the United States only twelve years to go from a culture where 90 percent of the population rode horses and only 10 percent drove cars to the reverse? Twelve years. We switched from using leaded gasoline in our vehicles and unregulated emissions to catalytic converters in the same span of time. In fifteen years, we replaced vacuum tubes in our radios with transistors and discarded black and white televisions for color. It took twenty-six years for air travel to become the dominant means of transportation among cities. And it took forty years for us to shift from a wood-burning civilization to coal, and then from coal to petroleum and nuclear. We forget how fast our culture can change, how fast our culture is always in the process of changing. Often in the blink of an eye. And part of our forgetfulness has to do with the strange aptitude we have for assuming that the way things are *now* is the way things have been for a very long time.

I find this to be true in my own work as well. I look at Melaver, Inc. today, a company that still has its challenges to be sure, but one that is made up of a passionate group of people, working as a team that is collectively and effectively focused on being a thought and product leader in sustainable real estate. They have been working cohesively toward realizing that vision for so long now that it feels as if that has always been the case. It hasn't. The company's evolution, which seems so smooth and effortless in hindsight, has been a painstaking process of managing change internally.

Managing change internally—a multi-year focus on practices and processes to build community—has enabled the company to promote changes externally in the community at large. Becoming a company focused on a green bottom line is based on building a strong culture of community inside the company. For more than ten years, my involvement with Melaver, Inc. has focused on shaping that strong culture of community and on creating the green glue that enables sustainable values to stick. This is a story of that green glue.

Consultants typically begin by listening and observing and making assessments. As I go back through my files, I am reminded of many of the problems—in HR-speak we call these problems "opportunities"—facing the company in the mid-1990s. If real estate is all about "location, location, location," then the problems facing this real estate company over a decade ago were all about "communication, communication, communication":

- Lack of a well-articulated vision commonly shared by all members of the company.
- A disconnect between the CEO's ostensible desire to have everyone share in the vision and a sense among staff members that at the end of the day, the CEO really was intent on doing what he wanted.

■ Fear of change within the organization, and an inability for staff members to articulate those fears in an open and trustful environment.

■ Lack of capacity among all staff members to brainstorm in ways that would facilitate critical, innovative thinking, without the process being intimidating.

■ Lack of alignment of some staff members with the sustainable direction of the company, with the prospect of some members of the leadership team leaving the company as a result.

■ Lack of clear strategic and tactical planning to realize a common vision and to hold individuals accountable for desired objectives.

■ Confusion as to roles and responsibilities, given the ethos of shared leadership within the organization.

■ Chronic frustration, among many at the company, of the bar being constantly raised, with no end in sight. (Such raising of the bar was also occurring in the middle of projects, making it challenging and costly to keep up with evolving standards.)

■ A sense that the shared leadership ethos was making decision making tedious and inefficient and confusing, and confusion about who really got to make which decisions.

■ Presence of significant personnel "noise," with staff members talking *about* each other, not *to* each other.

While the company was imbued with a strong set of values (shared leadership, a sense of meaning and purpose, etc.), those values lacked a set of practices and processes so that they could be effectively deployed. *Where is the stick-to-itiveness, the direction, the team alignment?* I thought. After facilitating my first Melaver leadership team meeting, I had the impression that there was little agreement, only unclear direction, and definitely no process on bringing team members together to accomplish their goals.

That first meeting also made me aware that something great was happening. I could tell these people wanted to be pioneers to lead the way as a unique and different company that wanted to do things right, sustainably. However, people on the team did not have the skills to communicate effectively, accept each other's different perspectives, and handle conflicts. Barriers were keeping this group of people from being all that a team could be. I wanted to help them figure out who they are and who they want to be, assist them in behaving professionally with one another, and influence them toward process to get results. I wanted them to have fun as a fast, innovative-thinking group. I intuitively knew they could become a real team of people working together to make a difference. With that assessment in mind, my role was fairly clear, if somewhat daunting:

■ Clarify the design of the organization to ensure better alignment and accountability.
■ Coach the leaders toward personal growth and improved people management skills.
■ Put HR processes into place to facilitate practice of key company values.
■ Develop the individuals into a high-performance team.
■ Support people through change and transition as the company evolved.
■ Facilitate conflict situations to be collectively managed constructively.
■ Enable the staff members to find joy in the purposeful work they were engaged in.
■ Connect the future changes team members would institute to continuity with history.
■ Improve communications internally.

CULTURE BUILDING

Addressing these issues amounts to culture building. Anthropologists tend to define a culture as a group of people with a shared sense of history, a language they hold in common, rituals that the group engages in, and rules of engagement that provide an overall governance structure. Those are the ties that make a community from a loose confederation of individuals. These were the same elements a business such as Melaver, Inc. needed to bind together to create a lasting community. Let's consider each element in greater detail.

Issues of Governance

The *Oxford English Dictionary* defines governance as proceedings or doings in business having to do with one's mode of living, behavior, and demeanor. In everyday parlance, we think of governance as the basic rules of the road: those ethical codes that guide wise behavior, the way a system is ordered, the various roles we all play in that organization, and the delegation of rights, responsibilities, and authority. All of these various facets of a system help shape the way a culture looks and feels and behaves. In the case of a values-oriented company focused on a green bottom line, how such facets of a system are constructed is indeed critical. Governance includes four key elements: values, organizational structure, leadership, and decision making.

VALUES

If you check the website of Melaver, Inc.[1] you will find wonderful stories that tell how it evolved into the sustainable development company it is today. One of my favorites addresses the values of Melaver, Inc.'s predecessor company, M&M Supermarkets (1940–1985):

> During the first forty-five years in business as a supermarket operation, the term core values didn't exist at Melaver. Values were simply the way we all worked and lived. "Do the Right Thing" was the litmus test for everything. The rest just fell into place.
>
> A few years back, we sat down as a company over a period of six months and asked each individual employee what mattered most. We distilled the long list to four values that matter most to us: ethical behavior, learning, service, and profitability. What is amazing is how precisely these values match how the family has managed its business for the past sixty years.
>
> Melaver, Inc. is in the service business. We always have been. We've always pushed ourselves to innovate, learn, and of course to profit by constantly improving upon the services we provide. But we also believe that our community profits from the ways we give back to it. Giving back is all about doing the right thing. This leads back to being a service company. Do Right. Learn. Serve. Profit. It's a closed loop.

There's a lot I find appealing in this story. It gives a strong nod to the past, and in so doing it links the changes the company is undergoing to an underlying foundation that speaks to continuity. It recognizes that values are an intrinsic part of the way this company has always conducted business, the air it breathes. And it speaks to the inclusionary nature of the process of articulating values for the current generation of staff members.

Because the issue of values has been addressed in great detail in the previous chapter, I don't feel there is a need to say much more about the topic here. I would like to point out that these values—and the democratic processes by which there were derived—inform everything else the company does. Hires are made on the basis of fit with these core values. So too are decisions regarding which vendors to use, other companies that Melaver, Inc. might consider engaging in joint ventures with, even such seemingly matter-of-fact business decisions as choosing a debt provider or an insurance underwriter. Price takes a back seat to values. I have often heard team members say, "The most expensive course of action we can take is simply choosing the cheapest option." Values, by contrast, are priceless.

ORGANIZATIONAL STRUCTURE

What does a values-driven organization look like? How is it organized? Does it look like and function as a typical business? Or should it be structured differently? Strange questions for a business to be asking itself. At Melaver, Inc., the management team of approximately ten staff members consists of C-level leaders (CEO, COO, CFO) plus various divisional heads. One of my first steps after working with the entire company on core values was to work specifically with this management team (called "Sanskrit," the origin of which is explained later on) on how the company should be organized.

Circling the Organizational Square

In early meetings to create an organizational chart, the management team drew all kinds of wild and wacky designs. Sure, there was the inevitable tree-branched hierarchy, starting with the CEO and trailing down to the COO, then the Sanskrit management team, and downward through the various reporting relationships. But most of the designs sketched out by the management team shared a basic common denominator: everyone reported first and foremost to the company's core values. From there, accountability was toward the company's various stakeholders (customers, shareholders, vendors, other colleagues, etc.). At the very bottom of the hierarchy, it was felt, was Sanskrit—the management team—itself. The end result was a concentric circle organizational diagram, best viewed with 3-D glasses (see Figure 2.1). The result was so unusual that the team was not sure they could present it to the board of the company!

As you can see from the new universe created, Melaver is a leading edge company that wants to place the company's values in the center (or core) of the company universe. To accomplish outcomes in the business units, the employees flow through the company's universe led by the values. Most companies have organization structures

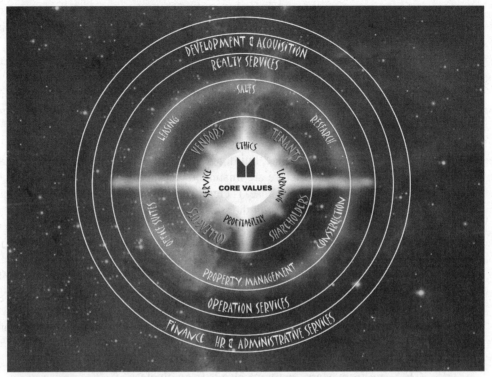

Figure 2.1   Melaver, Inc.'s 3-D organizational diagram. The company refers to the area beyond the outer circle as "The Sanskrit Galaxy."

that are linear in design, with a clear hierarchy. Melaver chose to avoid the traditional structure to make everyone accountable for outcomes.

The company would be flat, with everyone operating as equals. The team developed clear lines where necessary, but decided everyone would be "NTs" (No Titles). This decision about no titles was one that did not stick in application, as it confused clients and outsiders, but the concept still applies internally in the ongoing operations of the company. Titles exist and people are in charge of projects, but people at Melaver, Inc. have equal input into decisions. But having a non-linear, flat, non-hierarchical structure creates other challenges, such as communicating effectively and deciding who is responsible for what. Does having equal input into decisions mean that leadership is spread evenly across an organization?

LEADERSHIP

Fortune magazine, in its 1998 roundup of America's most admired companies,[2] identified leadership as an important characteristic of exemplary organizations. Companies being noticed and making a difference have leaders who are willing to find a way of conducting business through processes that are aligned with their true belief systems. Having the *right* people in leadership positions is key. Having *alignment* among those leaders takes work.[3]

A leader's job is to develop people and create a real team that accomplishes the mission of the company. Therefore, as a coach, I asked the Melaver leadership team to assess their beliefs, leadership styles, and the impacts they were having on each person on the team. Some decided they were not aligned with the new direction, some were unaware of the impact they had on others. It's not uncommon. You have a visionary leadership executive or team and a superb group of dedicated people working with that leadership team to execute, build, and maintain what's being envisioned. Also not uncommon in such a structure are misunderstandings.

The CEO thinks he (in this case) is clearly articulating a vision for the company over and over, *ad nauseam*. The rest of the staff just doesn't get it. Or do they? The staff may think the CEO is focusing too much on the long-term future and missing out on the critical problems in the here and now. The CEO just doesn't seem to get it. Or does he? Such misunderstandings are compounded in a values-centric company where there is a considerable amount of shared leadership. The CEO throws ideas out on the table in the interest of brainstorming about the long-term direction of the company. Others around the table seem hesitant. The CEO thinks the team is being resistant to change and gets frustrated. By contrast, the team wonders what in the hell their CEO wants from them. He keeps throwing out various ideas. *Sure wish he would make up his mind. If he really wants our input, why does he get so frustrated when people challenge him?*

CEOs are from Mars, everyone else is from clueless.

The CEO's Vision

Early in my work with Melaver, Inc., I asked the management team to think about a vision for the company and to try to articulate it at our next get-together. I got one response, from CEO Martin Melaver. What he wrote—almost ten years ago—and the responses from his staff are revealing.

There once was an amazing child of boundless curiosity. S/he was always asking "why" and "why not," always challenging, learning, and growing. This child was also wise.

And s/he knew that, to others, s/he would not always look like a child of boundless curiosity. S/he might look like an astronaut one day, or a teacher, or a businessman. And that was OK. But s/he wanted to take something along on this journey into becoming something else, so that s/he would always be reminded of the child of boundless curiosity that s/he really was.

The CEO's Vision (*continued*)

S/he thought for a long time. S/he was open to what other people thought and advised. S/he listened to her dreams. And then one day, seemingly out of nowhere s/he was given the gift of a kaleidoscope. And, of course, like most wondrous and magical gifts, this was not an ordinary kaleidoscope.

It was a kaleidoscope of extraordinary beauty and exquisite craftsmanship, though it looked quite plain when you first looked at it. The materials were simple, but seemed unusual and atypical. There were only four distinctive colors of glass that made up the inside soul of this object. And yet, when these four colors moved together in an ever-changing display of unique designs, it was almost impossible to say where one color stopped and another began. It was as if the colors were each distinct—and yet blended together in some indescribable harmony.

But as s/he began to use this kaleidoscope, s/he began to notice and feel things s/he had missed before. S/he noticed how much others benefited, when they too had a chance to put the kaleidoscope to use. S/he noticed that the community around her seemed to become richer as they all shared in the passion s/he had. And it struck her: How s/he used this gift became as important as what s/he could do with it.

It was all so strange at first. From the beginning, s/he could see things with the kaleidoscope that s/he felt no one else could see. And yet, when neighbors and friends and even chance acquaintances picked up the kaleidoscope, they seemed to see things differently. And so, using the kaleidoscope, s/he began to commit to understanding how others saw the world. And s/he began to commit to serving their needs as best s/he could.

The child's desire to take what s/he knew and what s/he learned and what s/he saw and apply it to the service of others was overwhelming. People came from all over to be a part of what the child had built.

As word spread and people came, the child began to experience things s/he had not seen before. One stranger advised that the child replace the expensive and rare glass pieces of the kaleidoscope with materials that did not need any care or maintenance. Another stranger came to her and suggested that s/he make many more, similar kaleidoscopes, so that many others might profit as well. And then there were so many people that seemed to expect that the child service their needs, no matter what they were.

And at such times, the child would gaze into the special lens s/he had been given, thinking about the numberless patterns it showed, until s/he knew deep down the right thing to do.

The CEO's Vision (*continued*)

Interpreting the Story

The story provides a glimpse of Martin's visionary leadership style, his way of communicating. He tends to convey meaning through storytelling, often speaking in metaphors and allegories. His conceptual communication is not always appreciated or understood.

I understood that the four colors of this magical kaleidoscope are the company's four core values—ethical behavior, learning, service, and profitability. Martin's vision for the company, told very obliquely, is to see the world through these four lenses with the wonderment and idealism of the child, to not be fearful of the capacity that we have as children (and often lose as adults) to ask *why* and *why not*. Still, how effective is this as a communication strategy? I often watched the rolling eyes of Sanskrit team members as Martin leaped into his visionary world while his teammates asked him practical questions: *What on earth do you mean by that? How are we going to achieve this? What are we going to do tomorrow?* Or made statements like, *We have real work to do, Martin.*

My role as facilitator was clear. I needed to bring this team together to appreciate both visionary and administrative leadership styles. Both were needed, but neither spoke the same language or understood the other. As we continued to work together, this strange company and I, the need for development of people became even more apparent to me. So I began coaching Martin and his leadership team, using assessment tools to create awareness, communication tools to create a better working environment for all, and evaluation tools to track how they were doing.

Leadership Development

A critical first step in the process was that each person on the management team undergo leadership development training, to assist with both self-understanding and understanding of others. An important feature of this leadership training would be learning how to seek out and accept feedback from colleagues. Other important features of this training included:

- **Desire to grow and mature.** This is easier said than done, since most of us will pay lip service to the desire to develop. But how truly open are we to this, even within a company that has learning as one of its core values?
- **Seek to understand before seeking to be understood.** The concept comes from Stephen R. Covey's *The Seven Habits of Highly Effective People*, a book the entire management team read and discussed.

- **Perception is reality.** Learning to listen and accept feedback in others required each person on the management team to comprehend this concept deeply, which is not an easy thing to do.
- **The perfect is the enemy of the good, even in visionary companies.** A vision that seems well-shy of ideal but has 100 percent buy-in by the entire management team is likely to be much more effective than an ideal vision that no one can get his or her hands around.
- **Process can be more important than outcome.** How a company's management team approaches various decisions, whom it includes, the respect it accords disparate voices, the time it devotes to thoroughly analyze all aspects of a question—these say volumes about what the management team truly stands for, much more so than the rhetoric of a mission or vision statement.

Individually, the members of the management team underwent leadership training. They followed up this training with a 360-degree survey of the entire company that focused on questions of management and leadership. Then the management team spent three days together looking at the feedback and working on a specific action plan of continues–starts–stops that enumerated the various things each person needed to continue doing, start doing, and/or stop doing to become more effective as a team and as a business.

As the management team began to understand each other better through dialogue and assessments, they also began to understand that this was not a one-off exercise but a paradigm change in the way they would work together from this point on. The work environment at Melaver, Inc. was always going to be fast-moving with high expectations. Resilience and adaptation were necessary. So, too, was a collective vision statement that everyone in the company could say he or she had a hand in, could agree upon, and could support. And so, within a year of when Martin first drafted his vision statement for the rest of the company, the management group met as a team to envision things together.

Collective Visioning

Envisioning is the hardest step of all in strategic planning because it requires one to journey away from day-to-day routines and into a world of possibilities without judgment of what will or won't work. Dreaming about what you could be means you have to allow your right brain to envision something you may never achieve, even as it is what wakes you every morning as you strive toward that vision. It calls for a willingness to be vulnerable with one another, a willingness that was probably in short supply when the management team first began to meet.

The exercise of envisioning also brings out differences among team members. It shows where people are focused, the direction in which they want to go, and the differences of those views. My job was to keep them focused until they could develop a consensus on a vision for the future. The way to successful envisioning is to value all opinions, honor the past, and move toward a future together. Their work together for the better part of a year learning to listen to each other is borne out in the company's Statement of Purpose:

> We aim to become a vertically integrated, truly sustainable real estate company.
>
> Our definition of sustainability focuses on the triple bottom line of economic performance, environmental footprint, and social engagement with the community. We view ourselves as an Enveloper®—enveloping our community in a fabric of innovative, sustainable, inspiring practices.

Collective visioning was the green glue that ultimately resulted in a statement of purpose that everyone felt ownership of and could understand and support. But the green glue that made this collective visioning stick was empathic leadership.

Empathic Leadership

Goleman, Boyatzis, and McKee, in their book *Primal Leadership*, state the business case for empathy very clearly:

> When leaders are able to grasp other people's feelings and perspectives, they access a potent emotional guidance system that keeps what they say and do on track. As such, empathy is the *sine qua non* of all social effectiveness in working life. Empathetic people are superb at recognizing and meeting the needs of clients and customers, or subordinates. They seem approachable, wanting to hear what people have to say. They listen carefully, picking up on what people are truly concerned about, and they respond on the mark. Accordingly, empathy is key to retaining talent. Leaders have always needed empathy to develop and keep good people, but when there is a war for talent, the stakes are higher.[4]

Most business writings on leadership focus on the competition for top talent and the value of retention as the underlying rationale for empathic leadership. I certainly wouldn't dispute that. Research has proven that people want to be recognized and valued. Showing empathy is part of giving the right kind of recognition, an aspect of an intangible compensation program, if you will. It motivates people to go the extra mile. And so on.

But I think that such a take on empathic leadership misses something more fundamental. Being an empathic leader, particularly in the context of shaping a vision for a company, makes a statement about ownership. It says *we need to understand your feelings and beliefs because without that, there is no strategic vision for the company.* It also sets the stage for a different approach to day-to-day management of the company, stating in effect that this leadership team intends to seek out and listen to everyone on matters that are of the utmost importance, all the way down to the most mundane. If the leadership team seeks to understand the needs of all its staff members at a deep psychological level in order to shape the direction of the company, imagine how it will govern relationships and practices on a daily basis.

Setting the tone for empathy from the top leadership and making empathy the topmost agenda item for that leadership team get the green glue flowing broadly and deeply throughout the company.

DECISION MAKING

A company that shapes its values from the bottom up, shapes its organizational design as a series of concentric circles built around values, and fosters a notion of empathic leadership is bound to have a different approach toward how various issues are deliberated and how decisions are made. On the one hand, there was a strong meritocratic element running through the company, a desire to structure decision-making processes in such a way that the very best thinking floated to the top. On the other hand, there was also a strong democratic element running through the company, a belief that the best thinking floated to the top only when everyone was empowered to be part of the debate. Too strong an emphasis on meritocracy could mean that the decision making fell to the strongest or most dominant influencers at the company. Too strong an emphasis on democracy could mean that decisions were reached when the lowest common denominator was found. The challenge is to find a golden mean between these two extremes.

Three primary concepts are drawn upon by Melaver, Inc. to help shape this golden mean: interdependence, subsidiarity, and consensus. Interdependence, used in the context of this particular company, tends to mean that a great deal of independence is afforded each and every individual at the company to make decisions, *but* the price of this independence is the strong expectation that individuals will draw upon other eyes and ears among colleagues to obtain different perspectives on a given issue. Subsidiarity holds that decision making is pushed downward to the level closest to the actors most intimately involved in a particular issue or project or policy. Taken together, interdependence and subsidiarity are used to leverage the small, entrepreneurial nature of the company and to empower all staff members to shape the direction of the company. Balanced against these two concepts is the practice of consensual decision making, which called for (and still requires) considerable practice.

Consensus is often misunderstood. It means we can all agree to implement a decision. That does not mean everyone agrees 100 percent, but that everyone can live with the team's decision. If anyone is not in agreement, then the discussion continues until some type of agreement is reached. Process is important in gaining a consensus decision. Most leaders would rather decide and direct as opposed to build consensus. The experience at Melaver, Inc. was no exception. It was fun to watch the transition of the team trying to work toward consensus, as people began to open up and trust one another.

One trick of the trade in our consensus-building processes was the use of green–yellow–red cards. Each member of the management team was provided with a set of these cards. As discussion progressed on a particular issue, the facilitator would check with the group from time to time to see whether consensus was being reached. A green card meant all was OK, a yellow card indicated concerns that needed to be addressed, and a red card signaled major resistance. While physical use of these cards has diminished over time, they are continuously deployed virtually in most every decision at the company. The real estate task force, for instance, cannot make a recommendation to develop a project if someone has "voiced" a red card. A recommendation with a yellow card indicates that a member of a task force feels caution needs to be exercised before a particular decision is made. The senior leadership team at Melaver, Inc. (the CEO, COO, and CFO) has agreed that no major decision will be made unless all three flash the green OK. They have agreed to continue discussing an issue whenever one of the three has flashed a yellow sign. A red vote by any of the three is tantamount to a veto and means that a potential deal is dead.

Decision making, which is never an easy thing for any business, is particularly challenging for the values-centric company focused on a green bottom line. Even seemingly mundane issues can unexpectedly raise subtle larger-scale issues that touch powerfully felt beliefs. For instance, a debate on whether or not to develop in a suburban or agricultural greenfield location (something Melaver, Inc. does not do) raises the issue of providing jobs and a better quality of life for an otherwise impoverished community. Individual values have the capacity to clash with one another from time to time. Moreover, decision making by consensus can lead to confusion and uncertainty. With multiple viewpoints all being aired in a marketplace of ideas, which perspective prevails and why?

Melaver, Inc. is constantly considering and reconsidering at what level of the company various decisions are made. You obviously don't want to tie up every decision in lengthy and unwieldy consensual debate. On the other hand, you want to make sure that a decision made by a small task force has sufficiently considered the larger, value-laden context perhaps embedded in the issue at hand. Leaders at Melaver were initially stricken with analysis paralysis as they fought making decisions and watched their team wander around looking for direction from the top. It took time for them to understand that getting input and having a consensus does not mean every person in the company must participate in every decision. They had to learn what decisions needed input from others, what decisions had consequences for team members.

Facilitative leaders (as opposed to directive leaders) have to believe in communication processes. Facilitative leadership does not mean letting go entirely of the role or the responsibility to lead. These processes involve inclusiveness and balance communications, problem solving, and decision making. Interdependence, subsidiarity, and consensus, integrated with each other, facilitate productive entrepreneurial activity and values-centric reflection. Using these three decision-making concepts certainly does not guarantee smooth sailing. But that is the choice of the values-centric company focused on a green bottom line: to have business-as-usual defer to business-unusual.

GOVERNANCE AND SHARED LEADERSHIP

Another aspect of governance embraced early on by Melaver, Inc. was that of shared leadership. It fits with the aspects of governance discussed above: an organizational design built around values, a collectively articulated company vision, leaders who devote considerable time and energy to engaging empathetically with all others in the company, and decision-making that engages most everyone at the company. These aspects of governance add up to an organization designed from the get-go to leverage everyone's leadership skills.

Perhaps the best example of this shared leadership is the large number of task forces one finds at the company—ridiculously large from an outsider's perspective. There seem to be almost as many task forces as there are employees: real estate, investment, personnel, philanthropy, communications, and cross-divisional are all standing committees at the company. And then there are the task forces that are created *ad hoc*. The proliferation of these task forces sometimes gives rise to disgruntlement, as a few feel the entrepreneurial spirit of the company is stifled by such team-oriented behavior.

Beyond the sheer number of these task forces, however, is the more germane fact that each group is cross-functional in its makeup. Operations, finance, brokerage, and senior management are represented on each task force. These cross-functional teams enhance the sense that there are no silos within the company, that everything the company does is integrated with everything else. Such integration improves communication and facilitates understanding among staff members who, because of their specific job functions, tend to speak different languages and analyze projects differently. As such, these cross-functional teams also help ensure that issues and projects are analyzed from multiple perspectives. Fundamentally, shared leadership is the means by which Melaver's governance structure shapes all the various rituals at the company, to which we now turn.

Rituals

All cultures engage in rituals in the form of ceremonies, customs, and regular practices that provide a sense of continuity and stability and order and even certitude. We tend to place our faith in these activities to ground us and keep us from drifting. I particularly like a specialized, psychological definition of ritual, cited in the *Oxford English Dictionary* as "a series of actions compulsively performed under certain circumstances, the non-performance of which results in tension and anxiety." This definition—compulsively performed actions that reduce stress—may sound a bit strong. Sometimes these rituals evolve naturally, such as the company's annual potluck Thanksgiving celebration and the practice of ringing a ship captain's brass bell when the company has cause for celebration.

Part of my role entails deliberately shaping rituals that bring everyone together in a fusion of work and play. The most prominent example of this is the company's annual retreat. Occurring each spring (the time of renewal!) over the course of two days and involving everyone at the company, the annual retreat is a time for team building, strategy development, and focusing on core values. Each year a different theme, usually based on the core values, is adopted. The retreat committee plans the activities of the event. I assist with the facilitation of process.

This time of work and fun includes perhaps the most moving ritual of the year: the company's annual core values awards dinner, which kicks off the retreat. During the dinner, Martin reads aloud stories colleagues have written about each other as they recall their teammates' exemplary efforts over the preceding year. You can almost hear the voices of individual staff members as they recall each other's small exploits of courage, determination, and leadership. Watching the face of an individual staff member as a particular story is read aloud about him or her is almost indescribable. It says *They noticed,* or *They see me for the best that I can be.* Mostly, though, it is a public recognition that this community cares deeply for what an individual has done and what he or she stands for.

The retreats are learning opportunities, not only on the theme for the retreat (e.g., a specific core value, metrics for a triple bottom line, the meaning of sustainability), but for learning about each other. Time is spent in teams and thinking sessions. The individuals see how they react to one another and see other team members in a setting outside the office. Retreats also attempt to give staff members time for relaxation, recreation, and visiting with one another.

Several retreats have been intense in the realms of interaction and focus on the company's identity and culture. Each retreat sets in motion an action plan for the ensuing year, largely to develop processes that address problems identified during the retreat. I have learned to go with the flow, as much of what I plan requires adapting to the moment or the situation. When staff members begin to feel freer to express some of their concerns, unexpected issues arise, and deeper systemic challenges are identified. Typically, the ritual of the annual retreat opens pathways into processes that need to be developed or refined over the coming year.

The follow-up processes we develop over the course of the year are deep-structure rituals that shape the company's green bottom line. They are rituals in the sense that they provide standards of conduct and behavior that shape the culture of the company. They indeed assist in ratcheting down tension and anxiety, and they also assist in making a community of individuals a high-performance culture. The rituals tend to fall into five conceptual categories: valuing multiple perspectives, transparency and straight talk, accountability, mentoring, and alignment. Let's consider each category and the particular rituals implemented for each one.

VALUING MULTIPLE PERSPECTIVES

In my work with Melaver, Inc., I decided that the best process for people to learn about each other, to improve communications, and to build a real team would be through self-assessment tools like personality assessments, leadership style assessments, skills assessments, and values-clarification exercises. As part of the "green glue" of culture creation at Melaver, Inc. each staff member is provided the opportunity to take the Myers-Briggs Type Indicator® (MBTI®) to learn about themselves, to learn about others at the company, and to learn how all the disparate styles within an organization can be leveraged effectively. The MBTI® has been an insightful tool for improving communications, working relationships, and coaching styles.

Isabel Briggs Myers and her mother, Katharine Cook Briggs, developed the MBTI® in the mid-twentieth century based on the work of Swiss psychoanalyst Carl Gustav Jung, who wrote *Psychological Types* in 1922.[5] The MBTI® identifies sixteen different personality types by examining our preferences for how we relate to the world (as Extraverts or Introverts), how we learn about the world (Sensing by using our concrete senses or by Intuiting), how we make decisions based on what we learn (Thinking logically or Feeling), and our attitudes about the world (Judging in an orderly way that leads to closure or Perceiving in ways that leave us spontaneous and open to possibilities). The MBTI® deals with four sets of opposites that, once we understand them, can increase our self-awareness and help us figure out why some behaviors seem more natural and easier than others. However, we do have a natural preference for how we see the world and make decisions based on our personality type preferences.

Psychological type theory is rich with examples of the *ah-ha's* people have when they learn how they like to see the world and make decisions. It is not unusual for people to comment that they thought they were very different from other managers or leaders but did not know that the difference was rooted in personality preference. They report a sense of relief and self-acceptance once they understand their type preferences as right for them. People often spend a great deal of energy trying to be what others expect instead of using their strengths as a benefit when adapting to an organizational culture.

I like to use the analogy of our handedness when facilitating learning about personality type. I ask people to write their name, and then I note that they used their preferred hand to do so. I then ask them to write their name again with their opposite hand. I hear all sorts of groans and comments on how hard it is to do that exercise. Because it is awkward—it does not feel natural—they do not feel good using their non-preferred hand. The analogy to personality type preferences is that we are asked every day to do tasks that may not use our natural preferences for how we see the world and make decisions. Thus, we may avoid those unpleasant tasks or find ourselves stressed when forced to do them. Using the information gleaned from our knowledge of personality enables us to be more cognizant of our own personality type and those of others, empathic of differences among colleagues, savvy as to how to leverage those differences, and more adaptive in our strategies to get the job done.

To know our core preference is helpful so that we know when the world is asking us to balance or use our opposite. Our goal in learning about ourselves through type is to know our personality preferences and to learn ways to balance and adapt. We do not use our type as an excuse for not performing, but we do know when something seems natural for us and when we have to give a little extra energy and focus to a task. We also have to learn the value of differences when talking about the diversity of type. A team with different types is a more productive team due to the different perspectives; however, there may be more conflict on that team as a result of different perceptions and needs based on different personalities.

For example, a person working at a financial institution who has a Feeling (values-oriented) preference may hide his concerns about how a decision might impact people in the organization. However, that very input might save the company time and money if it were expressed at a decision-making meeting by asking a question that heightens awareness of the impact. Once a team understands its members benefit from the various type differences of individuals, few reject values-oriented comments that balance the logic of a decision.

The glue that holds a team together is often awareness and understanding based on type difference. The rituals we established at Melaver, Inc. include conducting an MBTI® analysis on all staff members, following up this analysis with one-on-one sessions so the individual understands his or her type preference, collating all of the staff members' type preferences into a single table so that everyone can see the diversity of personalities that go into the construction of a team, and working with teams to leverage each other's type preferences. An overview of the Melaver team's preferences is provided in Figure 2.2.

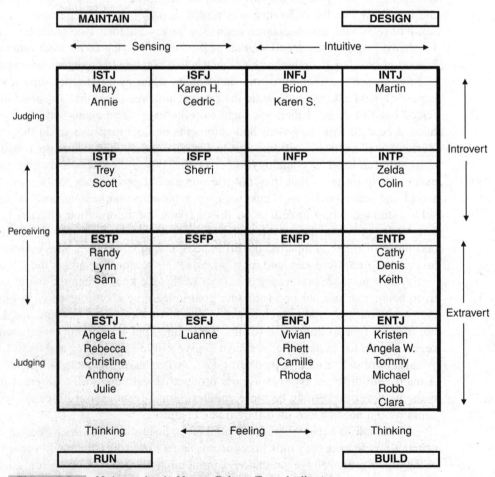

Figure 2.2   Melaver, Inc.'s Myers-Briggs Type Indicators.

A number of interesting things are worth pointing out about this company view of type preferences. When I first starting working with Melaver, Inc., there were more reported introverts than extraverts. With additional hires, the profile has changed. This may be owing to the company's consensual hiring processes, where a team interviews all potential hires. One issue I have with an inclusive interviewing process is that it might screen out introverts. Not only is the process intimidating for people who like to think before they respond to bombardment of questions by this team, it also creates a situation where the interviewee is asked questions on the spot about the company. A person who wants to gather facts quietly and then respond does not interview well in this open forum. The company likes this unique process and is aware that it might need tweaking to be sure that people of all type preferences are considered when hiring.

Another interesting insight about the chart in Figure 2.2 is the strong preponderance of TJs (Thinking and Judging, the four corners), indicating that this company as a group tends to make decisions emphasizing logical Thinking as opposed to Feeling, and it tends to move toward closure in its overall orientation rather than remain open and spontaneous to various possibilities. Given this bias, I try to remind team members that they need to pay particular attention to those whose preferences are for Feeling and Perceiving.

A third insight comes if you look at the chart as an amalgam of four quadrants having to do with Design, Build, Run, and Maintain. Melaver, Inc.'s culture is primarily constructed of builders and runners—those who construct ideas and those who execute these ideas. There isn't a surfeit of people who maintain the ideas once they are executed, so care needs to be exercised to ensure that personalities are in place to address day-to-day managerial issues. Moreover, there is not much excess capacity in terms of visionary designers. The company may not need much more capacity today, but it needs to keep this in mind as it plans for the long-term leadership needs of the company.

Another excellent tool used by some of the Melaver leadership for developing self-awareness is FIRO-B®, Fundamental Interpersonal Relations Orientation-Behavior® assessment, which helps us understand our behavior and that of others in interpersonal situations. Dr. Will Schutz developed FIRO-B® to examine three dimensions of interpersonal relations considered to be necessary and sufficient to explain most human interaction.[6] The FIRO-B® instrument consists of scales that measure how you tend to behave with others and what you seek from them. FIRO-B® measures and reports on three fundamental interpersonal needs: inclusion, control, and acceptance. Inclusion measures our need for recognition, participation, and contact with others. Control measures our need for influence, responsibility, and decision making. Acceptance measures openness, closeness, and personal warmth.

Within each of these needs, FIRO-B® also measures whether we prefer to initiate the behavior to achieve the need (expressed behavior) or whether we prefer others to do so (wanted behavior). For example, we may want inclusion but prefer others to initiate the interaction that results in our feeling included. When we use a tool like FIRO-B® to understand our interpersonal needs, we become more aware of why we react—and behave—the way we do. This level of understanding is necessary before we develop the ability to monitor and manage ourselves.

In an organization, interpersonal relations of managers can make the difference between failure and success. Using the MBTI® and FIRO-B® with managers and leaders in an organization can bring awareness regarding how others see us and react to us. By using assessments, we enhance our self-awareness and self-understanding. Since we all want to be understood, utilizing these instruments creates a language we can draw upon to explain needs, connect with others, and better understand actions and behaviors that differ from our own.

TRANSPARENCY AND STRAIGHT TALK

Zero politics. That was (and continues to be) a goal of this green bottom line company. Have you ever heard of something that wild and unreasonable, a business without the usual back-of-house politicking going on? Yet that was a clear value to this company. The logic is simple, if somewhat idealistic: Bring a group of talented, passionate folks to the table, enable them all to vocalize their individual beliefs and perspectives in a free and open exchange of ideas, and let the best ideas float to the top, irrespective of where and from whom they emanated. But from conception to implementation—talk about a firestorm!

The concept of zero politics, based on ideals of transparency and straight talk, was a belief lacking a set of clear practices and processes. Everything was shared on the management team level: recruiting and hiring, performance issues, development issues, financial data, even compensation. From the outset, I saw that people were exhausted from trying to keep up with the day-to-day workload while transitioning into new ways of getting the work done. Management meetings were long. In an attempt to create an open atmosphere, Martin used communication processes that were painfully honest and direct. One objective was to create greater accountability for one's actions, but there was a general fear of exposure in this environment. Another objective was to talk about issues directly with the person involved, as opposed to triangulating issues through a third party. People cringed when the agenda included time for discussion of behaviors that impacted team members.

This team was close. Some had more honest communications at work than they had in their personal lives. But this closeness was also an impediment. The effort at honest and direct communication was failing. Each person struggled to receive feedback from team members without being defensive. Team members felt more comfortable handling conflict indirectly through a sympathetic listener, than by addressing an issue directly, head-on. The most difficult part of this process was getting all members to a high level of comfort and trust with one another.

The first step in ratcheting up this level of trust was to work in smaller increments than the larger management team. Team members were divided into coaching pairs, with the objective of creating a safer environment in which to provide one another feedback. Pairs were expected to meet on a regular basis and proactively seek out feedback.

A practice that, over time, became standard among coaching pairs was the continues–starts–stops exercise. The practice calls for directed feedback on what a staff member needs to continue doing in order to be effective, what he or she needs to start doing in order to be effective, and what the person must stop doing in order to be an effective member of the team. I oversaw the work in large part by guiding various team members on how to elicit and receive feedback in a non-defensive manner. In the smaller, paired settings, this proved much more effective.

A third ritual in this overall approach to transparency and straight talk also involved small groups of individuals: 360-degree feedback. The practice is so widespread among companies today that it needs little explanation. Two to four times a year (initially more frequently, later less so), team members receive feedback from a supervisor, a direct report, and a peer. The focus of these 360-feedback sessions was on personal as well as professional growth, something of an anomaly in the use of this human resources mechanism but well suited to a values-centric company.

One final ritual that bears mentioning in our discussion of transparency and straight talk involves the company's anti-triangulation policy. Melaver, Inc., like most any other company, has had its share of indirect talk among staff members. Some of that was motivated by an unwillingness for parties to address each other directly, a natural fear of confrontation one finds in many social contexts. Another, more harmful motivating factor was that people created affinities for one another by bad-mouthing a third and absent party. Needless to say, such behavior can be extremely disruptive and unproductive. The company, after much discussion, instituted an anti-triangulation policy that basically holds staff members responsible for speaking directly to one another rather than via an intermediary. If one does use the resources of an intermediary, it is expected that direct discussion between parties will occur within twenty-four hours of the mediated conversation.

As a foursome of rituals, peer coaching, continues–starts–stops, 360-degree reviews, and anti-triangulation all focused on the complex of communication skills that have to do with self-assessment and feedback. Melaver employees may be focused and passionate, but they have the same issues I find in most companies. People talk, they do not feel heard. People hear what supports their own point of view. The most important part of leadership and management is *how* we say what we say, for it impacts every aspect of performance at work. People often spend their time and energy in reaction mode. We have a difficult time producing if our communications are not productive. And that challenge of communicating productively leads to the critical art of soliciting feedback.

FEEDBACK

In our personal relationships we receive feedback from those we care most about. Often those are the people to whom we listen poorly. Why is that? Perhaps we have heard it before, or perhaps we are threatened by the message.

It could be that we are defensive when told how we impact others. We want to be understood, and we want to be right. We hate being told when we do something wrong (in another person's view). In these situations, it's important to remember that all we need to do is listen and acknowledge the other point of view, even if we disagree with what is said. When receiving feedback, we can simply say, *Thank you for sharing with me how you feel* or *Thank you for sharing your thoughts.* Easy to say, hard to do—and hardest when we feel judged by the other person's comments. At those times, we need to breathe deeply, monitor ourselves in the present moment, and listen attentively. Receiving feedback is a skill that takes considerable practice. We might try telling ourselves this feedback may be a gift as we try to determine if there is some truth to what is being said. We can try to look for concrete advice on ways to improve our behavior or reaction. We often react in a righteous manner that says, *Who is he to tell me that?* Or, *She is not so perfect; she has faults too.* The truth is that every person has something to say in response to what we do and say.

At Melaver, I spend considerable time coaching people to ask for feedback, to accept what they hear as someone else's perspective, and to try to listen and learn rather than be defensive. Coaching can help people understand that knowing how others perceive our behaviors is helpful because it tells us how we impact others. Being coached to invite and accept feedback is a good learning tool that models receptiveness and opens the door to data that allow us to decide if we will change to more productive behaviors. We want our leaders to be consistent in the way they lead and treat others. Feedback points out our inconsistencies in how we behave with different people. We can look at the root cause underlying some of those behaviors. When there is a gap in perception between what you and your colleagues think, it is time to understand how to close that gap.

I find that some leaders resist this process because hearing what others have to say is risky business for their egos. What if someone has issues with your style of management, or what if someone rates you as a poor listener? If someone is rated lower than they expect on an assessment, that person might ask who provided the feedback so he or she can argue that the results were skewed for some rational reason related to that individual.

I find myself telling leaders they cannot talk their way out of perceptions others have formed about them based on their behaviors. Feedback certainly brings this home. It also tells you where you measure up in positive ways. Thus the goal is to look forward to hearing all the positives people have to say. When working with Melaver's management team, I facilitated exercises that led to the team understanding the barriers to listening that arise when we are more focused on wanting to be heard. Were team members listening with openness to learn what others were thinking, or were they listening to defend their position or with a closed view or perception? I find it helpful to ask team members to try to monitor their thought processes in real time, according to the examples in Table 2.1.

TABLE 2.1 EXAMPLES OF LISTENING	
OPEN	**CLOSED**
I can hear what he is saying.	How dare he think that way?
Perhaps her ideas have merit.	She has no credibility with me.

So basic and fundamental and obvious. And yet, even after years of practice, even high-performance teams need to go back to such "blocking and tackling" fundamentals. After all, most of us learned these basics in kindergarten: We talk when it is our time to talk, and we listen when it is our time to listen.

ACCOUNTABILITY

The most important aspect of feedback rituals is the interpretation and coaching that provides an individual action plan based on the results. We don't want just to engage in feedback sessions—we want to see efforts to leverage that feedback for positive results. We want to hold each other accountable by way of an action plan that measures and evaluates growth and development. That is where another tool is used, the balanced scorecard.

A balanced scorecard is used to integrate the feedback a staff member receives into a professional/personal action plan that dovetails with the overall strategic direction of the company. The company has a balanced scorecard each year, focusing on initiatives in four main areas: learning, internal processes, customer relations, and finances. The four areas are interrelated through these questions:

- What does the company need to learn in order to improve upon its internal processes?
- What does the company need to do with its internal processes in order to improve its delivery of services to customers (and other stakeholders)?
- What delivery of services does the company need to focus on in order to improve the financial performance of the company?

Typically two to four initiatives are assigned to each area of the company's strategic focus.

An individual's balanced scorecard functions precisely the same way, looking at initiatives for learning, internal processes, service, and financial results. Personal growth and development (learning and internal processes) are linked to professional growth and performance. It is worth noting that Melaver, Inc. determines year-end bonuses on the basis of these individual balanced scorecards, emphasizing the strong link it sees between personal growth and business productivity.

MENTORING

Melaver, Inc. wants to be a mentoring company, one that can guide others both within the company and outside the company. But where do mentors come from? Mentors are not always the older, more experienced people in the company. A mentor is someone you choose and ask to guide and advise you. A mentor is different from a coach, but a mentor can also play that role. And herein lies a quandary: While coaches can be assigned, mentors typically evolve out of a self-selective process. A senior leader of a company might say, "I want so-and-so to mentor this person." But unless there is a strong desire on the part of both people to mentor and be mentored, the desired program fails before it ever begins.

Mentoring, then, works when the desire to be mentored is present. It has its greatest chance for success when an organization exhibits these characteristics:

- A strong orientation toward learning among individuals within a company.
- Significant trust among colleagues to enable staff members to feel comfortable opening up to each other (openness as opposed to feeling vulnerable).
- Mutual respect for the varied strengths staff members bring to the table.
- Modeling of behavior among the leadership team, as they show themselves open to being mentored by others.

My own role as a consultant in facilitating the practice of mentoring is largely two-fold: drawing upon other rituals within the company to enhance the general sense of trust and mutual respect, and coaching particularly the senior leadership team to open themselves up to mentoring from others. The tactic that is perhaps most helpful in this regard is to open up the leadership team to using the phrase: "I need help with . . ." Oftentimes, this is simply a rhetorical gesture, one that feigns collaboration while being authoritarian. How many times, for instance, have we heard a leader say "I need help understanding . . ." which, when translated to our ears means, "I disagree with you but I'm going to do so in a polite way. The phrase, "I need help," needs to be authentic, a true admission of needing the guidance and direction of another. Like so many other aspects of green glue, it's easily said, but not so easily practiced.

The other helpful piece of the puzzle in facilitating the practice of mentoring entails hiring people who have the desire to learn and grow. As a company that sees itself as a learning company, Melaver, Inc. would not seem to face much of a challenge in this regard. And it doesn't. Every staff member has a line item in the budget each year for continuing education. At any given time, several are working on additional accreditation or advanced degrees. The calendar is filled with lunch-and-learn sessions, where knowledge and practices are shared among colleagues and facilitated by either internal or external experts. In any given month, a team member either is attending a conference to further professional/personal growth or is lecturing at a conference—often both. The company is replete with rituals that comprise formal and informal learning. But these rituals among a staff of learners raise a more fundamental question: How does a company fill its organization with a bunch of intellectually curious people without becoming a nonprofit think tank? The question leads us to our final category of ritual-shaping at the green bottom line company: alignment.

ALIGNMENT

One word that prevails in values-centric companies like Melaver, Inc. is "alignment" or "fit." That shouldn't come as much of a surprise. These companies devote considerable amounts of time and energy to culture-building rituals and, as a result, they want to make certain everyone on staff is aligned with the company's values, beliefs, and mode of governance. What may be surprising is the degree of diversity one finds at such companies, despite what would seem to be a fairly narrow set of hiring criteria. In the case of Melaver, Inc., the key to managing both alignment and diversity, ensuring clear focus on core values while also ensuring the staff does not lapse into group think, is its consensual group hiring practice.

The practice, as mentioned earlier, is not without problems. First, the group setting for interviews, in which numerous staff members all participate in the conversation, seems to favor extroverted candidates over introverted ones. It's a bias the company needs to be careful to counteract, to the extent that it can. Secondly, this group hiring process is very time consuming. Hiring for a new position literally takes months and, if that weren't enough, there have been times when, because of the fast-moving nature of the company, the needs for that particular new position changed during the course of the interview process. Moreover, given the lengthy nature of a hire, there's strong collective pressure to make a hire from the short list of candidates that eventually makes it to the group interview process, since no one wants to start the search from scratch. That pressure at times creates added tension and anxiety within the company, as the search committee wants to finish the process once and for all, while others on staff may prefer to wait for a better fit to come along. Third, it is very rare for staff members to reach total agreement on a hire: consensus, yes; total agreement, no. In fact, occasionally, the company has hired a new staff member when that person's direct supervisor actually preferred another candidate. Fourth, because these group interviews are structured like informal conversations among team members and don't follow a standard script of questions, interviews differ dramatically. Team members often tell me candidates get the interview they themselves script, which I can understand. But the HR policy side of me feels a certain discomfort with comparisons among candidates that appear to be objective but, in fact, contain systemic subjective elements. No one said hiring was a science. But this particular company seems to view it as almost pure art.

Having said that, there are strong positives running in favor of this approach to hiring. First, there is no arguing with success. Since instituting this ritual about five years ago, the company has retained virtually every single person it has hired. Second, it gives the prospective candidate a clear, composite picture of what this company is about. The short list of candidates who make it to the final round of a hiring process all have the hard skill set to do the job. At this point, the company wants to know primarily if its culture is one that the candidate would not only feel comfortable in, but one that would enable the person to be at his or her best. In this sense, the notion of fit and alignment is somewhat misunderstood. The company presents its culture to the candidate, and the candidate actually evaluates his or her fit with it. Of course, the obverse is also true: the company is evaluating the candidate's fit. Having participated in a number of these interviews, I can say that active listening will oftentimes tell you whether a prospective candidate gels with the culture or not.

I do have questions regarding the future feasibility of such an approach, such as whether, as the company continues to grow, it can afford to devote the time this process

TABLE 2.2 OVERVIEW OF GREEN GLUE RITUALS

Practice	Processes
General espirit de corps	Annual retreat Potluck Thanksgiving Ringing the bell celebrations
Valuing multiple perspectives	MBTI® FIRO-B®
Transparency and straight talk	Feedback through coaching pairs Continues–starts–stops 360-degree sessions Anti-triangulation policy
Accountability	Balanced scorecards
Shared leadership	Organizational chart Decision making Cross-functional task forces
Mentoring	Formalized continuing education Lunch-and-learns Conferences Executive and peer coaching
Alignment	Consensual hiring

requires and devote such a heavy degree of human capital to the effort. Some on the leadership team feel this is a drag on growth, while others, the CEO included, feel that it is precisely rituals like this that assist the company's deliberately paced growth.

Various rituals comprising the green glue of a values-centric company help create a rich and complex culture, one that takes time to feel at home with and comfortable in. An overview of the main components of ritual at Melaver, Inc. is provided in Table 2.2.

Language

Every culture has a distinct language. Businesses are no different. Most of us probably don't even think about it—it's a part of the environment we've grown accustomed to. But the newcomer to a company is often overwhelmed and confused by the acronyms and code words and general shorthand people use to communicate with one another.

Melaver, Inc. is no different in this regard. Team members talk about "going to the HILS," referring to the High Impact Leadership Seminar many attend to learn about feedback and leadership types. Staff members have LEGs, not the kind that hold up our bodies, but an internal program for the Leadership Enhancement Group. They call their senior leadership team Froot Loops (because they are graduates of the Kellogg

School of Management at Northwestern University), which has its monthly DMD sessions (Directors and Managers Discussion) with family shareholders. And so on.

There is, however, a type of language resonating throughout the company that comprises part of its green glue, a type of language that has to do with deliberateness of thought and action. Paradoxically, this aspect of the company culture happened somewhat by accident.

Martin and I were one day discussing the need to give a name to the management team that would lead the way for the new Melaver, Inc. I opened a dictionary and placed my hand on a word at random: *Sanskrit.* Wikipedia explains that the verbal adjective *samskrta-* may be translated as "put together, well or completely formed, refined, highly elaborated." It is derived from the root *sam(s)kar-* "to put together, compose, arrange, prepare" (*sam-* "together" and *(s)kar-* "do, make"). Sanskrit also is the root language for all Indo-European languages, a basis upon which other cultures built their own thoughts and dreams.

We both thought it appropriate for a team that would develop a new language for who they will become to be named after the old language of Sanskrit. It stuck. Today, the leadership team of Melaver is still called Sanskrit. Its staff is called *Hrdaya,* the Sanskrit term for heart, center, core of something, essence. Other areas of the company have Sanskrit-based names as well.

To many, perhaps, this might seem silly. But naming can be significant, and the Melaver team began to take the naming of various activities and projects seriously. The seriousness reflected a different kind of deliberateness rather than simply cobbling together acronyms. As task forces were created (for company philanthropy, for personnel policies, etc.) and projects were developed, the question of naming provided the team with a moment to pause and reflect upon the meaning behind a particular effort. *What are we trying to do?* was a question immediately connected to the query, *How do we want to communicate what we are doing by the name we select for this activity?*

Of course, behind the deliberate act of naming is the more fundamental language of values itself. I do not mean the more concrete language of the company's core values. Instead, what I have in mind is the fact that a value system is itself a language. Beginning in the mid-1990s (and continuing today), Melaver, Inc. began to speak a language that was not recognizable to most outsiders, a language about a sustainable ethos. Part of that language of sustainability focused on the uniqueness of place, and it made references to things like the company's environmental footprint, the triple bottom line, LEED accredited professionals, LEED certified buildings, carbon emissions, and the 2030 Challenge. Another part of that language of sustainability focused on people, making reference to the personality types of colleagues, behavioral guidelines, 360-degree feedback, anti-triangulation policies, etc. This language of sustainability, involving people and place, is another aspect of a holistic value system.

How you act, what you do, and the choices you make all speak volumes about what you believe and who you are. *Under promise and over deliver* is a maxim I often hear team members proclaim. Part of that orientation is a serious concern regarding "greenwash," an issue that is discussed in detail in Chapter 10. But the gist of this maxim has to do with making sure that all staff members express their values primarily through their actions, every day, in the respect they accord one another, the degree to which they engage in active listening, the extent to which they seek to understand first, the manner in which they keep themselves open to new and different ideas.

Each concept calls for processes and metrics to ensure that staff members are indeed walking the talk because in this language of a green culture, its walk is its talk. Granted, this emphasis on speaking softly and letting your actions do the talking has its challenges, particularly when it comes to issues related to marketing. But fundamentally, the language of deeds-over-rhetoric is an important building block of the values-centric company. Without it, the authenticity of a green bottom line company becomes questionable.

History

Every culture has a history. Business cultures do, too—even start-ups that have only recently received a business license. We all have stories to tell of the diverse roads that led us to this present moment. A collection of individual stories from the past helps comprise a collective history that is updated continually. We sometimes forget this in our race to get ahead or move forward. The old adage, *You can't know where you are going if you don't know where you came from,* is apt.

Melaver, Inc. is particularly fortunate in that it can draw upon a seventy-year company history that is steeped in values lessons. Annie Melaver, who started the family business with a corner grocery store in Savannah, Georgia, in 1940, used to close the store on occasion to bring soup to a sick neighbor. Norton Melaver, who took over leadership of the grocery business from his mother Annie, was noted for, among other things, taking a principled stance for racial integration that resulted in a boycott. The predecessor business to Melaver, Inc., M&M Supermarkets, was renowned for cutting-edge innovations and a strong customer service ethos, practices that enabled a small company to buck national trends as an independent operator that out-competed national chains. In the late 1970s, a new M&M store was developed around dozens of centuries-old live oaks, the first of many environmental practices engaged in by the company and its shareholders. Various family members involved in the business over the years managed to find time to devote to numerous community causes. Melaver, Inc. has a business history of a company in the community serving that community. Sound familiar?

Interestingly enough, while this rich company history was known to staff members of Melaver, Inc., it was more of a vague background than something everyone knew and referenced. Intent on its metamorphosis into a sustainable real estate company, the leadership team generally—and Martin Melaver particularly—seemed more focused on the road ahead than on what had brought them this far. So part of my work with the team was delving into company history, asking about the connections between the past and the present, digging up archived pictures from the early days, encouraging the leadership team to invite retired family members to talk about their days in the grocery business, and so on.

Part of this deep dive into history entailed delving in to the personal histories of the staff members at Melaver, Inc. I believe this to be a critical part of the green glue of a sustainable business culture, oftentimes overlooked. Too often we think of a business as having a history without considering the degree to which the business itself is made up of the multiple histories of the folks who have come to work with one another. Some teasing out of individual histories occurs naturally and informally, as staff members get to know each other better, but some calls for more formal practices and processes. I recall one exercise, for example, when team members were brainstorming about objects they would like to place in a time capsule at the first LEED shopping center in the country. Team members knew this was something of a watershed moment for themselves and the company. And so the general brainstorming exercise took on a very personal tone, as they began to reflect upon monumental objects in their own lives, about how objects can take on a quasi-sacred or special meaning, and what from their own lives today might speak to a future generation that might open the time capsule.

Stephen Covey is well known for popularizing the aphorism, "Begin with the end in mind." The idea is as simple as imagining how others might eulogize you once you have died, and what you would like this eulogy to include. Few of us, I think, would doubt the powerful effectiveness of such an approach. But part of what is often left out of such thinking is the recognition that for people to begin with an end in mind, it is critical to have a strong sense of where one began. A vision for the future, whether in terms of a values-centric organization or in terms of individuals working within such an organization, is intimately connected to the hopes and dreams and aspirations one grew up with. If a vision for the values-centric company needs to begin with the end in mind, it also needs to end with a beginning in mind.

There are many specific practices that help elucidate personal stories: having a website that showcases individual histories, creating a book of personal stories that any and all can refer to, using coaching pairs to do a deeper dive into one another, and so on. The key glue in all of these processes is empathy. Empathy touches virtually every aspect of the green bottom line company. It goes to the heart of how leaders conduct (or don't conduct) themselves. It is the fundamental basis behind much of a company's practices and processes, and of feedback work specifically. It is an essential component in all communication and, therefore, a company's language system. And it informs how a company approaches its own historical context.

What's your story? one staff member asks another. In so doing, she is inquiring into a person's background: where someone was born and grew up, the backgrounds of parents and grandparents, where a person studied and has worked, what the person's interests and hobbies are—the list goes on and on. Both the effort of inquiry and the genuineness of interest are critical here: where both are strong, so too is empathy, where both are weak or lacking, the effort seems superficial and disingenuous. One thing seems fairly clear: The interest a company takes in the deep context of staff members cannot be faked. It is there or it is not. When it is present, one finds a tapestry of stories and histories that deepen the value sensibilities of a company. A business culture becomes all the richer for it.

The Other Green in Green Glue

So what are the costs of all this green glue involved in culture creation? Does it pay off?

It is hard to sell the return on investment of soft practices in business. What seems to happen as I work with leaders of companies is that, as they begin to internalize the skills and practices introduced, they begin to see people act and perform differently on the individual and team level. They begin to train people to implement the human resource practices that improve people performance. They begin to see improved financial performance as people get better at leading and managing themselves and others. Thus, the strategic plan includes goals for development. Then, the leaders support the budget to implement the plan.

Sustainable companies get it. They know they have to devote dollars to ongoing practices to ensure long-term success. Just as we learn that sustainable farming is not a fast business, we know that sustainable business cannot plant a seed, ignore it, and expect to see results. Thus, managing the people side of your business is one of the most important strategies you can implement. Remember that you have to have the right people on the farm working together, aligned in values and following a vision. They need to be going in the right direction together to create alignment in order to see their vision become a reality. They need the right skills and knowledge to perform. The organic fertilizer for a sustainable company is interpersonal skills.

Having such a long, consistent tenure with Melaver, Inc. does enable us to put some hard numbers in place to analyze the costs and benefits of the green glue that goes into shaping a green culture. Over the most recent seven years, I have devoted about fifty-four hours annually (or just over a half-day each month) to working with this company. Of those fifty-four hours, two full days are allocated to the annual retreat with all staff members present, and the balance is equally divided between my work with the Sanskrit team (about ten people) and one-on-one coaching sessions. Roughly speaking, I devote about a third of my time to the entire staff, a third to the leadership team, and a third to individual discussions. The time devoted to Melaver, Inc. is not precisely the same year after year. Work at an annual retreat can and often does set off a flurry of human resources work focused on a particular aspect of culture building. As the work progresses and various practices are implemented, my time tends to trail off until a new set of challenges presents itself and the cycle begins anew. And there are times when fiscal restraint or the sheer volume of work facing the staff limits the amount of money and time Melaver, Inc. is able to devote to outside consultants, myself included. A breakdown of my time from 2000 through 2007 can be seen in Table 2.3.

About five hundred staff hours are spent annually working with me, on average. Most of that time is spent working with entire staff. About 25 percent of the hours are with the Sanskrit management team; a small amount of time is spent in one-on-one coaching, as shown in Table 2.4.

So what are the costs and benefits involved in the time devoted to culture building? There is, of course, the cost of my time, which—on average—amounts to $10,725 each year. There is also the lost opportunity of revenue that comes from diverting

TABLE 2.3   BREAKDOWN OF CONSULTANT'S TIME, 2001–2007

YEARS	TOTAL HOURS
2001	78.61
2002	90.73
2003	24.07
2004	40.98
2005	11.68
2006	67.74
2007	61.50
Total	375.31
Average per year	53.61

"productive" work to culture-building (green glue) practices, which is basically a 1.25 percent reduction in potential revenue enhancement. We've used these figures to create a synopsis of these costs, totaling about $1 million over ten years, in the analysis of Green, Inc., the fictionalized company we introduced in Chapter 1, which is intended to be a reasonable approximation of Melaver, Inc. The costs per year and totals for lost opportunity and facilitator's time can be seen in Figure 2.3, the Updated Cash Flow Analysis for Green, Inc.

The benefit side of the ledger is obviously trickier to calculate. As part of our analysis of Green, Inc. in Chapter 1, we considered the issue of retention and the cost to the company of losing just any worker. Deliberately left out of that calculation was the retention of a good worker. Bob Willard, in his book *The Sustainability Advantage,*[7] estimates this premium to be around $159,000 or two to three times an annual salary when one takes into account these factors: loss of productivity while employee considers pros and cons of leaving, cost of company's time trying to keep the person from leaving, cost of payroll/benefits administration, cost of separation allowance, cost of training, cost of

TABLE 2.4 STAFF HOURS SPENT ANNUALLY WORKING WITH CONSULTANT

Entire staff	357 hours
Sanskrit team	125 hours
One-on-one	18 hours
Annual total	500 hours

Years	0	1	2	3
REVENUES/SAVINGS				
Revenues from Chapter 1		21,300	21,300	21,300
Revenues from Chapter 2				
Retaining key staff members		79,000	79,000	79,000
Total revenues from Chapter 2		79,000	79,000	79,000
Sub-total of all revenues (Chapters 1 & 2)		100,300	100,300	100,300
COSTS				
Costs from Chapter 1	(166,000)	(114,500)	(91,875)	(97,286)
Costs from Chapter 2				
Lost Opportunity from green glue		(46,875)	(53,906)	(61,992)
Cost of facilitator's time		(10,725)	(10,725)	(10,725)
Total costs from Chapter 2		(57,600)	(64,631)	(72,717)
Sub-total of all costs (Chapters 1 & 2)	(166,000)	(172,100)	(156,506)	(170,003)
Total Cashflow	(166,000)	(71,800)	(56,206)	(69,703)
Discount Factor	1.000	0.909	0.826	0.751
PV Cashflow	(166,000)	(65,273)	(46,451)	(52,369)
NPV	335,955			
IRR	20.85%			

Figure 2.3   Updated Cash Flow Analysis for Green, Inc.

lost future customer revenue from this lost employee, impact on departmental productivity, cost of backfilling, and cost of lost knowledge, experience, and contacts.

It is my belief that the type of culture we have been discussing specifically enhances the retention rate of middle to top management. To what degree? That's hard to say. If the typical large company turns over its entire workforce, including its good workers, every four years, let's assume that a conventional company the size of Green, Inc. turns over its good management team twice as slowly, or every eight years. Much of Melaver, Inc.'s management team has been in place already for eight years. It has added people to the team as the company has grown, but it has lost only two people during that time. Average tenure with the company among the Melaver management team is now about seven years and increasing annually.

With this real data in mind, let's assume the following:

■ Green, Inc. pays $80,000 (on average) for a member of its management team.
■ Green, Inc. turns over its senior management team every twelve years.
■ Green, Inc. pays a headhunter a 30 percent fee to hire a new manager.

The benefits, then, to constructing a green culture for Green, Inc. thus amount to approximately $79,000 a year, as seen in Table 2.5. For the sake of smoothing the analysis, I've held this savings statically consistent over ten years. In actuality, the numbers are

4	5	6	7	8	9	10	TOTALS
21,300	21,300	471,300	538,800	548,925	550,444	550,672	2,766,640
79,000	79,000	79,000	79,000	79,000	79,000	79,000	790,000
79,000	79,000	79,000	79,000	79,000	79,000	79,000	790,000
100,300	100,300	550,300	617,800	627,925	629,444	629,672	**3,556,640**
(103,490)	(110,605)	(118,765)	(128,125)	(138,866)	(151,190)	(165,335)	(1,386,037)
(71,291)	(81,985)	(94,282)	(108,425)	(124,688)	(143,392)	(164,900)	(951,737)
(10,725)	(10,725)	(10,725)	(10,725)	(10,725)	(10,725)	(10,725)	(107,250)
(82,016)	(92,710)	(105,007)	(119,150)	(135,413)	(154,117)	(175,625)	(1,058,987)
(185,506)	(203,315)	(223,772)	(247,275)	(274,279)	(305,307)	(340,960)	**(2,445,024)**
(85,206)	(103,015)	326,528	370,525	353,646	324,137	288,711	
0.683	0.621	0.564	0.513	0.467	0.424	0.386	0
(58,197)	(63,964)	184,316	190,138	164,978	137,466	111,311	0

likely to be much more volatile. In the early years of a major transition such as Melaver, Inc. underwent, there is a likelihood of losing a few good people early on who are simply not aligned with the direction of the company (hence the loss of two people from Melaver's management team). However, in later years, the retention rate becomes extraordinarily strong, well beyond what my analysis is projecting.

TABLE 2.5   FINANCIAL BENEFIT OF RETAINING GOOD MANAGEMENT

ITEM	COST PER PERSON	STANDARD SMALL COMPANY	GREEN, INC.	ANNUAL SAVINGS
Turnover of good personnel (years)		8	12	
Annual loss of good personnel per year		1.25	0.8333	
Cost of recruitment	$ 7,000	$ 8,750	$ 5,833	$ 2,917
Headhunting	$ 23,850	$ 29,813	$ 19,875	$ 9,938
Cost premium for losing a good worker	$159,000	$198,750	$132,500	$66,250
Total				$79,105

The costs of culture building for Green, Inc. and the benefits derived by a stronger retention rate among the management team result in costs over ten years of around $1,058,000 and benefits over the same period of $790,000, for a loss over ten years of $268,000 or $27,000 per year. It's not a significant number in the scheme of things, when one considers that Green, Inc.'s annual gross revenue is around $12 million. Secondly, I think that in the interest of being very conservative, the benefits are probably understated and the costs (particularly lost opportunity of revenue) may be overstated. For example, in Chapter 1, the assumption was made that Green, Inc. would benefit by doing two $15 million development projects in Years 6 through 10 of the analysis. If just one of those projects each year increased in scope from $15 million to $16 million, the $27,000 projected shortfall would be covered.

Having said all of that, the numbers for Green, Inc. bear out the fact that the culture-building investments discussed in this chapter still make sense. If we take the discounted cash flow analysis from Chapter 1 and update it with the figures we have been discussing here, we find that despite a conservative analysis that probably overstates costs and understates savings and income, Green, Inc. is still projecting a positive net present value of its various soft investments of $336,000 and an internal return rate of 20.85 percent—still above the company's targeted return of 15 percent. The updated cash flow analysis can be seen in Figure 2.3.

Concluding Remarks

A conventional real estate development firm works in a linear fashion. A company ties up a piece of land, and the development team creates a vision for what the project will look like. A design team, comprised of architects and engineers, develops a plan for how the project will look, while a financial team develops a plan for how the project will be profitable. A construction team executes the plan, constructing the conceived buildings, while a leasing or sales team works on filling the project with tenants and/or owners. Finally, a management team steps in to maintain the project to desired standards. From envisioning to building to running to maintaining, the process is linear. A conventional business functions similarly. A small cadre of leaders conceives of a vision for the company. The vision is delegated to an upper management group that develops a multi-year plan for building the vision. Middle management is charged with executing the plan. And the day-to-day troops are tasked with maintaining the organization.

By contrast, the green bottom line company is shaped through integrated design, where envisioning, building, executing, and maintaining come together. The vision for such a company is fundamentally about the values of its people and about community, the community of people within the organization and the external community of which this organization will be a steward. People within the green bottom line organization, guided by an empathic leadership team, collectively draw upon their values to construct a culture that is the physical manifestation of the company's vision—a culture comprised of governance structures, rituals, language, and history that enables the organization to cohere. Empathic leadership, self-awareness and respectful understanding of others, refined communication skills, mentoring—these practices shape the culture and make it stick. If people in an organization are the bricks of a culture, then the culture-building practices of an organization serve as the mortar—the green glue—that holds it together. The vision of a green bottom line company is indistinct from the way that company's culture is constructed and practiced. The vision of a green bottom line company is its culture, and the culture, in many ways, is its vision.

Businesses typically work from a vision that gets translated into a multi-year strategic plan that, in turn, leads to various tactical initiatives. The business has a blueprint for where it wants to go and how it intends to achieve its objectives. Creating a vision, defining shared values, assessing the struggles impeding the creation of a high-performance team, developing initiatives and processes for improvement, measuring the company's progress—these standard elements of a company's blueprint determine how its culture is shaped and the critical role people play in creating it. The green bottom line company, instead, has what I like to call a greenprint. A greenprint begins with a central question: *How can we build a sustainable business model with processes that will hold it together?* The fundamental response to that question—a company's strategic greenprint for operating sustainably over the long term—is all about people.

What inspires people in an organization? I believe that the shared values and beliefs of a leadership team can trigger intellectual activity, excitement about direction, and greater commitment. A leadership team that can inspire, develop, and empower people within an organization will create a well-run machine, but studies report that most people in organizations use only 5 to 10 percent of their abilities at work. But if people are using only 5 to 10 percent of their abilities on the job, they must be applying the rest to other activities that bring meaning to their lives.

After a decade working with Melaver, Inc., it is clear that the company's focus on sustainability inspires people to give well beyond 10 percent of their abilities. They feel good about their reduction in carbon emissions, their reduced consumption of energy and water, and their general stewardship of the environment. They also feel good about giving back to the community through a strong sense of civic responsibility. Perhaps most importantly, Melaver, Inc. staff members feel good about the sense of belonging they have within a culture they helped shape.

The company's belief in what it does, this green glue, does more than hold the company together. It impacts the community such that others start getting stuck to it as well. The company discovers a network of people beyond its walls watching what it does, and what they find is indeed inspiring. This company "gets it." But it took considerable time and effort for this passionate and talented group of people to pull it together.

It takes a significant investment for an organization of people to open themselves up to feedback from others. As a coach, I tell leaders that they will likely receive feedback that they could improve their communication style. After seminars and coaching sessions, they still struggle. At Melaver, Inc., some changes are visible only after constant coaching and practicing of skills until they become internalized. It takes time to plant seeds and watch them grow. It is rare to find leaders who actually implement the changes that are necessary to make a difference.

It takes perhaps even a greater investment for an organization to manage the change that open communication sets in motion. Managing change and assisting employees with their personal transitions through change are valuable skills. Change happens; it is the transition to a new way that we resist. Some people are ready, others are not. Some resist, and some lead the way. All go through stages that include letting go of something, being in a zone where we do not have clear outcomes and moving towards a comfort level with the new way of operating. In fundamental ways, the green bottom line company, by dint of undergoing changes within its own business culture, reveals the nature of changes our civilization as a whole must soon undertake.

NOTES

[1] www.melaver.com

[2] Thomas A. Stewart, "America's Most Admired Companies: Why Leadership Matters," *Fortune,* March 2, 1998.

[3] Warren Bennis, "The Leadership Advantage," *Leader to Leader* 12 (Spring 1999), pp. 18–23.

[4] Daniel Goleman, Richard E. Boyatzis, and Annie McKee, *Primal Leadership: Learning to Lead with Emotional Intelligence* (Boston: Harvard Business School Press, 2004), p. 50.

[5] For more information on the MBTI, see Isabel Briggs Myers with Peter B. Myers, *Gifts Differing* (Palo Alto, Calif.: Consulting Psychologists Press, Inc., 1980); Isabel Briggs Myers and Mary H. McCaulley, *MBTI Manual: A Guide to the Development and Use of the Myers-Briggs Type Indicator, Third Edition* (Palo Alto, Calif: Consulting Psychologists Press, 1998). The Myers-Briggs Type Indictor and MBTI are registered trademarks of the Myers-Briggs Type Indicator Trust.

[6] For more information on FIRO-B, see Will Schutz, *FIRO: A Three-Dimensional Theory of Interpersonal Behavior* (New York: Holt, Rinehart & Winston, 1958); Will Schutz, *The FIRO Awareness Scales Manual* (Palo Alto, Calif.: Consulting Psychologists Press, 1978); J. A. Waterman and J. Rogers, *Introduction to the FIRO-B, Third Edition* (Palo Alto, Calif.: Consulting Psychologists Press, 1996). FIRO-B is a registered trademark of Consulting Psychologists Press.

[7] Bob Willard, *The Sustainability Advantage* (New York: Henry Holt & Co., 2002), p. 34.

3

GREEN FROM THE INSIDE OUT

TOMMY LINSTROTH

SUMMARY

Once a company shapes its organization around a core set of values and then devotes time and resources to developing a culture that is in synch with those values, it needs to consider the extent to which its own practices serve to "walk the talk." Potential hires, staff members, outside vendors and, increasingly, customers and clients are seeking out organizations that are authentic in the products and services they offer as well as in their organizational practices. As head of sustainable initiatives for Melaver, Inc., Tommy Linstroth is charged with ensuring that the company "walks" its green values every day and is authentic from the inside out.

Linstroth establishes a framework for organizations seeking to increase the authenticity of their sustainability message. This chapter first considers the general challenge of trying to be authentic against the moveable benchmark of sustainability. The issue of authenticity then leads Linstroth to what sustainability means to the organization and how that concept is determined. Next, the metrics of sustainability are considered. Much as financial performance is measured by looking at net operating income and return on investment, a company's sustainability metrics can be measured, analyzed, and improved. The chapter then considers what metrics to focus on, how to measure them, and how to improve upon them. Also considered are the many resources available to help create a sustainable organization. The chapter ends with a general discussion about where to go next, including how to evaluate partners and vendors to further improve the authenticity of sustainability.

> This chapter is not intended to define absolutely—nor could it define absolutely—what sustainability is and how it should impact a company. What it hopefully will do is identify how to develop a sense of sustainability that is authentic to a particular organization, one that an entire firm can describe, affirm, and embrace through its day-to-day practices.

Sustainability and Authenticity

Think for a moment about bamboo. In the last few years, this quick-growing grass (it's not a wood, as is commonly thought) has become a material of choice in many green buildings. It is most often used as a substitute for hardwood flooring, but also appears in plywood, cabinets, countertops, bedding, and window treatments. The *moso* species of bamboo, which can be used to make these products, can grow up to thirty feet in one year, and needs only five years to mature. The same plant can also be cut and harvested four or five times before it must be replanted. These characteristics make bamboo a sought-after material in the green building community. Because bamboo is rapidly renewable, easy to grow, cost competitive, and able to be reharvested, it is considered a very sustainable building material.

Or is it? Nearly 100 percent of the bamboo products entering the United States and used in green buildings are grown and manufactured in China or Southeast Asia and shipped halfway around the globe to reach the U.S. marketplace. Hmmmm. Perhaps that bamboo flooring is not looking as green as it seemed initially. But the benefits of not cutting down domestic forests still outweigh the impacts of transportation, right? Maybe—but let's look at the other factors that come into play. Were vast swaths of native forests and vegetation slashed in China to make room for bamboo plantations? Were fertilizers and pesticides applied to ensure quick growth, and where did that runoff go? What are the labor conditions? Are workers being exposed to toxic fumes while they glue the layers of bamboo together? Are they being paid fair wages? Are the manufacturing plants meeting specific emission standards?

In asking these questions, I'm not condemning bamboo. It's a fantastic material, with near-endless potential. Many bamboo products may be grown and harvested in a sustainable manner, with workers receiving a fair wage and working in healthy conditions. That's not the point of this narrative. The point is that any product or business claiming to be green or sustainable is (and should be) subject to an ever-increasing level of scrutiny. It is no longer enough to offer a sustainable product. Consumers, partners, employees, and investors are looking for firms that raise the bar on product or service offerings by embracing wholeheartedly the concepts their products purportedly advance or support. Firms must be authentic in their actions to have their products or services fully embraced in today's marketplace.

The ramifications of authenticity (or lack thereof) can destroy a company's ability to market a sustainable product or service. Would you, for example, buy that new recycled-content countertop if the firm that manufactures it were releasing ozone-depleting gases with each slab of material it produced? Or would you look for an alternative material that would serve the same purpose but is not made by a company that is destroying the ozone layer? (This is a fictional example, at least to the best of my knowledge.)

There are numerous examples of companies facing challenges to their authenticity:

■ Toyota, seen as an environmental innovator for the development of the fuel-sipping hybrid Prius, joined other auto manufacturers in a lawsuit against the state of California to challenge the state's stringent vehicle emission requirements.

■ General Electric's Eco-imagination campaign highlights efforts to become a more sustainable company by focusing on the research and production of clean technologies and green products. Will consumers buy this or will they remember the GE that dumped millions of pounds of PCBs into the Hudson River or the fact that it is responsible for the largest number of Superfund clean-up sites in the country?[1]

■ British Petroleum (BP) has spent hundreds of millions of dollars since 2000 re-branding itself "Beyond Petroleum," while its leaking oil pipes in Alaska in 2006 caused the largest-ever North Slope oil spill[2] and after announcing a $3.1 billion deal to pursue extracting oil from the Canadian tar sands in upcoming years. Does that sound like moving beyond petroleum?

CAST STUDIES IN AUTHENTICITY

What follows are some real, personal experiences involving green development. I will provide the framework, and you can assess the degree of authenticity for each one.

Case One

Scenario: A meeting with a fellow green developer with a clever master plan to create a large-scale sustainable development. This development would rekindle interest in a forsaken part of town, with hundreds of thousands of square feet of green buildings being constructed over the next few years. During the course of the meeting and a tour of the developer's main office, I noticed there were no recycling bins available, and that indeed, numerous aluminum cans were being thrown away in the trash. Could a green developer whose office lacks the easiest, most basic recycling effort be considered authentic?

Lesson: In the green world, first impressions last. Recycling is often used as a first sniff test for organizations—it is a basic tenet of sustainability, is cheap and easy to do, and is visible—if your organization does not have a recycling program for paper, plastic, glass, and cans—get one!

Case Two

Scenario: An open call for architectural/design proposals to win a commission to build a new green building for a public agency. This process involves sitting before a group of decision makers and giving your best case for why your firm should be selected to design the building. As this was a green building, with sustainable aspects to the building referenced throughout the request for proposals, the agency charged with awarding the contract assumed that the competing architectural firms that were submitting proposals were already either green or had a true commitment to going green. Naturally, during the question-and-answer portion of each firm's presentation, the topic of sustainability came up. One firm, which had expressed a strong desire to focus its work around designing green buildings (and thus *this* green building), was asked two questions: First, does your firm have a recycling program? No, it did not have a recycling program. Second, what is your definition of sustainability? You could see the collective gasp as the representatives of the firm stumbled for what they thought the correct answer should be, obviously conjuring up a response on the spur of the moment. After conferring among themselves about who should answer the question, the response came from an outside consultant (not a member of the firm), who spoke vaguely about the importance of energy efficiency. Everyone around the table could sense this idea had not been discussed previously. There is no right or wrong answer to defining sustainability. Part of the challenge with the notion of sustainability—as well as its potential—rests with the fact that it has a different meaning to each person, firm, and organization. The important thing is that the topic has been broached within an organization and deliberated, so that a company indeed has *some* definition it feels is meaningful and provides purpose and direction. Not having a thoughtful response creates an authenticity gap for a firm hoping to broadcast its desire to design sustainable buildings.

Lesson: When you advocate your desire to design sustainable buildings, create a sustainable product, or offer a sustainable service, you need to have a well thought out concept of what sustainability signifies, and everyone in your organization should be able to articulate it and/or provide a personal perspective on the company's general philosophy. (This concept will be covered in more detail later in the chapter.)

Case Three

Scenario: A meeting with another architecture firm with two green projects in its portfolio. A casual conversation ensued about the firm's other current projects—a building here, a parking deck there—lots in the hopper. Knowing the firm had worked with green technology before for clients who required it, the representatives were asked if they were incorporating a few basic green features, such as waterless urinals or ultra-low flow toilets, into the projects they had started after the completion of the aforementioned green building projects. The answer was, no, they had not thought about using waterless urinals or low-flow fixtures. Incorporating sustainability only when an owner/client/consumer/investor asks you to is certainly not authentic.

Lesson: Sustainability must be fully embraced through all aspects of an organization and its product and service lines. If it is not, the finicky marketplace will move on to a company that does. To be seen as giving lip service to sustainability—providing it only when it is convenient or when a client specifically requests it—or to be found to be greenwashing can be a fatal flaw within the ever-growing sustainability community.

So how does one avoid being labeled well-intentioned but lacking authenticity? While the degree at which the bar is set varies with each partner/vendor/client/investor, the rest of this chapter will suggest a framework for moving an organization from being a company that has a green product to one that embodies sustainability in its everyday business practices.

What Does Sustainability Mean to Your Organization?

In the past few years, sustainability has become a buzzword that's tossed around on every topic. Many organizations issue an annual sustainability report. In fact, if you Google the word sustainability, you will get over 13.5 million hits!

One early and oft-cited definition of sustainability comes from the Brundtland Commission. Its *Report of the World Commission on Environment and Development*[3] defined sustainability as that which "meets the needs of the present without compromising the ability of future generations to meet their own needs." Sufficiently vague? Whose needs are to be met? Those who need a three-car garage to keep their boat out of the weather, or those who are on the street without food and shelter? This definition is often one's first exposure to sustainability, and while it conveys the right message—we need to balance our needs against those of future generations—a company's definition of sustainability needs to be much more specific. The details will vary from organization to organization and person to person. This is OK—what it means to a real estate company will play out much differently than what sustainability means to a large manufacturing company.

At Melaver, Inc., there are several components to our definition of sustainability. First, we aim to have a profound, positive impact on the communities in which we live and work and a notable lack of impact on the environment in those communities. Second, we examine every decision we make according to a triple bottom line: Decisions made must not only be financially profitable, but environmentally and socially beneficial as well. The decisions we make fall into the sweet spot identified by the confluence of all three components, seen in Figure 3.1.

As a general concept, triple bottom line sustainability is easy to understand. It means that we are turning a profit (so we can stay in business), but not harming our environment, all the while bettering our communities and society. For us, the meaning of sustainability comes through many actions we take and even more through actions we *do not* take. But what about for your organization?

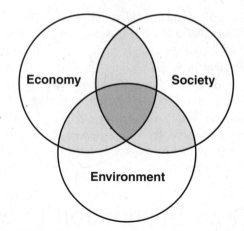

Figure 3.1   Sustainability as triple bottom line.

According to Daniel Esty and Andrew Winston, authors of *Green to Gold*, there are two driving forces behind the push for sustainability. "First, the limits of the natural world could constrain business operations, realign markets, and perhaps even threaten the planet's well-being. Second, companies face a growing spectrum of stakeholders who are concerned about the environment."[4] This would lead one to believe that the path to sustainability is charted through a narrow strait: avoiding constraints posed by the natural world while also avoiding pressure from various stakeholders. Flashback to the authenticity question: Imagine a reporter asking you why your firm has pursued sustainable initiatives. What do you really want to be the rationale, and what do you feel is the true explanation? Consider which of the following explanations seems most compelling:

- "Well, we didn't necessarily want to, but boy, a few board members sure did."
- "Our CEO made us go in that direction."
- "We had to make that move in order to get a certain client's business."
- "Our customers were asking for it."
- "We felt it was an opportunity to create greater brand value for the company."
- "It fit our overall set of core values and beliefs."
- "Our staff members felt it was an important step to make."
- "The market was moving in that direction."
- "We were·facing regulatory pressures that pushed us in that direction."
- "We felt a moral responsibility toward addressing a global need to reduce our footprint."

Usually an organization develops an overarching sense of authentic sustainability either as a top-down decision by key executives or as an upwelling of support from employees. Each approach has advantages and disadvantages. It often takes a combination of both approaches to truly create a sense of buy-in for any new idea. Sustainability is no different. For example, having the CEO unilaterally decide that the company will pursue sustainability in all business operations can be seen as just another directive coming from the top, lacking employee buy-in. This can lead to a range of feelings, from resentment about additional work that obviously might arise from this decision, to a what-the-heck-does-he-know? mentality, to a sense of relief that upper management is finally moving in the direction of employees' personal beliefs. In the bottom-up scenario, employees can incorporate sustainability concepts into everyday business practices, from ordering recycled paper to changing to compact fluorescent light bulbs. But this approach, too, has pitfalls—it can have limited reach, and often gets stalled as it progresses up through middle and upper management.

In a thirty-person firm such as Melaver, Inc., the organization's CEO espoused the need for sustainability in all aspects of the organization. However, this transition did not come overnight as a mandate. Rather, concepts were identified, and a two-year odyssey followed to determine what sustainability really meant to the company and what it meant to each individual within the organization. This process culminated at a company retreat, with each employee identifying traits that the concept of sustainability must embody, and resulting in a sense of ownership for all employees. Every single person had input into what sustainability meant to the organization and a hand in shaping the direction of the company. It also meant that the employees' vision had full support of the company executives, ensuring the role of sustainability as a primary factor in all business decisions as a successful, lasting concept for the company.

Will this exact approach work for a firm with five thousand people? Five hundred? Even fifty? Each company has a unique methodology for decision making and its own logistical processes. (Imagine one thousand employees crowded into one room trying to define what sustainability means to their organization!) The main point is that it takes a comprehensive approach to determine what sustainability means for each company, an approach that should include not just the executives but all the folks on the ground who will be implementing the decision day to day.

If sustainability is part of the fabric of the organization (as it is at Melaver, Inc.), someone can walk into a company's office and ask anyone, from the CEO to the receptionist, about what sustainability means to the company and what the company is doing about it, and the person will receive a fairly consistent answer. Part of this is the comprehensive process by which a company determines what sustainability means, which translates to genuine buy-in to the overall vision. Part is the continual reinforcement coming from the sustainability department, and part is the continued support from upper management.

Sustainability affects all components of an organization, from finance to operations to legal and human resources. To help span this diverse spectrum within the organization chart, many companies (Melaver, Inc. included) designate a sustainability director or green officer to oversee sustainability initiatives and to help ensure that sustainability is incorporated into business operations. However, there can be pitfalls when designating the responsibility for incorporating a sustainability program throughout the company to one person or department. This was a concern at Melaver, Inc. when the position was first created—that sustainability would simply be the responsibility of the "green guy," rather than the responsibility of everyone in the company. Having this position should not absolve individual employees of the responsibility to continue to implement the sustainable vision as well as to come up with new ideas for furthering the overall goal.

Esty and Winston identify three issues that can arise by isolating sustainability responsibilities to one person or department: 1) lack of support outside of the sustainability department, 2) insufficient budgets, and 3) lack of communication between disparate groups within the company.[5] Let's consider each of these challenges in somewhat greater detail.

Lack of support. Widespread support for sustainability initiatives often comes from three mechanisms: that those affected were part of the idea, that the idea is so fantastic that everyone benefits, or that the new idea doesn't affect anyone's day-to-day activities so there is no reason not to support the initiative. However, if an idea that arises from an isolated green team does not receive broad organizational support, or if there is resistance to change or no enforcement mechanism, a new program may not last, especially if it necessitates changes in employee behavior or routines.

At Melaver, Inc., we have faced this issue multiple times. As part of the evaluation of our environmental footprint (discussed in greater detail later) we examined our levels of waste and the amounts of items we were purchasing. A little digging turned up some interesting numbers. Our Savannah office provided liquid refreshments to employees at the company's expense. Soda, bottled water, tea, coffee, and juices were provided in the break room for employees to consume at their leisure. In fact, the office was purchasing $1,800 worth of soda annually (over six thousand cans!), creating both procurement and disposal issues. Even when soft drink cans are made from recycled material, they still require a large amount of energy to clean, manufacture, fill, and distribute.

Multiple solutions were suggested to reduce the impact of offering beverage alternatives to employees. The sustainability department suggested eliminating soda cans altogether and installing a soda fountain in the kitchen. This led to discussions regarding maintenance, freshness, space, etc., and represented a potential significant change of habit for most employees. Instead of being able to grab a can of soda, one would have to grab a glass, syrup bags would have to be changed, spilled soda would need to be cleaned from the fountain, the ice maker would have to be kept running. After a course of meetings with all the employees to discuss alternatives that would provide refreshments while lowering the environmental and financial impacts, a compromise was soon reached, with the support of upper management. We eliminated the purchase of canned soda, but provided each employee with a refillable glass that he or she could take to the sandwich shop downstairs and fill, with the cost of the soda absorbed by the company. Employees still did not pay for soda, but instead of popping a can in the kitchen, they had to walk downstairs and fill their glasses themselves. This was a significant change, one that would have been more difficult to implement if it had been limited to a directive from the sustainability department. Because the discussion involved all the employees and senior management, the solution was eventually acceptable to everyone.

Often the sustainability department will suggest ideas that push the envelope, knowing they may be scaled back in the course of debate and implementation. That's what a sustainability department is for—to drive the debate and expand the organization's sustainability practices.

Insufficient budgets and intangible advantages. This issue is nicely illustrated by the unfunded government mandate. In Savannah, Georgia, the county jail is required to house state inmates, but the county is not fully compensated for costs associated with feeding, housing, and monitoring them. In the absence of compensation, the requirement becomes an unfunded mandate, and the financial burden is passed to the county, and thus to the taxpayers. The county was not provided with the resources to ensure compliance with the state requirement, but was mandated to act without adequate funding.

This might seem an odd comparison to creating a sustainability program (though some CFOs could be compared to wardens). However, if the mandate to create sustainability programs is issued, whether the mandate is simple (operate greener) or more complex (become a net generator of electricity), the program will not succeed if it is not backed by an adequate budget. Nothing hampers progress more than creating a directive but not providing adequate resources to implement it.

Many sustainable initiatives deal with energy and water efficiency and offer an attractive payback for the initial capital expenditure. Replacing old, inefficient lights in an office with new, high-efficiency fixtures may cost a few thousand dollars up front, but with a payback of a year or so based in energy cost savings as well as the environmental benefits, this type of project is usually green-lighted immediately. For example, a building on which we performed a lighting retrofit in Birmingham, Alabama required a capital investment of around $35,000. The annual savings in electricity is at least $30,000. This is just over a one-year payback, with $30,000 of operational savings every year after (even more if the price of electricity goes up) and the additional benefit of reduced carbon dioxide emissions.

Similar low-hanging fruit involves water savings. Typical faucet aerators use 2.2 gallons per minute. By spending less than three dollars per faucet, water consumption can be cut to half a gallon per minute. This is a water use reduction of over 75 percent, with a capital cost of a few dollars per fixture.

However, not every sustainable measure yields immediate financial savings, and at times, a direct payback is not necessarily realized by the company. For example, creating an extensive recycling program may involve a monthly collection fee as well as additional time and effort on the part of staff, but rarely offers any financial payback.

In short, all three projects discussed above—changing out lighting fixtures, installing faucet aerators, and recycling—further the organization's goal of sustainability. But if every single program pursued must offer an immediate financial payback in order to be implemented, the organization would soon run out of initiatives to pursue.

While an organization does not have to budget hundreds of thousands of dollars toward these initiatives (though some firms have budgets in the millions), funding is often necessary to truly implement a sustainable vision. The upside is that organizations can start small and begin to implement low-cost, low-hanging fruit programs suited to their organization before moving forward with more costly items (alternative energy solutions such as photovoltaic panels, cogeneration of heat and electricity, vegetated roofs), while beginning to accrue the differentiation, marketing benefits, and goodwill that such actions provide. Oftentimes, a multi-year approach to continuous process improvement leads to this happy and unexpected discovery: previous years' low-hanging fruit gives way to other easy pickings in subsequent years.[6]

Lack of and inconsistent communication. Focusing sustainability initiatives in just one department can lead to a significant gap in understanding the role of sustainability in an organization's business model. Toyota, usually considered the greenest of the major automakers, joined other auto manufacturers in a lawsuit against the state of California to challenge vehicle emissions requirements. It also lobbied against increasing federal corporate average fuel economy (CAFE) standards. Do you think its sustainability department was happy to explain this to Prius owners? McDshell packaging and opening the world's first CFC-free restaurant in Vejle, Denmark, released its first "Report on Corporate Social Responsibility" with great fanfare, only to be strongly criticized by Paul Hawken for such unsustainable practices as using 600 gallons of water for every quarter-pounder and using ten calories of energy for every calorie of food it produces.[7] Senior executives at Pepsico have publicly expressed concern over the increasing amount of refined sugar in the American diet, a long-term trend linked to increased levels of obesity in the United States, but are nevertheless caught up in the paradox of a business model built around peddling soda.[8] Over two thousand CEOs of major global corporations have signed on to the United Nations Global Compact, emphasizing basic principles concerned with human rights, labor standards, the environment, and anti-corruption.[9] All but a small fraction, perhaps, live up to those principles. Inconsistency in a company can be exacerbated by isolating sustainability in one department, rather than diffusing it throughout the organization.

How do you overcome these three challenges? One solution is integrating "the concerns, needs, and incentives of those on the company's operational front lines into the game plan."[10]

As building owners and managers, Melaver, Inc. implements what we call Mark of a Difference sustainable property management practices. One practice is a green cleaning program. Rather than taking a one-size-fits-all, top-down approach, we meet with the janitorial crew at each property. We ask about problem areas, recurring issues, and hard-to-clean spots. We strive to involve all the parties at the beginning, especially those who are directly impacted by sustainable practices. In the case of green cleaners, we've learned that the same cleaning products are not always available in different markets. Different buildings and tenant types require different green cleaning approaches. And some green cleaning products just don't work well. We have learned a lot by involving, from the early stages, the front line operators—getting their buy-in and support and creating a sense of ownership in the results.

The same approach holds true for creating an effective sustainability program company-wide. It requires communication and buy-in from all aspects of the organization. Early and frequent discussions with financial officers will help secure an adequate budget for new initiatives. Dialogue with those who actually implement new policies

or practices, from purchasing recycled paper to green cleaning to choosing new partners or vendors, helps ensure a thorough understanding of the rationale, and provides
a broader spectrum of ideas and support. Consistent communication and interaction
fosters a sense of sustainability into all facets of the organization. Having a values-
centric organization and a set of cultural practices built around those values—issues
discussed in the prior two chapters—are critical foundation stones that support an
ethos of consistent communication and staff-wide interaction.

How Sustainable Are We?

The decision has been made to push an organization in a direction that embodies sustainability as a core component. Support is company wide, from the C-level positions
(CEO, COO, CFO) down to the front line workers. There might even be someone
specifically tasked with overseeing these initiatives and armed with a budget to boot.
How do you measure success?

A primary hindrance in the drive toward corporate sustainability is the lack of
clearly defined metrics. Financial metrics are fairly standard for most industries. You
can measure success based on rate of return, return on equity, gross sales, or increase
in profits. But how do you measure the sustainability of an organization's day-to-day
business operations?

The most common metric organizations consider is the company's carbon footprint.
A carbon footprint refers to the amount of carbon dioxide (and its equivalents, as
explained below) generated through business operations. (The term "environmental
footprint" is often used as well, though an environmental footprint may be more inclusive than activities that result in carbon emissions, such as procurement and recycling.)
While not by any means a complete snapshot of an organization's environmental
impact, a carbon footprint gives a company a starting point for analyzing energy consumption, travel, water use, and waste. Methods of determining your carbon footprint
vary widely by industry.

DETERMINING YOUR CARBON FOOTPRINT

Measuring your company's carbon footprint (and reducing your contribution to climate
change and global warming, a large-scale environmental threat) starts with greenhouse
gases. The World Resources Institute defines global climate change as "the desta-
bilization of the earth's climate system caused by an increase in the concentration of
greenhouse gases (GHGs) in the atmosphere."[11] In a nutshell, our atmosphere contains
a variety of heat-trapping gases called greenhouse gases, so named because they blanket the earth, functioning in a similar fashion to a farmer's greenhouse or a car sitting
in the sun. In both cases, the glass lets light pass through, but traps the heat energy.
This results in temperatures within the greenhouse or car being significantly warmer
than temperatures outside. These greenhouse gases form a blanket around the planet
that works like the glass of the greenhouse or the car, keeping Earth's surface about

54 degrees Fahrenheit warmer than it would be without an atmosphere.[12] GHGs are necessary for life on Earth—otherwise our planet would be freezing. The problem occurs when the levels of greenhouse gases sharply increase, as has been occurring at an accelerated rate over the past two hundred years. This dramatic increase in GHGs throws off the delicate atmospheric balance and contributes to a host of climatic changes, the most significant being an ever-warming planet, which in turn contributes to the melting of the polar ice caps, rising sea levels, and a dramatic shift in weather patterns.

Carbon dioxide (CO_2) is the primary greenhouse gas of interest, as human activities are greatly increasing its concentration in the atmosphere, though there are six main greenhouse gases that scientists measure.[13] In the past century, the concentration of carbon dioxide in the atmosphere has risen almost 30 percent, methane levels have more than doubled, and nitrous oxide concentrations have increased by 15 percent.[14] Furthermore, the growth in atmospheric greenhouse gas concentrations has been progressing at a more rapid rate over time. For example, concentrations of CO_2 have increased during the last forty years at a rate 50 percent higher than seen in the previous two hundred years (54 parts per million [ppm] increase from 1960 to 2000 vs. the

TABLE 3.1   COMMON GREENHOUSE GASES

GAS	ATMOSPHERIC LIFETIME (YEARS)	GLOBAL WARMING POTENTIAL (100 YEAR)
Carbon Dioxide (CO_2)	50–200	1
Methane (CH_4)	9–15	21
Nitrous Oxide (N_2O)	120	310
Hydrofluorocarbons (HFCs)		
HFC-23	260	11,700
HFC-125	29	2,800
HFC-134a	13.8	1,300
HFC-143a	52	3,800
HFC-152a	1.4	140
HFC-227ea	33	2,900
HFC-236fa	22	6,300
Perfluorocarbons		
Perfluoromethane (CF_4)	50,000	6,500
Perfluoroethane (C_2F_6)	10,000	9,200
Sulfur Hexafluoride (SF_6)	3,200	23,900

Source: Intergovernmental Panel on Climate Change.[16]

36 ppm increase from 1760 to 1960).[15] What this means is that humans are releasing a variety of greenhouse gases at much higher rates than has happened historically. Additionally, all of these gases differ in their ability to trap heat. As Table 3.1 shows, the six primary greenhouse gases have varying lives in the atmosphere, and vary in their ability to trap heat. This makes comparing the gases difficult, so GHG emissions are often reported by each gas's global warming potential. For example, methane has a global warming potential of 21, meaning that one molecule of methane has twenty-one times more heat-trapping ability than one molecule of carbon dioxide. Thus, while carbon dioxide is the gas we hear most about, even minor releases of other gases have significant impacts. Some of the gases are over 20,000 times more potent than carbon dioxide, though they are released in such small quantities that they are not likely to trigger widespread warming on their own.

When reporting emissions for an organization's carbon footprint, all the gases are represented as carbon dioxide equivalents, or eCO_2, which translate the global warming potential for each gas into one common denominator so it is possible to compare total emissions, regardless of which gas is emitted. For example, if a company is responsible for one thousand units of carbon dioxide (with a global warming potential of 1) and one hundred units of methane (with a global warming potential of 21) the total GHG emissions would be expressed as 12,100 units of eCO_2. (1,000 $CO_2 \times 1 +$ 100 methane $\times$ 21 = 12,100).

So Where Do All These Gases Come from?

Carbon dioxide is by far the most significant GHG, making up 77 percent of global GHG emissions. Nearly all CO_2 is released through our burning of fossil fuels for energy generation (coal for electricity, natural gas for heating, or oil for transportation). Methane is the result of decomposition of organic matter in landfills, but it is also associated with leaking natural gas, as well as rice and livestock production. Nitrous oxide is released through fossil fuels combustion (especially automotive) as well as fertilizer production and use. Hydrofluorocarbons, perfluorocarbons, and sulfur hexafluoride are complex compounds that are produced for industrial and manufacturing processes, as refrigerants, and as replacements for ozone-depleting substances. Typically, service-based organizations do not emit these gases unless they have a refrigerant leak. These gases are associated with mining, refining, and electricity distribution.

Now that you know what the greenhouse gases are and how they are expressed, your organization can get around to measuring its own carbon footprint. This will not require the services of a dedicated climate scientist, greenhouse gas specialist, or environmental engineer. This is a process that can be done in-house.

What Resources Can Help Me with Carbon Footprinting?

A challenge to implementing a successful sustainability strategy is the lack of available resources. While a Fortune 500 company may have the assets to hire a specialist to focus on greenhouse gas emissions and other sustainability indicators, smaller service-based organizations may not have such resources or expertise available. Fortunately, there are a host of free resources available to help organizations large and small implement sustainability programs. Two in particular that help organizations reduce their carbon footprint are the Environmental Protection Agency (EPA) Climate Leaders program and the World Resources Institute (WRI) publication, *A Service Sector Guide to Greenhouse Gas Management.*

The EPA Climate Leaders program is a partnership between the EPA and industry to develop comprehensive climate change strategies for organizations.[16] Companies that join this voluntary program commit to audit and reduce their corporate greenhouse gas emissions over an agreed-upon timeframe and to report progress annually to the EPA. In return, the EPA provides extensive technical support, including site visits, software, and phone support from contracted firms, which are individually assigned to each partner. Additional benefits accrue to industry partners through EPA recognition, access to resources and case studies, and credibility of participating in an externally validated greenhouse gas reduction campaign.

The organizations comprising the Climate Leaders program represent a broad range of industry sectors including cement, forest products, pharmaceuticals, utilities, information technology, and retail, with operations in all fifty states. The ever-growing number of partners range from thirty-person organizations such as Melaver, Inc. to Fortune 100 companies such as 3M, Anheuser-Busch, and Intel. No matter how large or small the organization, each partner receives the same level of ongoing technical support to facilitate the reporting and reduction process. In recent years, increased participation in the program seems to augur a general tipping point of awareness in the business sector of global warming and the need to address this problem proactively. (See Figure 3.2).

The Climate Leaders Greenhouse Gas Inventory Guidance is based on an existing protocol developed by the World Resources Institute and the World Business Council for Sustainable Development. Climate Leaders Partners develop their GHG emissions inventory using the Climate Leaders GHG Inventory Guidance. Companies are required to document emissions of the six major GHGs (CO_2, CH_4, N_2O, HFCs, PFCs, and SF_6) on a company-wide basis (including at least all domestic facilities) associated with:

- On-site fuel consumption and energy use.
- Industrial process-related emissions (as applicable).
- On-site waste disposal.
- On-site air conditioning/refrigeration use.
- Indirect emissions from electricity/steam purchases.
- Mobile sources.

For many smaller firms, especially service-based organizations, this may seem like an arduous task. Fortunately for these firms (Melaver, Inc. included), very little data needs to be collected. The Climate Leaders program is very user friendly. The free technical support will help each organization establish what is included in the inventory, how to collect the data, how to input data into the software, and finally, how to interpret the results. This program is a great way to begin to audit your greenhouse gas

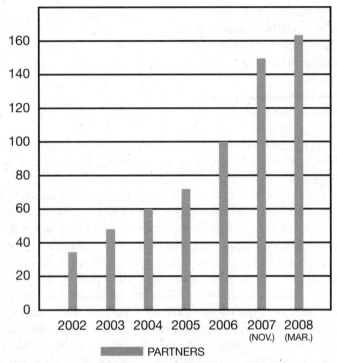

Figure 3.2   **Companies in the EPA Climate Leaders Program.**

emissions, though it does require yearly follow-up with the EPA and a strong commitment to accurately include and audit all emission sources.

The WRI publication *A Service Sector Guide to Greenhouse Gas Management* is a do-it-yourself manual for service firms to create their own greenhouse gas reduction plan. The easy-to-read publication provides a program similar to Climate Leaders but is written more for internal use. Unlike Climate Leaders, it does not provide publicity or technical support (and it doesn't have reporting requirements). WRI identifies heavy industry as a leading source of GHG emissions and acknowledges that many business-based GHG programs focus on manufacturers and utilities.[17] However, climate change is a broad issue with many contributing factors, including emissions from service sector companies such as banks, law firms, retailers, and real estate managers.[18] In fact, buildings are responsible for over half of U.S. greenhouse gas emissions, and over 76 percent of all power plant electricity is used to operate buildings. Thus there are substantial GHG reductions available through changes in service sector energy use and the products and services offered.

The WRI guide identifies a process similar to the Climate Leaders process, but goes further into building the business case internally. The guide also gives detailed analysis on establishing the process for a carbon footprint, identifying the following seven steps:

1 Assign resources.
2 Design a GHG inventory.
3 Collect data.
4 Calculate emissions.
5 Set target.
6 Reduce emissions.
7 Report results.

WRI also has free software tools (available at www.ghgprotocol.org) that help quantify the emissions from fuel used in facilities, electricity purchases, business travel, and employee commuting.

The main difference between the two programs is that the WRI program has a do-it-yourself structure, which eases some of the external reporting requirements and gives a bit more flexibility. The Climate Leaders program is a bit more demanding, calling for a more significant allocation of a company's time, money, and human resources, but it offers personal tech support, recognition, and bi-annual conferences for sharing best practices and networking. Both programs are helpful tools to begin the process of establishing a carbon footprint and establishing reduction goals.

At Melaver, Inc., we chose to join the Climate Leaders Partnership to help with our footprint analysis. We then took a hard look at our business operations to see where emissions came from, which ones we could measure, and which ones we were responsible for. Our analysis showed we were emitting greenhouse gases from three primary areas: electricity used to power our offices, business travel (both ground and air), and the staff's commutes to the office. While the company does not dictate where employees live or how they get to work, we did claim responsibility for commuting as part of our comprehensive carbon footprint.

To determine our footprint based on these three areas required collecting some basic information:

- For electricity, the total annual electric bill for each of our offices.
- For business travel, the total miles driven on company business, total miles flown, and vehicle fuel efficiency.
- For commuting, the frequency, distance of commute, and fuel efficiency for each employee's vehicle.

Many organizations have this information already. And after establishing a benchmarking process, systems can easily be put into place to facilitate data collection on an ongoing basis. For example, Melaver, Inc.'s electric bills were already on file through accounts payable. This held true for business travel as well, since in order to be reimbursed, mileage and flight information must be turned in. To assemble commute information, a brief survey asking employees to list their vehicle, fuel efficiency, round trip commute in miles, and how many days per week this trip is made. (Carpooling and public transportation are excluded from the calculations.) Larger organizations may wish to take a representative sample rather than creating an extensive spreadsheet charting each employee. For accuracy, Melaver, Inc. records every employee's exact commute distance and each vehicle's fuel efficiency. This was done through a simple e-mail survey where employees listed the pertinent information, which was recorded and is updated annually.

Once this data is collected, it is placed into a spreadsheet (provided by Climate Leaders for Melaver, Inc.) that has various emissions factors associated with each fuel type based on region. The region where your business is located affects the fuel mix used to produce electricity. More emissions, for example, are generated in parts of the country that rely more extensively on coal for electricity generation. See Figure 3.3 for regional breakouts and Table 3.2 for emission factors assigned to each region.

Our Calculations

While the software provides emissions data for other pollutants, such as sulfur dioxide and carbon monoxide, our company focused on eCO_2 as a benchmark. These results led to the next question: Now that we have computed 245 tons of carbon emissions, what does that mean? Is this good or bad? How does that rank against other

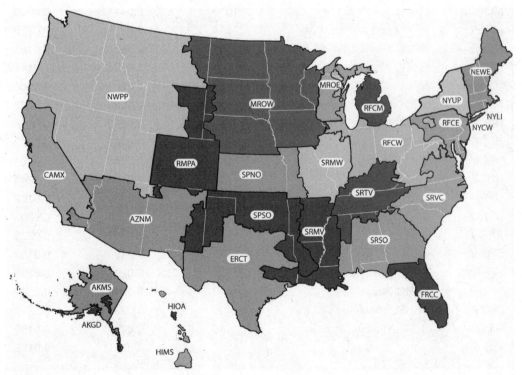

Figure 3.3   Electricity generation regional map.[19]

firms? These questions do not have clear-cut answers (there is, for example, no requirement saying real estate firms with under thirty people should emit 300 tons or less of eCO_2). But knowing your eCO_2 is a start, a baseline. The result of our company's emissions audit was aggregated into Table 3.3.

TABLE 3.2   ELECTRICITY GENERATION EMISSIONS BY REGION[20]

EGRID	SUBREGION	LB CO_2/MWH	LB CH_4/MWH	LB N_2O/MWH
AKGD	ASCC Alaska Grid	1257	0.0266	0.0064
AKMS	ASCC Miscellaneous	480	0.0239	0.0044
AZNM	WECC Southwest	1254	0.0175	0.0148
CAMX	WECC California	879	0.0359	0.0084
ERCT	ERCOT All	1421	0.0214	0.0148
FRCC	FRCC All	1328	0.0541	0.0160
HIMS	HICC Miscellaneous	1456	0.1006	0.0183
HIOA	HICC Oahu	1728	0.0911	0.0212
MORE	MRO East	1859	0.0411	0.0303
MROW	MRO West	1814	0.0275	0.0289
NEWE	NPCC New England	909	0.0798	0.0153
NWPP	WECC Northwest	921	0.0223	0.0141
NYCW	NPCC NYC/Westchester	922	0.0384	0.0060
NYLI	NPCC Long Island	1412	0.1020	0.0162
NYUP	NPCC Upstate NY	820	0.0240	0.0114
RFCE	RFC East	1096	0.0276	0.0172
RFCM	RFC Michigan	1641	0.0348	0.0254
RFCW	RFC West	1556	0.0195	0.0244
RMPA	WECC Rockies	2036	0.0244	0.0302
SPNO	SPP North	1971	0.0236	0.0303
SPSO	SPP South	1761	0.0303	0.0230
SRMV	SERC Mississippi Valley	1135	0.0420	0.0133
SRMW	SERC Midwest	1844	0.0214	0.0288
SRSO	SERC South	1490	0.0395	0.0249
SRTV	SERC Tennessee Valley	1495	0.0233	0.0237
SRVC	SERC Virginia/Carolina	1146	0.0294	0.0192
U.S. total		1363	0.0305	0.0198

Year 2004 eGRID Subregion Emission Factors
(Source: eGRID2006 Version 2.1, April 2007)

TABLE 3.3   MELAVER, INC. 2006 GROSS EMISSIONS FROM OPERATIONS

SOURCE	eCO$_2$ (TONS)	SO$_2$ (LBS)	NOx (LBS)	CO (LBS)	PM10 (LBS)	VOC (LBS)
Electricity	91.1	1,178.1	431.1			
Employee Commutes	71.1	36.0	810.0	7,739.7	16.2	828.0
Business Travel—Ground	29.3	15.2	341.1	3,259.4	6.8	348.7
Business Travel—Air	53.4	235.0	5,874.0	8,811.0	0	4,272.0
Totals	244.9	1,464.3	7,456.2	19,810.1	23.0	5,448.7

As with any metric, looking at trends over a multi-year period gives a clearer picture of the data, providing a sense of total emissions as well as emissions per staff member—a good indicator of how a company's growth affects emissions. At first blush, the company was actually responsible for more emissions in 2006 than it was in 2004. However, the company has grown in the past three years—more employees mean more commutes, more electricity, more travel. But by analyzing emissions per employee, the company saw an almost 15 percent decrease in greenhouse gas emissions, even while expanding our real estate holdings regionally during this period caused an increase in business travel, as shown in Table 3.4.

TABLE 3.4   MELAVER, INC. ECO$_2$ EMISSIONS (IN TONS), 2004–2006

SOURCE	2004	2005	2006	% GAIN SINCE 2005
Electricity	95.0	87.0	91.1	4.7%
Employee Commutes	73.0	71.0	71.1	0.1%
Business Travel — Ground	26.0	28.3	29.3	3.5%
Business Travel — Air	35.3	38.6	53.4	38.3%
Totals	229.3	224.9	244.9	8.9%
Tons per employee	11.5	10.2	8.7	
% reduction per employee		−12.17%	−14.46%	

Each company will need to determine a methodology for normalizing emissions. It could be per employee, per square foot, per unit produced. Those sorts of goals can be worked out with whichever resource (e.g., Climate Leaders, WRI) your firm uses to audit emissions. However, the overriding factor is that greenhouse gas generation must be slowed and then stopped to prevent global climate change, so we will begin moving toward looking at absolute emissions, 245 tons of eCO_2 for 2006 vs. normalized emissions. In fact, Melaver, Inc. has pledged to become carbon neutral—meaning we will reduce and offset our emissions until they are balanced at zero, a goal we have achieved and will continue to achieve going forward. (More on this later.)

OK, Now What?

Once the baseline audit of emissions is completed, an organization can begin to establish a GHG reduction plan. It could be an absolute reduction (e.g., we will hold emissions to 2005 levels going forward). Or GHG emissions can be normalized, based on per employee, or per square foot occupancy. The audit also allows you to focus on the areas where conservation and efficiency will have the most impact. For Melaver, Inc. (see Figure 3.4), electricity consumption was the leading contributor to our carbon footprint, accounting for almost 40 percent. This became the logical place to start reducing emissions.

Fortunately, reducing electricity consumption offers the opportunity for a financial return. Investments made to reduce energy consumption result in lower operating expenses, which adds directly to the bottom line. At Melaver, Inc., over the past four years we have invested in many quick return opportunities to both save money and lower our environmental footprint. Such opportunities include lighting retrofits that reduced our lighting demand by 60 percent, switching to laptops and LCD monitors, installing low-flow aerators on our faucets (to reduce hot water demand), and installing window film on our office building in Atlanta to reduce solar heat gain and decrease our air conditioning demand. This resulted in a 5 percent reduction in absolute emissions and a 31 percent reduction per staff person. This is a win/win—saving money and reducing emissions. Also included in our overall strategy were the reno-

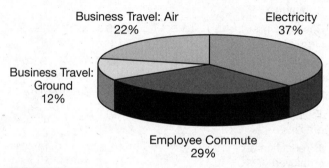

Figure 3.4   **Melaver, Inc. 2006 eCO_2 emissions.**

vation of our main office in Savannah, Georgia to LEED for Commercial Interiors standards and the build-out of a new office in Birmingham, Alabama, also to LEED Commercial Interiors criteria.

Every organization has unique opportunities to reduce emissions based on its footprint and operational procedures, from simple lighting retrofits such as Melaver, Inc. undertook, to system-wide computer energy usage monitoring software and building automation controls. Other opportunities exist in the commute and travel sectors, from encouraging and rewarding carpooling to investing in hybrid vehicles. These measures can have a significant impact on the carbon footprint of an organization, often with a financial payback (from reduced energy consumption and reduced fuel for vehicle fleets) as well.

What Comes Next?

Melaver, Inc. committed to achieve carbon neutrality when we joined the EPA Climate Leaders program in 2005 (Climate Leaders allows participants to set a range of goals, from holding emissions steady to achieving carbon neutrality), and we have made strides in reducing our emissions. However, the company is still emitting over 200 tons of eCO_2 annually. So how did we become carbon neutral? By involving ourselves in a variety of carbon offset programs. Carbon offsets are credits to your carbon emissions by claiming eCO_2 reductions elsewhere. This can be from reducing emissions from another, non-company related source or by investing in clean energy generation. Businesses realistically cannot immediately begin to operate without generating carbon emissions. Unless a company's staff does not drive or fly, unless all employees walk or bike or use mass transit to get to work, carbon dioxide is still going to be generated. This is where carbon offsets come in.

There are two forms of offsets—investing in actual emissions reduction projects and the purchase of green electricity or renewable energy credits (or green tags). We'll first look at investments in actual emissions reductions.

Emissions Reductions Projects

Here the possibilities are endless, though it may be harder justifying the expense if all you are concerned about is a payback, because there often is none. Melaver, Inc. created a multifaceted program to offset emissions. For example, the company partners with the Savannah Tree Foundation to plant hundreds of trees in the community each year, trees that begin to absorb carbon dioxide immediately and will continue to do so for the next fifty years. Planting trees helps sequester carbon dioxide from the environment, where it is used in photosynthesis to allow trees to grow, producing oxygen. Tree planting has taken its knocks by critics for carbon storage. After all, once the tree dies and decomposes, won't all that CO_2 be released back into the atmosphere? To an extent this is true. Also, many tree-planting programs are not beneficial to reforestation—planting one hundred acres of monoculture does not a forest re-grow! In Melaver, Inc.'s case, trees are planted throughout the community, not in one large area, which helps foster ecosystem diversity and reduce the heat island effect. If local tree planting is not an option, many national non-profit organizations partner with industries to plant trees in an environmentally beneficial manner and often can quantify how much CO_2 is being sequestered to help a company reduce its carbon footprint.

Other projects outside of the confines of the Melaver, Inc. office have resulted in actual emissions reductions. For example, in 2006 we purchased and distributed over one thousand compact fluorescent light bulbs (CFLs) in the Savannah community. Through a partnership with our lighting distributor, we were able to secure Philips CFLs for around three dollars apiece. These 27-watt bulbs replaced traditional incandescent 100-watt light bulbs, almost a 75 percent energy savings. Melaver, Inc. staff manned a table at the City of Savannah's 2006 Earth Day festivities and distributed the light bulbs, one per household, to anyone who wanted one. When these thousand CFLs were installed in houses around the community, it resulted in an annual reduction of over one hundred tons of eCO_2, almost half of Melaver, Inc.'s total footprint. This was a low-cost investment ($3,000) with far-reaching impact. The recipients benefited from lower energy bills, and we offset our own emissions by claiming credit for the reduction. Similarly, in 2007, we partnered with Goodwill Industries of Savannah, whose main office was in an older building with extremely inefficient lighting. Melaver, Inc. made an investment of around $7,000 and replaced eight hundred lights and two hundred ballasts with new, energy efficient lights, resulting in an ongoing annual reduction of twenty tons of eCO_2 and energy cost savings for Goodwill for years to come.

Many similar opportunities exist in communities all across the country. They may just take a little digging to find. For firms that would rather work through a third party, organizations such as Climate Trust can be of assistance. According to its website (www.climatetrust.org), Climate Trust is a non-profit organization dedicated to providing solutions to stabilize our rapidly changing climate. Climate Trust invests in specific projects that reduce GHG emissions, such as providing electricity at truck stops so truckers can plug in to operate their rigs rather than run on diesel to power the battery; optimizing traffic signals in cities to reduce traffic idling at stoplights; upgrading large facilities such as the Blue Heron Paper Plant in Oregon City, Oregon, which produces paper utilizing over 50 percent recovered fiber; and funding a program that retrofits inefficient boilers in schools.

Purchasing Green Energy

An alternative to offsetting emissions through third-party efficiency projects is the purchase of green energy directly through a utility company. The purchase of green electricity through a utility, if available, is a rather straightforward procedure. Many utilities offer (at a premium) the option to purchase green power (power generated from wind, solar, biomass, or small scale hydro) to customers. Depending on the amount purchased, this could effectively reduce the GHGs from a company's electricity consumption to zero. Often customers can elect to purchase blocks of green electricity, such as 100 kilowatt hours per block, or simply a percentage of use (e.g., 25, 50, or 100 percent). While this does not mean that electrons from a wind farm are directly entering your building, it does cause the utility to purchase more green power at a designated premium, helping more renewable energy to come online.

Purchasing Green Tags

In the same vein as purchasing green power directly from the utility is the purchase of green tags. Green tags are a decoupling of the environmental benefit from the actual power purchase. There are three basic premises behind green tags. First, renewable power still costs slightly more to generate than traditional fossil fuel derived electricity. Second, it is not possible to purchase green power in all markets. Finally, people are willing to pay extra for more environmentally benign power. Green tags separate the electricity purchased from the sale of the environmental benefits.

Green tags are sold as a commodity. Green energy generators such as wind farms put their electricity directly into the electrical grid. Companies wishing to procure green energy that are not within the service area of a utility that produces green electricity can buy the rights to the energy produced by the wind farm in the form of the green tag. This is done through third party brokers who work with the green power producers and the end users to transfer the green energy credits.

There are many brokers of green tags throughout the country. A starting point is with an independent third party such as Green-E, a nonprofit organization that certifies green tags. They ensure that the green energy is actually produced and that the product is not being double-sold to multiple end users. Purchasing green tags that have been certified by Green-E or similar groups helps ensure the green tags are from qualifying sources. Green tags can then be used to offset any amount of GHG emissions. To achieve carbon neutrality, Melaver, Inc. purchases green tags for wind energy produced in Kansas. The carbon that is offset from green electricity production in Kansas is applied to our total GHG responsibility to offset the amount of carbon our business operations generate.

So Why Not Just Offset Everything?

Unfortunately, green tags are becoming an easy way to pay penance for carbon emission sins. Many firms choose to simply purchase inexpensive green tags rather than take steps to actually reduce emissions. While this does help more renewable energy come on line, it does not reduce an organization's absolute carbon emissions. A commitment to sustainability requires that a company always first reduce emissions as much as possible, and then look at ways to offset. This tracks back to authenticity. Buying your way out is not what customers or partners view as authentic. This holds true at Melaver, Inc. as well. Many of our staff members voiced concern over the initial purchase of green tags. However, these purchases have been complemented by programs directly focused on reducing energy use. What cannot be done through these efforts is then offset through green tags. Many large organizations purchase green tags to offset electricity use, including Whole Foods, Coca-Cola, Duke University, FedEx Kinko's, Nike, the U.S. Army, and the World Bank.[21] Whether green tags are right for your organization depends on the size of your carbon footprint and your reduction alternatives.

One More Thing

It is worthwhile to add a postscript to this lengthy discussion of greenhouse gas emissions. As a result of opening two new offices and adding a number of staff members, our overall emissions went from 245 tons of eCO_2 to 270 tons—not encouraging considering the total emissions, but all in all OK given the fact that our overall emissions per employee have remained stable. What is particularly interesting in our ongoing effort to audit our footprint is the fact that one of our projects, Enterprise Mill in Augusta, Georgia, is entirely powered by hydroelectric (water) power from the Augusta canal that has flowed through this historic cotton mill since the mid-nineteenth century. The mill generates power well in excess of the needs of this mixed-use, 250,000-square-foot commercial and residential development, and we sell the excess power back to the grid at a green premium. How much excess? About 2,050 tons of eCO_2 are avoided as a result of power generation through alternative energy On-site. In short, given the fact that as a company we are directly responsible for 270 tons of eCO_2 while generating almost ten times that amount in clean energy, our company is not just carbon neutral but carbon negative *without the use of offsets*. At roughly nine tons of eCO_2 per staff member, we are essentially providing enough clean energy to the grid to account for growth of almost two hundred additional employees, while still remaining carbon neutral. This particular example of using a specific real estate project for energy generation indicates the direction our industry needs to go: on-site, alternative distributive energy.

So What Do We Get Out of All of This?

Calculating your company's carbon footprint and creating an emissions reductions program result in both tangible and intangible benefits. Tangible benefits include:

1 Financial savings through energy efficiency.
2 Financial savings from operational efficiencies.
3 Being a product leader and first to market with low-carbon goods and services.

The first two benefits are straightforward. If you use less electricity or fuel to generate energy, financial savings accrue straight to the bottom line. Lower annual operating costs also result in a higher valuation for buildings, which can have a significant impact. For example, a $10,000 decrease in electricity costs can translate into an increased property value of $100,000 using a capitalization rate of 10 percent. That is one heck of a return on investment. Not only does this money go straight to the bottom line, you are also reducing your greenhouse gas emissions and improving air and water quality. That is the win-win decision the triple bottom line intends: Your firm benefits financially, the environment benefits from less GHG emissions, and society benefits from cleaner air. Carbon offsets, regardless of which means an organization pursues, come at a cost. Green tags cost money, but the less carbon an organization generates, the fewer offsets are needed. This translates into cost savings for organizations looking to reduce their footprint.

Another way of looking at this, from the perspective of an environmental economist, is to internalize the externalities. Organizations seeking to reduce their carbon footprint can internally assign a cost to carbon or other negative environmental pollutants, putting a value on emissions that traditionally is not accounted for in profit and loss statements. This type of accounting can change the financial outcome of decision making and quickly lead to more sustainable thinking across the organization. Assigning a value to the emission or reduction of greenhouse gases makes it a tangible liability or asset instead of a concept. Many organizations, from Dow Corning to Bank of America to King County, Washington, participate in the Chicago Climate Exchange (CCX). The CCX creates a market value for carbon and makes it a tradable commodity, similar to a stock. Companies that become members and make carbon reductions realize a market value that can be sold to other companies. The growing CCX and its membership put a true dollar value on carbon dioxide emissions and reductions. Thus, the externality of carbon dioxide production is either a financial penalty for emitting companies or an asset for companies that find innovative ways to reduce emissions.

The third benefit is a bit more esoteric, having to do the value creation from new-to-the-world product and service offerings. First-mover advantage from anticipating a new need and creating a new carbon-reducing product—for example, a carbon scrubber—can be highly profitable and facilitate a company's long-term competitive advantage. A consistent message being communicated by business leaders across industry sectors is not only that a company can do well by doing good, but that energy-efficient products and services represent a financial windfall to those companies that are early entrants into this soon-to-be-burgeoning market.[22]

Intangible benefits include:

1 Market differentiation.
2 Corporate social responsibility to shareholders.
3 Enhanced employee satisfaction and retention.

Intangible benefits, by definition, are hard to quantify. However, there is true value to be realized by market differentiation and a commitment to sustainability that organizations can realize. For example, if there is a choice between a product from Company A or a product from Company B, but Company B has a program to reduce carbon emissions, an aware consumer will be more apt to purchase the product from Company B. In a tough, competitive business climate, with consumers five times more apt to think that a company's reputation on environmental issues is an important factor in purchasing decisions, "environmental compatibility breaks ties at the shelf."[23] Becoming the first in your industry to go green provides first-mover advantage—companies that later adopt such measures will be seen as following the trend of the market leader. Can you assign a value to this? As we see throughout this book, realizing value from intangible benefits, particularly brand enhancement and reputation, is tricky. Nevertheless, it certainly earns an organization its fair share of free publicity through news coverage, demonstrated market leadership, and product differentiation. It also facilitates business development as clients actively seek to associate with a first-mover, values- centric company.

Corporate social responsibility to shareholders relates to the growing awareness of sustainability in investor decisions. Socially responsible investment (SRI) funds are growing rapidly, and Dow Jones, among other organizations, has created a Sustainability Index (DJSI) devoted to socially responsible investing. As shareholders increasingly hold organizations accountable for their GHG emissions and environmental performance, firms that willingly adopt measures to reduce emissions and improve environmental performance (rather than being forced, through regulation or other outside pressures) are seen in a better light by investors.

Finally, there are the numerous employee-related benefits discussed in Chapters 1 and 2. These include less employee turnover, greater productivity, higher retention rates, and increased job satisfaction. While quantifying these benefits is difficult, there are certainly real dollars attributable to each.

Beyond Carbon

A sustainability analysis can go far beyond carbon emissions. At Melaver, Inc., we continually try to expand the scope of our environmental footprint reduction, each year analyzing more criteria. For example, we reviewed 100 percent of our purchases for the past year and were surprised by the results. Earlier in the chapter we discussed the amount of soda purchased and the issue of recycling aluminum cans. The company found the same held true for bottled water. An analysis of bottled water purchases revealed that we were purchasing almost fifty gallons of bottled water at a cost of nearly $11 per gallon. This compared to the $0.007 per gallon we pay for municipal water, which we filter and chill for immediate consumption. The comparison is shown in Table 3.5. We based the calculation on our water bills in 2006 (33,552.35 gallons at a cost of $221.64).

The true cost of bottled water is more than 1,500 times the cost of municipal water plus the embodied energy associated with petroleum-based plastic bottles. True, we recycled bottles that were discarded at the office, but we could not guarantee that 100 percent were recycled. If, for example, guests took bottles with them, we did not know where the bottles would end up. Also, energy is associated with trucking bottled water around the country and in recycling the bottle material. Not buying bottled water resulted in a win-win, both financially and environmentally, for the company. We are saving money by not paying a 1,500 percent upcharge on each gallon of water, and at the same time we are reducing the environmental impact associated with producing, shipping, and recycling plastic bottles.

However, best intentions do not always yield optimal results. Our audit also showed we use almost 150,000 sheets of paper annually (even with double-sided printing). While this reinvigorated our efforts to reduce paper use, we also moved to switch from 30 percent recycled content paper to 100 percent recycled content. What looked good

TABLE 3.5 MELAVER, INC. 2006 WATER CONSUMPTION COMPARISON			
WATER CONSUMPTION	GALLONS	COST	COST/GALLON
City	48	$ 0.32	0.0066
Bottles	48	$516.00	$10.75

on paper (no pun intended!)—fully recycled content paper with only a 5 percent cost premium over standard paper—did not work in application. The 100 percent recycled content paper would not run through our copy machines. Indeed, if the copy job was more than a few pages, half the copies would be ruined, resulting in more paper being needed to reprint! This trial and error led us back to the 30 percent recycled content paper we had been using, though we are still exploring paper options.

While some procurement results surprised us (each employee goes through one pen per month, on average), we were right on track for others. Our use of 100 percent recycled toilet paper, napkins, and facial tissues was the norm, as well as green cleaners, recycled content trash bags and writing paper, and a significant reduction in battery purchases since switching to rechargeables. We ended up with a fantastic baseline from which to measure our progress in the following years. Decreasing paper consumption, reducing per-person pen use to eight per year instead of twelve, etc., all result from accumulating the relevant data and taking the incremental step to find more sustainable solutions. Don't think clients won't look to see if you use recycled materials— not only on your photocopies, but also on stationery, marketing materials, and mailers. In fact, one early change we made was to print our business cards on 100 percent recycled content paper with soy-based ink. (You better believe that almost everyone we hand a card to notices.)

MOVING FORWARD

You've audited your emissions, greened your procurement, started carpooling, cut energy use and cost—what comes next? As mentioned, this chapter is neither a rigid protocol nor an all-inclusive solution. In fact, it is intentionally an overview to help organizations take the first steps toward a more sustainable future. We have not walked far enough down the green brick road at Melaver, Inc. to even begin trying to offer definitive advice for every situation.

We hope other steps we have taken regarding our partners and vendors will help increase the ripple effect sustainability seems to have. Much like throwing a pebble into a still lake causes waves to ripple outward to the shore, authentic acts of sustainability often have far-reaching effects. For example, getting recycled content business cards and letterhead meant working alongside our marketing partners and printing companies to find solutions that conveyed our message in an environmentally responsible manner. Our Savannah-based printer, Spanish Moss, now uses recycled content paper and soy ink as the norm, instead of as a special order. That is just one example of the ripple effect of sustainability.

If a small, thirty-person firm such as Melaver, Inc. can have this sort of impact on partners, how does the ripple effect work for larger firms? Wal-Mart is redefining many aspects of retail shopping—from how products are packaged and how fish are caught to how goods reach stores and how supply chains are managed. Starbucks' use of recycled sleeves in lieu of double cups provides a quiet but persistent conservation message. Do those ripples have an effect? You bet they do. In fact, they can change whole industries.

You don't have to be an environmental nonprofit to incorporate sustainable practices, and you don't have to be as big as Wal-Mart to have a far-reaching impact in creating a more sustainable future for the planet and for generations to come. What it takes is an authentic commitment to approach business decisions with a different mindset—one that considers a triple bottom line. The first step is establishing what sustainability means to your organization. The next step is determining what the appropriate metrics are. After all, how can you measure success if you have nothing to compare it to? Small steps can go a long way—get the framework in place and then look to rapidly expand and grow upon early successes. Continue to push the boundaries, and most importantly, re-evaluate where you are and where you came from. Sustainability is a dynamic concept. You can only determine where you are going by knowing where you have been.

Putting a Price Tag on All These Efforts

SO WHAT DOES ALL OF THIS COST?

In trying to be conservative, we estimate that about 5 percent of a sustainability associate's time is devoted to auditing the company's overall environmental footprint and overseeing various programs that help reduce that footprint. This means that 5 percent of overall opportunity cost for revenue generation is lost to planning and implement-

Years		0	1	2	3
REVENUES/SAVINGS					
	Energy Savings		1,000	1,000	1,000
	Water Savings		500	500	500
	Soda reduction program		1,800	1,800	1,800
	Sub-total Revenues/Savings		3,300	3,300	3,300
COSTS					
	Purchase of green tags		(1,600)	(1,600)	(1,600)
	Tree purchases for carbon sequestration		(4,000)	(4,000)	(4,000)
	Purchase of CFLs for the community		(3,000)	(6,000)	(5,000)
	Lighting retrofits		(1,500)		
	Low-flow aerators		(15)		
	LEED expenses for Savannah office		(39,000)		
	LEED expenses for Birmingham office		(20,000)		
	Value of time of sustainability associate		(9,375)	(10,781)	(12,398)
	Sub-total Costs		(78,490)	(22,381)	(22,998)
Total Cashflow		0	(75,190)	(19,081)	(19,698)
Discount Factor		1.000	0.909	0.826	0.751
PV Cashflow		0	(68,355)	(15,770)	(14,800)
NPV		(203,828)			

Figure 3.5   Costs and benefits of greening a company's operations.

ing internal company programs. In the case of Melaver, Inc., such programs have included making our various offices LEED certified under the Commercial Interiors program, retrofitting our lighting, installing low-flow aerators on sinks, partnering with the Savannah Tree Foundation on a carbon sequestration tree-planting program, and partnering with various community organizations to replace incandescent lighting with compact fluorescents.

AND THE BENEFITS?

There are certainly savings in energy and water consumption that occur as part of this overall program. For the sake of this analysis, though, we are assuming that the savings, while a benefit, are negligible: annual savings of about $1,000 in electricity annually from our LEED Commercial Interiors offices, water savings of $500, and savings from our alternative approach to soda provisioning of about $1,800 annually. Cost savings through ease of attracting and retaining talented staff might also be factored into our analysis, but we have already considered those line items in Chapters 1 and 2. Finally, there are intangible benefits relating to brand and reputation enhancement, benefits that we will consider at the end of Chapter 4.

All in all, we estimate total costs of our internal green program over ten years to amount to $356,000, with total benefits of approximately $33,000. Discounted back to Year 0, total investment in our green program amounts to nearly $204,000 (see Figure 3.5) or approximately $22,000 per year.

4	5	6	7	8	9	10	TOTALS
1,000	1,000	1,000	1,000	1,000	1,000	1,000	10,000
500	500	500	500	500	500	500	5,000
1,800	1,800	1,800	1,800	1,800	1,800	1,800	18,000
3,300	3,300	3,300	3,300	3,300	3,300	3,300	33,000
(1,600)	(1,600)	(1,600)	(1,600)	(1,600)	(1,600)	(1,600)	(16,000)
(4,000)	(4,000)	(4,000)	(4,000)	(4,000)	(4,000)	(4,000)	(40,000)
(5,000)	(5,000)	(5,000)	(5,000)	(5,000)	(5,000)	(5,000)	(49,000)
							(1,500)
							(15)
							(39,000)
							(20,000)
(14,258)	(16,397)	(18,856)	(21,685)	(24,938)	(28,678)	(32,980)	(190,347)
(24,858)	(26,997)	(29,456)	(32,285)	(35,538)	(39,278)	(43,580)	(355,862)
(21,558)	(23,697)	(26,156)	(28,985)	(32,238)	(35,978)	(40,280)	
0.683	0.621	0.564	0.513	0.467	0.424	0.386	0
(14,725)	(14,714)	(14,765)	(14,874)	(15,039)	(15,258)	(15,530)	0

Years	0	1	2	3
REVENUES				
Shaping values, chapter 1		21,300	21,300	21,300
Creating a culture of green glue, chapter 2		79,000	79,000	79,000
Green from the inside out, chapter 3	0	3,300	3,300	3,300
Sale of Green, Inc.				
Total Revenues		103,600	103,600	103,600
EXPENSES				
Shaping values, chapter 1	(166,000)	(114,500)	(91,875)	(97,286)
Creating a culture of green glue, chapter 2	0	(57,600)	(64,631)	(72,717)
Green from the inside out, chapter 3	0	(78,490)	(22,381)	(22,998)
Total Expenses	(166,000)	(250,590)	(178,888)	(193,002)
Total Cashflow	(166,000)	(146,990)	(75,288)	(89,402)
Discount Factor	1.000	0.909	0.826	0.751
PV Cashflow	(166,000)	(133,627)	(62,221)	(67,169)
NPV	132,127			
IRR	13.98%			

Figure 3.6   Amalgamated data for Green, Inc.

If we amalgamate the data from our discussion in this chapter with the costs and benefits analyzed in the previous two chapters, we find that the total investment in green—values, culture, and internal operations—still shows net positive discounted cash flow of $132,000 over ten years, with an internal rate of return of 13.98 percent (see Figure 3.6).

This return of 13.98 percent is now below our hurdle rate of 15 percent—not great news, of course, but hardly cause for much alarm. For one thing, it is the collective view of our writing team that we have, in fact, greatly understated the benefits of green in the interest of providing a conservative real-case study of a green bottom line. Moreover, the story is far from complete. We still need to consider the costs of creating a whole company of folks who are expert in green development—as well as the benefits that accrue to such an investment in working our way up the green learning curve—to which we turn our attention in Chapter 4.

NOTES

[1] Robert F. Kennedy, Jr., *Crimes Against Nature: How George W. Bush and His Corporate Pals Are Plundering the Country and Hijacking Our Democracy* (New York: HarperCollins, 2004), p. 184.

[2] http://money.cnn.com/2006/08/07/news/international/oil_alaska/index.htm accessed 12/28/07.

4	5	6	7	8	9	10	TOTALS
21,300	21,300	471,300	538,800	548,925	550,444	550,672	2,766,640
79,000	79,000	79,000	79,000	79,000	79,000	79,000	790,000
3,300	3,300	3,300	3,300	3,300	3,300	3,300	33,000
						0	0
103,600	103,600	553,600	621,100	631,225	632,744	632,972	3,589,640
(103,490)	(110,605)	(118,765)	(128,125)	(138,866)	(151,190)	(165,335)	(1,386,037)
(82,016)	(92,710)	(105,007)	(119,150)	(135,413)	(154,117)	(175,625)	(1,058,987)
(24,858)	(26,997)	(29,456)	(32,285)	(35,538)	(39,278)	(43,580)	(355,862)
(210,365)	(230,311)	(253,228)	(279,560)	(309,817)	(344,585)	(384,541)	(2,800,886)
(106,765)	(126,711)	300,372	341,540	321,408	288,159	248,431	
0.683	0.621	0.564	0.513	0.467	0.424	0.386	0
(72,922)	(78,678)	169,552	175,264	149,939	122,207	95,781	0

[3] "Our Common Future," *Report of the World Commission on Environment and Development*, published as an Annex to document 42/187, United Nations General Assembly, 96th Plenary Meeting, 11 December 1987.

[4] Daniel C. Esty and Andrew S. Winston, *Green to Gold: How Smart Companies Use Environmental Strategy to Innovate, Create Value, and Build Competitive Advantage* (New Haven: Yale University Press, 2006), p. 8.

[5] Ibid., p. 248.

[6] Joseph J. Romm, *Cool Companies: How the Best Businesses Boost Profits and Productivity by Cutting Greenhouse Gas Emissions* (Washington, D.C.: Island Press, 1999), pp. 144–5, 165–170.

[7] Jeffrey Hollender and Stephen Fenichell, *What Matters Most: How a Small Group of Pioneers Is Teaching Social Responsibility to Big Business, and Why Big Business is Listening* (New York: Basic Books, 2004), pp. 100–103.

[8] Marc Gunther, *Faith and Fortune: The Quiet Revolution to Reform American Business* (New York: Crown Publishing, 2004).

[9] Andrew Savitz and Karl Weber, *The Triple Bottom Line: How Today's Best-Run Companies Are Achieving Economic, Social, and Environmental Success—and How You Can Too* (San Francisco: Jossey-Bass, 2006), pp. 70–72.

[10] Esty and Winston, *Green to Gold: How Smart Companies Use Environmental Strategy to Innovate, Create Value, and Build Competitive Advantage* (New Haven: Yale University Press, 2006), p. 249.

[11] Samantha Putt Del Pino, Ryan Levinson, and John Larsen, *Hot Climate, Cool Commerce: A Service Sector Guide to Greenhouse Gas Management* (Washington, D.C.: World Resources Institute, 2006), p. 9.

[12] Roger A. Hinrichs and Merlin Kleinbach, *Energy: Its Use and The Environment* (New York: Thompson Learning, 2002).

[13] Concentrations of atmospheric CO_2 have risen from 280 parts per million (ppm) in 1850 to 364 ppm in the late 1990s, the highest concentration on earth in 20 million years.

[14] Roger A. Hinrichs and Merlin Kleinbach, *Energy: Its Use and The Environment* (New York: Thompson Learning, 2002).

[15] Lester R. Brown, *Eco-Economy: Building an Economy for the Earth* (New York: W.W. Norton and Company, 2001).

[16] Data is from Intergovernmental Panel on Climate Change, *Climate Change 1995: The Science of Climate Change* (Cambridge, U.K.: Cambridge University Press, 1996).

[17] See www.epa.gov/climateleaders

[18] Samantha Putt Del Pino, Ryan Levinson, and John Larsen, *Hot Climate, Cool Commerce: A Service Sector Guide to Greenhouse Gas Management* (Washington, D.C.: World Resources Institute, 2006), p. 1.

[19] Ibid.

[20] United States Environmental Protection Agency website.

[21] United States Environmental Protection Agency website, http://www.epa.gov/cleanenergy/energy-resources/egrid/index.html accessed 3/22/08.

[22] Putt Del Pino, et al., p. 9.

[23] Fred Krupp and Miriam Horn, *Earth: The Sequel: The Race to Reinvent Energy and Stop Global Warming* (New York: W. W. Norton & Company, 2008); Ted Nordhaus and Michael Shellenberger, *Break Through: From the Death of Environmentalism to the Politics of Possibility* (New York: Houghton Mifflin Company, 2007).

[24] Jacquelyn A. Ottman, *Green Marketing: Opportunity for Innovation* (New York: J. Ottman Consulting, Inc. 1998), p. 15.

GREEN IS THE COLOR OF MONEY

DENIS BLACKBURNE

SUMMARY

In this chapter, Denis Blackburne, CFO of Melaver, Inc., first considers the costs of gaining green expertise, looking at the cost of the learning curve and how a company can learn from some of the more expensive mistakes made by Melaver, Inc. From there, Blackburne examines the various ways a company could not only recoup its initial investment in going green but actually could profit handsomely from it. Section two of the chapter looks at the short-term, tangible cost reductions likely to be realized by a green company, primarily from reductions in operational expenses, insurance premiums, and loan underwriting.

The third section looks briefly at the longer-term, intangible cost reductions associated with reduction of risk and liability. The fourth section considers the short-term, tangible ways a green company can create revenue and value, with particular attention given to capturing market demand for green through additional green project development and consulting work. Also addressed is the enhanced value that accrues to a sustainable portfolio.

The fifth section contains an assembly all of the financial pieces of data into an analysis of the cost and benefits of developing green expertise. In the sixth section, all of the financial data on the fictionalized company Green, Inc. from this and previous chapters are assembled and synthesized to provide an overall (and quite healthy) financial picture of the green bottom line of a sustainable real estate company. In the concluding section, Blackburne considers the intangible value of a green company, concluding that perhaps such a company's "X" factor—its reputation, image and culture—comprises the greatest value of all.

Following the Money

The money aspect of sustainability is top of everybody's mind. This is what everybody wants to know first. Can I afford it? How much does green cost? Does it make economic sense? What is the value of green? What is the payback? Is it a fad or is this green stuff here to stay? Why is everybody jumping on the green bandwagon? Should I?

A remark often heard these days at conferences devoted to sustainable real estate practices is that while green has now arrived on Main Street, it hasn't become mainstream. Yet. That situation is changing rapidly, with financial considerations serving as the major driver. If the initial drive for the sustainability movement was initiated by the heart—a commitment and passion for doing the right thing at all costs—what will keep this movement sustained over the long haul is the wallet: market demand for green products and services, financial capital to address those needs, and business models that prove conclusively that a company not only can do well by doing good, but can do *better.*[1]

My role involves this last function, shaping our company's green strategy in ways that assist us to perform better in the marketplace. It is to help the company plan the resources it needs to reach its goals, track its evolution, keep score, and quickly determine any necessary changes. It is to challenge or at least question with a skeptical eye the proposals to make out-of-the-ordinary investments on the front end, monitor performance in real time, make adjustments on the fly, and assess results with hard-nosed objectivity on the back end. Yes, I've heard the unkind joke: finance people know the price of everything and the value of nothing. But if the business case for sustainability is to gain widespread traction, it has to pass muster with people like myself.

What Does It Cost to Become a Green Company?

Does it cost more to build green? Our basic response to this question, which may seem contrarian or off-putting, is one almost no one seems to consider: "More versus what?" There is no such thing as an absolute standard for building. A developer may choose to capture the rainwater on the roof and reuse it in the irrigation system instead of placing marble columns in the foyer. A company may decide to put in a more efficient HVAC system rather than fancy finishes in common areas. It can be somewhat difficult to answer this comparative question with a simple yes-or-no answer along with a percentage figure.

The answer also depends on how green a developer wants its project to be, what level of green certification is being sought, what aspects of an overall green program a developer wishes to emphasize, and how experienced the developer is with sustainable development practices. As a basic rule of thumb, our own experience, borne out by data provided by the U.S. Green Building Council, is that to achieve a LEED certified or LEED Silver certification, no additional cost premium is called for. To achieve Gold certification, there is a cost premium of around 2 percent, and a Platinum certification entails a cost premium of around 7 to 8 percent.

Those numbers, though, tell only a part of the story. Those familiar with the LEED program and its flexible point system are well aware that there are "cheap" points (such as bicycle racks) and considerably more expensive ones (a green roof), such that a developer simply chasing the least expensive points purely for certification can develop a LEED building more cost effectively than a developer who is more focused on the more profound impacts of a project on the environment and on future costs of operations. Adding to the complexity, any exploration of the first costs of green development needs to consider the overall objectives of a project. For instance, our company has recently signed on to the 2030 Challenge, involving a commitment to aggressive reductions in carbon emissions over the next two decades. The commitment impacts how we intend to allocate our overall expenditures on construction. While the LEED program provides for the tallying of points in five main areas of development (Site, Water, Indoor Air Quality, Materials & Resources, and Energy & Atmosphere), our company has decided to place particular emphasis on energy, which is a more expensive route to take—particularly given our twenty-year objective of developing zero net energy buildings. That commitment is likely to entail a lower LEED rating at a higher than average first cost—at least until higher performance energy systems (including use of alternative energy) become more price competitive owing to greater market demand and government subsidies in this area (desperately needed, by the way).

There is the learning curve associated with developing a green project. If a company develops its first LEED project and someone from that company says that it did not cost anything more, that person is probably not a finance guy, has found a clever way to "dress up" the numbers, is flat-out lying, or a combination of all three. As with learning anything new, there is most certainly a cost premium associated with adapting to a new set of processes. There is the additional time involved in doing things differently, including the time it takes to get one's hands around the LEED program. This is true even if a company hires a consultant to expedite matters. There will still be members of a company's development team that have to understand and implement what this consultant is talking about. There is also a learning curve associated with changing from a linear approach to development to an integrated approach. In a conventional development project, the developer creates an overall vision for the project and hands off the vision to the design team, which gives shape to the vision, which then gets handed off to finance for cost evaluation, back to design for refinements, and then to the contractor who builds the project. A leasing team is brought in to find paying tenants. And the whole thing gets handed off to the property management group to make things run smoothly. In a green development, all of these disparate parties are present from the get-go. Done well, siloed areas of expertise are jettisoned as all work collaboratively to achieve a high-performance building. Done poorly, turf battles emerge, various professionals lay claim to their own area of expertise, and collaboration seems well-nigh impossible. It takes time to manage this collaborative process both creatively and efficiently.

Finally, there is the thought process of determining the environmental objectives of a project. A manager who simply gives the charge to her subordinates *deliver me a LEED building* has given virtually no direction as to what the objectives of such a building are: Reduce energy or water consumption, by how much? Focus on those aspects of a building that will help improve operational efficiency, how? Get LEED certification, what level and for what purpose? The market demand for LEED buildings is growing exponentially, to the point where it will soon become the new baseline standard for construction and hence something of a commodity. The developer who takes environmental impacts seriously or the business person who is looking for a competitive advantage already needs to think beyond the label of LEED to the underlying purpose of what her company is trying to achieve. That thought process, like everything else in a green project, takes time before a smooth process takes hold.

It took our company probably three years to manage this learning curve to the point where we can now develop a green building with the same efficiency of time and money as a conventional project. The first aspect of this learning curve involved investing in resources so that a significant portion of our staff became well versed in the LEED program—to the degree that about 80 percent of our staff is now LEED accredited, with everyone taking the exam at least once. The investment includes providing for a one-day workshop for all staff, covering the cost of the LEED exam, paying a bonus to all who pass, and the lost opportunity for time spent at the workshop and/or taking the exam. Total cost of this investment was approximately $49,000, as indicated in Table 4.1.

A carryover part of this learning curve entailed the engagement of non-Melaver, Inc. partners in the LEED program. While this was never mandated, the fact that so many within our company were ramping up their knowledge of green development had a carryover effect with those who were working with us. Our legal counsel, IT vendor, and outside marketing firm all became LEED accredited, along with some two dozen other professionals (architects, landscape designers, contractors, sub-contractors) directly associated with our early LEED projects.

Yet another part of this learning curve involved the school of hard knocks, learning the ropes the hard way, directly in the field. We hired LEED consultants early on, only to

TABLE 4.1   COST OF INVESTING IN THE LEED PROGRAM

ITEM	COST PER PERSON	# OF STAFF MEMBERS	TOTAL COST
LEED course	$350	27	$ 9,450
LEED exam	$300	27	$ 8,100
LEED bonus for passing	$750	22	$16,500
Subtotal			$34,050
Lost opportunity			$14,800
Totals			$48,850

discover they were learning with us and knew little more than we did. We incurred premium pricing for materials and technologies either out of our own ignorance or because prices had yet to become competitive. We faced up-charges in architectural fees as a result of re-designing projects in the midst of construction because we were changing our environmental performance objectives as we were building. And, not surprisingly, we suffered from delayed receipt of rent because our projects were taking longer to complete.

We estimate that our first LEED project, the Whitaker Building, probably saw a 15 percent cost overrun as a result of our learning curve (approximately $150,000). It did not help that our first project involved renovation of a 19th century building on the National Historic Register, challenging us to meld historic constraints to stringent contemporary energy standards. Our learning curve on that development in terms of time and process was a significant part of the cost premium.

We estimate that our second project, Abercorn Common, also entailed a premium as a result of our learning the LEED program. Much of this premium is factored into our discussion of this project in Chapter 5. However, it's worth pointing out there that a significant portion of our premium came in the form of longer time delivery to market and with that delay, deferral of rental income (approximately $417,000). It's almost impossible to calculate how much of that delay was owing to making this building green and how much to other factors that many development projects face (threatened litigation from a neighboring retailer, changes in design, uncertainty in the retail market post-9/11, etc.). Since those first two projects (Whitaker and Abercorn), the premium we calculate paying for green development has declined to the point where we feel it is largely negligible. An overview of the cost of our learning curve on our first four LEED projects, amounting to roughly $1.5 million, can be seen in Table 4.2.

TABLE 4.2   COST OF LEARNING CURVE ON EARLY LEED PROJECTS

PROJECT NAME	PROJECT TYPE	LEED CATEGORY	PREMIUM (%)	PROJECT COST	LEED PREMIUM ($)	LOST INCOME	TOTALS
Whitaker	Office/ Retail	New Construction	15.00%	$ 1,211,479	$ 150,000		
Abercorn: In-line	Retail	Core & Shell	4.60%	$18,293,000	$ 840,000	$416,667	
McDonald's	Retail	Core & Shell	4.60%	$ 1,410,000	$ 50,000		
Building 600	Retail	Core & Shell	4.60%	$ 1,788,000	$ 57,000		
Learning Curve Totals					$1,097,000	$416,667	$1,513,667
Cost of LEED Education							$ 48,850
Total Cost							$1,562,517

All told, our initial three-year learning curve amounted to approximately $1.56 million ($1.513 million + $49,000 in LEED-related education), not an insignificant premium to pay but, as we will see by the end of this chapter, an investment that is worthwhile. What did this investment teach us? Lessons learned include:

- Be clear from the outset as to the green objectives of the project.
- Stick to those objectives as rigorously as possible.
- Make sure that everyone on the development team is strongly aligned with and has a passion for the environmental objectives of the project. Most professionals in real estate these days will profess strong interest in doing a green project. Many, unfortunately, are just paying lip service in order to get your business. Collaborative, integrative design work can be painful without strong alignment among all involved.
- Use the resources of an experienced consultant who thoroughly understands design and construction as well as the LEED process and who is capable of overseeing the details of the LEED submittal process.
- Start with a smaller-sized project (but not too small). There are certain fixed costs associated with the LEED program (developing a LEED scorecard, energy commissioning, the LEED submittal process) that are factored in more economically with a larger project. However, you don't want to bite off a project that is too big or too complex your first time out.
- Be present on the development site as much as possible. What can be agreed upon in an office or conference room may not be what is being implemented in the field.
- Realistically assess the capacity of your company to adopt different organizational processes than the ones you are currently using, and try not to overtax the staff with too many changes all at once. That's easier said than done, particularly in a values-centric organization such as ours that is passionately driven to do things differently.
- Expect mistakes and anticipate them financially by underwriting them as part of your company's overall financial analysis.
- Enjoy the journey. It's a feel-good process from beginning to end, despite the learning pains.

We began this section of our chapter with a basic question: Does green cost more? The general answer is that it depends on what your objectives are and the general maturity of the market for this type of development. The particular answer in the actual experience of Melaver, Inc. is, "Yes, it does cost more." We believe the $1.56 million initial three-year investment in green is an investment in our long-term future and should be amortized over the lifetime of the business. Analyzed in this way, the annual cost of green for Melaver, Inc. over a thirty-year period at a 7 percent interest carry is about $125,000. Placing this additional annual cost in the context of our fictionalized company Green, Inc. results in additional expenses of $125,000 (or 1.49 percent) and a reduction in net operating income from $3.75 million to $3.62 million. Overall return on Green, Inc.'s equity drops from 15 percent to 14.50 percent or a 50 basis point reduction, as shown in Table 4.3.

TABLE 4.3 GREEN, INC.'S INVESTMENT IN SUSTAINABLE DEVELOPMENT	
GREEN, INC.	
# of employees	20
Annual payroll	$ 2,400,000
Total assets	$100,000,000
Debt	$ 75,000
Equity	$ 25,000,000
Gross revenues	$ 12,000,000
Return on assets (ROA)	12.00%
Expenses	$ 8,250,000
Net income after debt	$ 3,750,000
Return on equity (ROE)	$ 15.00%
Cost of capital	7.00%
Cost of green	$ 1,562,667
Annual amortization	$ 124,758
Revised expenses to include amortization	$ 8,374,758
Revised net income to include cost of amortization	$ 3,625,242
Amortization as % of revised expenses	1.49%
Revised net income after amortization	14.50%

As a values-centric company, Melaver, Inc. is comfortable paying a 50 basis point premium for being a green developer. However, this thumbnail sketch of the costs of becoming a green company tells only a small part of the story, since we have considered the expenses of going green without factoring in any of the benefits. Obvious follow-on questions are: Has this first-cost premium been worth it? Does it pay for itself? Are there tangible ways to justify this initial investment in green? As mentioned briefly in Chapter 1, there are four main areas to consider when evaluating performance:

■ Short-term and more tangible cost reductions
■ Longer-term and more intangible cost reductions
■ Short-term and more tangible revenue/value creation
■ Longer-term and more intangible revenue/value creation[2]

The first three areas are relatively simple to document, and I will be considering them throughout the remainder of this chapter, focusing exclusively on the costs and benefits of a green bottom line that concern operations. Short-term and more tangible aspects of cost reductions include looking beyond first costs by considering the life cycle cost of utilizing newer technologies, operational savings, and savings from insurance and finance underwriting. Longer-term and more intangible cost reductions include reduced risk exposure from liability and regulation. Short-term and more tangible revenue creation entails the additional business that accrues from anticipating market demand for green, enhancement value of assets, easier access to capital, and increased productivity. The fourth area, longer-term and more intangible value creation—at least as far as our own company is concerned—is clearly relevant but is more challenging to document objectively. I will only briefly consider this fourth aspect of value creation.

Short-Term and More Tangible Cost Reductions

OPERATIONAL SAVINGS

As real estate developers, we are handicapped when it comes to recapturing the cost of our investment through operational savings. First, we are a service-based business and not a manufacturing company, and as such are not heavy direct consumers of energy and water. Second, most of our investment in green materials and technologies is directed at our tenants and accrues to their benefit, not ours. Because of the way the market traditionally structures leases, it is challenging to recapture a considerable amount of our investment in green through reduced costs. Let's consider this second issue in slightly more detail.

An industrial lease (for warehouses) is typically structured as a triple-net lease, which means that the tenant pays a base rent to the landlord and then the tenant is responsible for paying taxes, insurance, all operational charges including its utilities and cost of common area maintenance. Unless it is specifically covered in a lease, the landlord/owner's investment in high-performance materials and technologies is paid for by the owner, but the benefits accrue to the tenant through operational savings, higher productivity, etc.

A similar situation occurs in retail leases, where the tenant pays the landlord base rent and its pro-rata share of taxes, insurance, and common maintenance fees, but is directly responsible for paying its own utility bills. Here again, while owner investment in high-performance technologies will result in lower utility bills, the savings in operational expenses accrues directly to the tenant and not the owner. It is interesting to note that several of our national retail tenants are reporting that their stores in our green developments are among the highest performing stores in the nation, with their profitability owing directly to reduced operational expenses. You don't need an advanced degree in economics to realize that it is much easier for a business to increase its profitability by reducing its overhead than by increasing its sales. And that is precisely what our retail tenants are seeing—great for them and perhaps, eventually, great for owners as well, once standard leasing structures are modified to account for the growing attention to green development. But for now, the operational savings for retail development accrue to our tenants, not us.

Multi-family residential projects—condominiums and affordable housing developments—reflect the same constraints. In both product types, the investment in higher-performance technologies is made by the developer while the savings in utility costs accrue to the occupant. With affordable housing development underwritten by tax credit financing, there actually may be innovative ways to recapture some of the developer's initial investment in green, with the developer essentially serving as the utility intermediary/provider. But the market for green affordable housing is still too new and dynamic to document this potential.

Finally, there is the office lease, in which a tenant pays to the landlord a base rent plus its pro-rata share of taxes, insurance, and operating expenses above the base year of its occupancy. Say, for example, a tenant enters into a lease in 2010. In the initial year of the lease, the owner receives base rent of $20 per square foot and incurs expenses from taxes, insurance, and operating expenses of $6 per foot, for a net income of $14 per foot for this tenant. In future years, the tenant will pay to the landlord its base rent plus whatever amount in excess of $6 a square foot the landlord faces in increased operating expenses. Investing in high-performance technologies will enable the landlord/owner to reduce its operating expenses and hence recapture some of the value of its investment. But owing to the structure of office leases, new tenants and renewing ones will get a "free ride" as a result of the owner's investment.

Note that in all of these various product types, I have focused only on the limited recapture potential of investment in high-performance technologies. Not factored in at all are the other LEED-related investments having to do with materials and resources, indoor air quality, etc. In short, current market conditions and outdated leasing structures provide little opportunity for a landlord to participate in the reduced operating expenses of its tenants, even though the landlord has made a significant investment in its tenants' operations. As the market for green matures and greater numbers of institutional developers construct green buildings, I think this situation is likely to change.

To promote green buildings, we need an innovative green lease structure where the interests of the landlord and the tenant are aligned. In today's scenario the landlord is effectively penalized financially as improvements to the energy and/or water management in a property typically result in costs that are absorbed by the owner and benefits that are realized by the tenant. A green lease structure, where landlord and tenant share the costs and benefits of energy improvements would seem logical, as there are enough advantages to be shared by all.

Despite the current limitations, Melaver, Inc. does realize some small degree of savings through reduced operational expenses. Our offices in Savannah, Atlanta, and Birmingham are all in LEED certified quarters and, as such, realize energy savings annually of approximately $1,000 and water savings annually of around $500. Going to double-sided printing has resulted in a reduction of paper consumption amounting to $300 each year. Purchase of a company hybrid car is anticipated to reduce gas consumption and be financially better than reimbursing miles. And the decision to stop purchasing canned sodas for staff members and instead provide them with chits to purchase soft drinks from a neighboring retail tenant has resulted in a small but symbolic saving of $1,800 annually. Most of these financial benefits have already been considered in the previous chapters.

REDUCTION IN INSURANCE COSTS

For years, my colleagues and I had been buttonholing our insurance underwriter, contending that our insurance premiums should be reduced because of our green portfolio. Our emphasis on craftsmanship and durability, we argued, meant that our buildings would hold up better over the long term. Our investment in more energy- and water-efficient systems would similarly result in more durable properties. We argued that our buildings, with fewer volatile organic compounds and tighter building envelopes, made them healthier than others and thus less exposed to lawsuits having to do with sick building syndrome and the like. To no avail. We simply could not make any headway. Until August 29, 2005, when Hurricane Katrina hit New Orleans.

As most in the real estate world are well aware, insurance rates since Katrina have skyrocketed. Particularly hard-hit in terms of price increases has been the southeastern United States, both along the Atlantic shoreline and the Gulf. In the last four years, insurance rates in our region have increased by approximately 11.5 percent a year, which means our own insurance premiums during that time period should have jumped from $0.83 (per 1,000) to $1.28. The key words here are *should have*. The fact of the matter is that during that same time period, our insurance rates have increased more modestly, going from $0.83 in 2004 to $0.99 in 2008. This equates to $29,000 in annual savings ($0.29/1,000 × $100 million).

The underlying cause of this stability in our insurance rates can be linked to our specific values/green orientation—but not necessarily our LEED orientation. We need to be careful here. With certain caveats in mind, a developer can build a LEED certified building in most places: in a desert environment, along our coastal marshlands and hammocks, even in the middle of an old-growth longleaf forest. Our own company standards are more stringent than those set forth by the U.S. Green Building Council in that we voluntarily choose to build in urban core settings well away from the environmentally sensitive areas of our coastal wetlands. It's part of our particular ethos as a company. It is also part of our ethos to devote ourselves to projects that speak to the needs of a particular community. Hence, we do not replicate one type of product (for instance, condos) over and over again in market after market. Such a business model, typical for many companies, simply does not fit our belief system, which holds that every locale is unique and that is it incumbent upon us as developers to address the uniqueness of the communities within which we work.

As it turns out, our belief system and practices have also paid off for us financially (though that was hardly the intent). Because our developments are located well away from the liminal areas between marsh and high ground, and because we have a portfolio of properties diversified both geographically and in terms of product type, we did not match the risk profile that resulted in so many real estate firms having to pay significantly higher insurance premiums after Katrina.

REDUCTION IN FINANCING COSTS

Financial markets are changing so radically in their overall approach to green development that it is almost certain that in the nine-month time lag between the writing of this chapter and its publication, much of what I will say will have become passé. A green asset has a greater value/risk relationship than a conventional property and should be considered as such by the financial community. Its operating expenses are lower, its overall exposure to liability is less than for a conventional development, its exposure to volatile energy prices is less, the longevity of its capital equipment is typically greater. And there is the co-branding opportunity, enabling a financial institution to capitalize on its underwriting of cutting-edge projects that do the right thing. All these elements, among others, make a green project ideal for discounted underwriting.

It wasn't that way in 2002 when we began our first LEED project (the Whitaker Building), and it wasn't that way almost five years later in 2006 when we initially began looking for a financial institution to purchase a majority interest in a portfolio of sustainable properties we had developed. However, the financial landscape is changing quickly. Over fifty international banks, including ABN Amro, HSBC, and Citigroup have signed on to the Equator Principles, which require environmental assessments for major loans. In 2005, J. P. Morgan (now JPMorgan Chase) announced that it was restricting its lending and underwriting policies for projects likely to have an environmental impact. The Paris Bourse requires companies listed on its exchange to provide clear information quarterly on their environmental practices. PNC Financial Services Group has made a strong commitment to developing its branch network according to LEED criteria. Goldman Sachs is building its new headquarters in New York with the goal of achieving LEED Platinum certification. Bank of America is building offices to LEED standards and has also announced a specific fund dedicated to green developments. Other financial institutions are quickly jumping on the bandwagon.

So how does all of this activity in the financial sector affect the underwriting of green projects? Favorably. We have had three experiences within the past year that bear this out. First, the sustainable portfolio we initially looked to sell to an institutional investor (more about this later) we eventually determined would be better managed if we simply refinanced and retained ownership. The two banks that ended up doing the deal with us, RBS Greenwich and IXIS Capital, as well as our financial broker, The First Fidelity Companies, understood the value of our green assets and provided non-recourse financing, 80 percent loan to market value, interest only for five years, no cross collateralization with an all-in price of about 5.7 percent for the ensuing ten years. Given the volatility of the cost of capital at the time of this writing, it is likely to be difficult for the reader to compare this underwriting to financial conditions at the time this is being read. It is hard to say with any certainty that we saw a discount in underwriting as a result of our green orientation. There was a very strong interest in our portfolio. Why? The market liked the solidity of our offering, some of which is probably owing to its green orientation. Suffice it to say that we probably realized a discount of 25 to 30 basis points compared to what a more conventional portfolio would have likely received at the time. As a small privately owned real estate development company out of Savannah, Georgia, we did not expect to command the small credit risk spread we were offered. We simply did not anticipate achieving such a rate. If we equated that discount to Green, Inc.'s portfolio of $100 million (with $75 million in debt), this would amount to an annual savings of $187,500.

A second case study is our development of Sustainable Fellwood, a mixed-use, mixed-income development project involving a remarkable public-private partnership that includes the Housing Authority of Savannah and the City of Savannah. The development, which is part of the LEED pilot program for Neighborhood Development and looks to be a prototype for how green affordable housing can work, is funded by tax credit financing provided by the state of Georgia. Without getting into too much detail here, a developer typically receives tax credits on such projects, which it then sells on the open market at a commodity price. At the time we approached the market, pricing was ninety cents on the dollar for federal tax credits and approximately twenty-five cents on the dollar for state tax credits. The interest among various institutional bidders for Sustainable Fellwood was so strong that these tax credits were eventually sold for a premium, amounting to a windfall of some $1 million on a $8 million deal. Part of this premium was owing to market rates moving upward at the time. However, being conservative, we feel that a full 50 percent of this premium, or $500,000, was owing to the unique green affordable housing nature of the project. We feel that this is a rather exceptional event having as much to do with the branding opportunities for the financial institution that landed this deal as it does with the green nature of the project. Nevertheless, the lesson seems to be clear: A significant discount is to be had for companies willing to invest in projects that deeply integrate social and environmental justice issues. In this light, perhaps some radical re-thinking of triple bottom line practices is called for: Rather than focus first on the financial bottom line of a project and then layer it with social and environmental considerations, perhaps reversing the order is—at least sometimes—called for.

One final case story involves a local community bank that, as part of a deal to underwrite one of our projects, agreed to create a new policy offering a 25 basis point discount to any green commercial or residential project in the area. This is a case study we feel hopeful about, since we leveraged our own work to seed favorable financing for any green developer in our community. As the saying goes, a rising tide lifts all boats. At least that's our intent.

HIRING AND RETENTION OF PERSONNEL

In Chapters 1 and 2 we noted that a values-centric company improves upon a company's capacity to hire and retain staff members and that a culture of shared leadership more particularly facilitates the hiring and retention of key managerial personnel. I simply would like to reiterate this notion, without accounting for this value-add twice. It is a value-add worth mentioning.

The current image of green is cool and trendy, but more importantly it stands for values that many of us espouse today, especially the younger, dynamic generation that will become the leaders of tomorrow. More than a small portion of the population wants to work for a company with sound ethical standards. The Enron debacle and other scandals have caused many people to reconsider working for corporations that only look to maximize financial profits and thus wealth creation. They would rather work for entities that have clearly articulated ethical values on social and environmental issues and walk their talk.

The image of green will help attract and retain key individuals within a company. Is this important from a financial aspect? Yes. Let me rephrase that...*yes!* A company's personnel cost outweighs the cost of property ownership and/or rental and operational building costs. By "outweigh," I mean personnel costs are usually significantly more than building costs, whether one owns or rents the properties in which business is carried out. If I look at the general management and administrative costs in our company, which are probably pretty typical for a large chunk of similar entities throughout the country, our human resources costs represent about 70 percent of these expenses, whereas building costs are between 5 and 10 percent. Being able to attract and retain good talent more successfully than other companies is a clear savings to the sustainable company, reducing churn rate, leveraging knowledge within the company over a greater span of years, and enhancing general *esprit de corps*, which clearly has a positive effect on productivity.

Longer-Term and More Intangible Cost Reductions

REDUCING EXPOSURE TO REGULATORY CENSURE AND LITIGATION

Longer-term, intangible cost reductions are often difficult to quantify since these line items tend to involve the projection of future risks. How does one reasonably quantify risks that we think will not happen, lawsuits we think we will likely avoid, penalties from regulatory restrictions that have yet to be codified? I'm not an actuary, skilled in risk appraisal work. Moreover, our company is a small, service-oriented business working in an industry that is much less exposed, say, to litigation related to toxic emissions—as a chemical company would be—or to regulatory issues such as carbon trading—as a manufacturing or energy company might be. Still, these issues do get factored into a company's bottom line. A green company that is diligent and attentive to its management of "sources and sinks" (where resources come from and how they are disposed of) is less likely to risk the same level of exposure (financially, legally, etc.) as a conventional company. These issues need to be considered, even if they tend to be rather remote and contingent as far as our own company is concerned.

Bob Willard, in his book *The Sustainability Advantage*, calculates the evaluation of risk exposure as follows: He assumes that 15 percent of a company's gross revenue accounts for a business' general and administrative (G&A) costs, that 5 percent of G&A costs are risk related, and that 5 percent of the amount delineated as risk expenses could be eliminated by a company taking a proactive, sustainable orientation. If we applied this formula to our fictionalized company, Green, Inc., we would realize a cost savings of $4,500 annually, as seen in Table 4.4.

Willard's rule-of-thumb calculation may be helpful when dealing with large, multinational companies (where the numbers start to look like international telephone numbers), but with smaller companies such as our own company or Green, Inc., it's

TABLE 4.4 GREEN, INC.'S SAVINGS FROM RISK REDUCTIONS	
Gross Revenue	$12,000,000
General & Administrative expenses (15% of GR)	$ 1,800,000
Risk related expenses (5% of G&A)	$ 90,000
Savings by sustainable orientation (5% of risk related expenses)	$ 4,500

relatively useless. In 2007, for example, we were notified by the state Department of Natural Resources (DNR) that a small property we had owned briefly fifteen years ago and that was once tenanted by a local dry-cleaning business contained ground contamination from the tenant's dry-cleaning fluids. It would cost about $400,000 to clean up the site, and DNR was looking at every owner of the property in the chain of title to determine whom they would ask to foot the bill.

My points are these:

- Oftentimes, irrespective of actual culpability, regulatory agencies such as DNR (and others) will look to the deepest pockets for remediation of point or non-point source pollution. A green company, one that is more sensitive to and knowledgeable about the processes and practices of its tenants, is much less likely to lease to a user who might present a potential liability risk—environmentally, socially, or financially.
- The benefit of being green as analyzed by Willard is probably significantly understated, especially when it comes to small to mid-sized real estate companies. There is an accounting gap between the relatively small amount of savings Willard's formula provides for (particularly in regard to small companies) and what actual exposure to risk might be. We sold the property with the dry-cleaning operation over a decade ago as part of our shift to a more sustainable orientation. Using Willard's formula, that decision would have reduced our exposure by $45,000 ($4,500 over ten years)—when in fact our actual potential exposure turned out to be almost ten times that amount.

In short, it is challenging to estimate with any accuracy the savings that come from reducing a company's risk to regulatory censure and litigation—particularly when it concerns companies such as ours. As such, I will include it in my overall calculation of the value of going green, but will try to understate this savings in the interest of being conservative. Suffice it to say that a company such as Green, Inc.—with total assets of $100 million in green properties—is much less prone to the type of clean-up penalties DNR was looking to collect than a conventional portfolio of properties of similar value. The savings for a green portfolio in terms of reduced risk exposure is a very relevant savings, much more significant than Willard is allowing for. A very conservative estimate of the reduced risk exposure of having a green portfolio would be to consider that once in ten years, one-half of a percent of a company's total assets may be at risk. In the case of Green, Inc., which has a total asset base of \$100 million, this would mean that it would have reduced liability risk of \$50,000 annually (\$100 million × 0.5% ÷ 10).

There is one other aspect of exposure to risk that, while possibly remote, has given us pause. We typically think of our sustainable ethos as one that reduces our exposure to regulatory censure and litigation. And to a large extent, I think that is correct. But what about the case where a company's sustainable ethos actually increases exposure to risk? This is certainly a possibility when it comes to the way in which a company markets its green credentials: should a company's talk exceed actual performance, it could face costly repercussions. Words to the sustainably oriented: under-promise, over-deliver, and document all performance claims. The sustainability movement suffers damage otherwise.

A related type of exposure—increased liability precisely from a company's sustainable ethos—comes from our own vault of strange but true business case studies. Early on in a potential development project we were bidding for, we booted off the team we had cobbled together one company that simply was not aligned with our sustainable beliefs and practices. We initially thought it was, but once we realized we were mistaken, we rectified the situation as swiftly as possible. This erstwhile team member sued us for breach of an agreement that had never existed. Lessons learned:

1 Sometimes it does seem that no good deed goes unpunished, and

2 Vet, vet, vet and continue to vet alignment of all businesses potentially involved in a joint project since just the least bit of friction can amount to a peck of pain.

Short-Term and More Tangible Revenue/Value Creation

ASSET VALUE ENHANCEMENT

It is our contention that the asset value of a green building is greater than that of an equivalent standard building simply built to code.

The value of a company share on the stock market can largely be summarized as the value of cash the company holds plus the discounted value of all the future cash flows. The real estate world works similarly. Buildings as such don't have any cash, so the asset value is simply the discounted value of projected future cash flows. How does green affect these future cash flows?

A green building, thermally more efficient, has a reduced cost maintenance structure. It is typically built to higher standards, calling for less reserves set aside for future maintenance. And productivity within a green building—a topic addressed in Chapter 7 by our COO, Colin Coyne—is superior to that of a conventional building, auguring stronger demand for this type of product in upcoming years.

Those, at least, are the main financially tangible arguments for maximizing the value of a green asset. We probably should have had those arguments foremost in our mind in 2006 as we sought to refinance a portfolio of six sustainable properties (retail, office, and residential in four markets in the Southeast). Our first approach was, however, clumsy and taught us the importance of correctly framing our pitch to the financial community: less in terms of our passion for developing a different way, more in terms of long-term value enhancement of the asset.

It was the spring of 2006. Martin (our CEO), Colin (our COO), and I put on our dark suits and fancy ties and traveled to New York to visit several banking institutions. At that time, our naïve if heartfelt pitch was along the lines of, "Hey, we have a portfolio of cool green properties for you to finance...." And they were cool: an early LEED office/retail building listed on the National Historic Register, the first LEED retail shopping center in the country, a mixed-use office/residential rehab of a 19th-century mill that runs entirely off hydroelectric power and sells the excess power back to the grid, an office building that became one of very few in the country to receive a LEED for Existing Buildings certification. The reaction from these prestigious banks was a mix of interest and caution. Mostly, they were skeptical about this green stuff. One very reputable bank essentially said, "Gee, thanks for coming guys, but please take the elevator and get off on the second floor. Go see our community banking team. I'm sure they could give you a $10,000 grant or so..." This was embarrassing, as we were actually trying to refinance a sizeable portfolio of properties with an overall occupancy rate of close to 95 percent and a high degree of credit-worthy tenants, providing solid returns in line with the markets in which they were located.

We realized we were not speaking the right language. At that time, green developers still conveyed a certain stereotypical image among financiers—tree-hugging, Birkenstock-wearing individuals willing to build something that cost too much as part of a goofy fad (although I must say the financial industry is now quickly catching up on the benefits of building sustainably). We realized that we needed to reframe our work as standard fare enhanced by green attributes, not as an exotic offering the financial community might have a hard time understanding.

And so we reframed the pitch. We started with the validity of the real estate, the types of properties, the prime locations, the quality of the tenants, the rent roll, and the appeal and look of the properties. Then, once the key values of our properties had been determined, we noted "Oh, and by the way, these properties happen to be high-performing buildings with better energy efficiency, and are eco-friendly." Bingo...doors started to open! In the financial and insurance sectors, green construction needs to be framed in terms of its high performance, reduced exposure to energy cost volatility, and longer-term retention of value than its environmental friendliness (or the values-centric orientation of a company). Once the value of the real estate was established using standard evaluative criteria, the value of green was seen as a very important add-on. It not only generated additional value but also provided a differentiation factor.

Once doors started to open, there was a keen interest in refinancing this portfolio. We worked on the future cash flow projections, taking into account the proven savings that had been realized over the previous year or so, thanks to the energy efficiency and water saving elements we'd put in place. We worked on every cash flow for every property for the ensuing five years to come. This resulted in some interesting discussions with the potential financing entities, the financial broker, and us. It was very inspiring as first the broker "got it" and finally so did the lenders. Everybody recognized the value of the future enhanced cash flows. True financial analysts don't simply consider the first year's net operating result to determine the value of the properties. They will also consider the projected cash flows in the years to come. In this case, the improved future cash flows enabled the lenders to recognize a greater value of the properties.

Since lenders provide loans on a loan-to-value basis, this enabled us to extract a lot more of our capital than we had originally anticipated. We were looking for $60 million in refinancing and ended up with $72 million. Some would say that this is due to being in the market at the right time with the right product. Maybe, but I sincerely believe that a good part of the additional recognized value of the assets was due to the projected improved cash flows resulting from our investment in green. We had realized a 20 percent value enhancement to our portfolio through our refinancing efforts, amounting to $15 million in value creation (see Table 4.5). Even if only a quarter of that value increase was owing to the green aspects of our sustainable portfolio, the value enhancement would have amounted to almost $4 million, not bad for those early years of climbing the learning curve of sustainable development.

ATTRACTING ADDITIONAL DEVELOPMENT WORK

In Chapter 1, we considered the additional business development work that began to come our way in light of our sustainable ethos. As noted in this earlier chapter, our company fields quite a number of third-party development projects, selecting only a few that we feel fit with our capabilities, interest, and expectations of partner alignment. It is our feeling that part of the reason others approach us is owing to our values-centric orientation. But part of the reason also is owing simply to the fact that we have experience developing green projects. It is reasonable to expect that other green real estate companies will also see additional business development work as the market demand for green expands (but the current supply of those with actual experience is still rather limited).

How much additional business development should a company expect? The answer depends a great deal on the size of one's company, the capacity the company

TABLE 4.5 VALUE ENHANCEMENT OF MELAVER, INC.'S SUSTAINABLE PORTFOLIO

	BUDGETED ASSUMPTIONS	ACTUAL PERFORMANCE	VARIANCE ($)	VARIANCE (%)
Value of sustainable portfolio	$75,000,000	$90,000,000	$15,000,000	20%
Loan-to-value	80%	80%		
Financing available	$60,000,000	$72,000,000	$12,000,000	20%

has for focusing on such third-party work, etc. A company such as ours should conservatively be able to take on at least two such development projects each year, of around $15 million each in scope Assuming a 4 percent development fee for each of these projects, assuming further that 25 percent of that fee will go for expenses leaving 3 percent of actual net income (4 percent less ¼ equals 3 percent), and assuming that one-half of the resulting net income of 3 percent is owing to our reputation (1.5 percent) and the other half owing to our experience building LEED (1.5 percent), it is safe to project additional business development income of 1.5 percent of $30 million (2 projects of $15 million each), or $450,000 annually. We will be using that figure as part of our overall analysis of our fictionalized company, Green, Inc.

ATTRACTING REVENUE THROUGH FEE-BASED CONSULTANCY

Depending on the capacity of a green bottom line company, there may be the opportunity to leverage knowledge and experience by providing green development consultancy services to other companies. We've done a good bit of this type of work over the years, ranging from pro-bono consulting to non-profit organizations interested in reducing their environmental footprint in various ways to rather sizeable fee-based work. In considering such work, my colleagues and I find that we are continually juggling among several competing initiatives: the desire to spread knowledge of sustainable practices as quickly and as broadly and as deeply as possible, irrespective of the potential generation of fee income (outreach and advocacy); the need to allocate human resources to our own development projects (attending to our own core business and avoiding spreading ourselves too thin); and the capacity to generate additional income while also spreading the word through fee-based consulting (additional income).

Every business will address this balancing act differently. Our own strategy for balancing these disparate initiatives is to consider fee-based consulting work 1) only when it does not impede the delivery of our own development work; 2) typically in locations that are in close proximity to our offices, to ensure good work/life balance for staff; 3) in locations outside our area only if local expertise is not available, and only if the nature of the project is likely to push our own learning further. Even with those fairly stringent criteria in mind, we still face the potential of consulting on numerous projects each year, providing us with the opportunity for significant additional revenue. The important thing to note is that the market demand for knowledge of and experience with sustainable projects is quite high, and the potential for additional income not insignificant.

PERSONNEL PRODUCTIVITY

Earlier in this chapter, we addressed the fact that a green company is likely to face reduced administrative costs as a result of being more successful than many companies at attracting and retaining good staff members. There is also the value creation that occurs through greater employee productivity. Because this is not something our company can document with any degree of precision (what would our benchmark be?), I am not including this value-add in my overall accounting of the benefits of being a green company. But just because I have not factored this into my analysis does not mean that the value does not exist. On the contrary. The U.S. Green Building Council cites numerous studies where employees in a LEED certified building have improved productivity, higher retention rates, less absenteeism, and less sick leave. This is money in the bank for employers, easily outweighing any perceived costs of owning or developing a green building. In an office environment, for example, productivity is reported to be 2 to 18 percent higher in a LEED building. Such enhanced productivity would allow companies to do more, take on more tasks, and/or operate with a smaller staff. Whichever way you look at it, such increased productivity represents value.

Pulling the Data Together

We began this chapter by discussing the learning curve faced by a company intent on becoming knowledgeable about and experienced in the practices of developing green—going so far as to put an overall price tag of $1.56 million on this learning curve over the course of three years.

We then considered the various ways a green development company might reduce its costs and add revenue as a result of its green ethos. By now, you will hopefully agree with me that there are indeed tangible benefits to going green. Among those benefits of cost reduction and revenue generation, we made the following conservative determinations:

- We can expect insurance savings of about $29,000 annually.
- There is a 50 basis point savings on loans on green properties. In the case of Green, Inc., with $75 million in debt, this would amount to an annual savings of $187,500 ($75 million × .0025).
- A very conservative estimate of the reduced liability exposure of having a green portfolio would be to consider that once in ten years one half of a percent of a typical company's total assets may be at risk to litigation of an extraordinary nature. A green company would simply be insulated from such risk by avoiding a range of practices that others would be less knowledgeable about. In the case of Green, Inc., this would amount to an annual savings of $50,000 ($100 million in assets × .005 ÷ 10).
- A sustainable company would conservatively see additional demand for green development work amounting to 1.5 percent of net income on two projects annually, each amounting to $15 million, for additional development income of $450,000 ($15 million × 2 × 1.5 percent).
- A sustainable company would see demand in the market for consultancy work very conservatively amounting to 0.5 percent net income on projects worth $30 million, for additional income of $150,000 annually.
- A company could realize a one-time value enhancement of its sustainable portfolio, an increase in value we have conservatively estimated at 5 percent of the equity portion of the company's holdings. We are not certain as to how long such an opportunity will

remain in the market. Certainly, as it becomes more common for developers to create green real estate portfolios, the value-enhancement opportunity diminishes—to the point where it becomes a liability if a development company owns a non-green real estate portfolio. In the case of Green, Inc., this enhancement amounts to $1.25 million ($25 million × 5 percent). Moreover, as a result of this value enhancement, a green company has the opportunity to reinvest this cash in the business, earning an added return of this initial value recapture in ensuing years.

■ If this enhanced value is realized through refinancing or sale, Green, Inc. can take its additional value enhancement of $1.25 million, leverage it with debt of approximately $3.75 million to develop an additional property worth $5 million. It earns 15 percent on the new development, pays 7 percent on the debt, and amortizes its principal payments over thirty years for additional revenue of $230,000.

Years	0	1	2
REVENUES/SAVINGS			
Savings in insurance costs		0	0
Savings in financing costs			
Reduced exposure to liability risks			
Reinvestment of revenue savings			
Additional development work because of LEED experience		0	0
Reinvestment of Added Revenue from yr 1			0
Reinvestment of Added Revenue from yr 2			0
Reinvestment of Added Revenue from yr 3			0
Reinvestment of Added Revenue from yr 4			0
Reinvestment of Added Revenue from yr 5			0
Reinvestment of Added Revenue from yr 6			0
Reinvestment of Added Revenue from yr 7			0
Reinvestment of Added Revenue from yr 8			0
Reinvestment of Added Revenue from yr 9			0
Consulting work as a result of LEED experience		0	0
Reinvestment of Consulting Revenue from yr 1			0
Reinvestment of Consulting Revenue from yr 2			
Reinvestment of Consulting Revenue from yr 3			
Reinvestment of Consulting Revenue from yr 4			
Reinvestment of Consulting Revenue from yr 5			
Reinvestment of Consulting Income from yr 6			
Reinvestment of Consulting Income from yr 7			
Reinvestment of Consulting Income from yr 8			
Reinvestment of Consulting Income from yr 9			
Asset value enhancement			
Asset value enhancement reinvested			
Total Revenues	0	0	0
COSTS			
LEED education of staff	(49,000)		
Learning curve associated with early projects		(150,000)	(417,000)
Total Costs	(49,000)	(150,000)	(417,000)
Total Cashflow	(49,000)	(150,000)	(417,000)
Discount Factor	1.000	0.909	0.826
PV Cashflow	(49,000)	(136,364)	(344,628)
NPV	2,168,868		
IRR	34.50%		

Figure 4.1   Costs and benefits of Green, Inc. becoming sustainable

Let's now assemble all of this data into a ten-year discounted cash flow statement to get an overall picture of whether our investment in becoming a green company has, in fact, paid off. In keeping with the conservative methodology of the previous chapters, we have restricted income and revenue production to the latter years of this ten-year analysis (Years 6 through 10), while front-ending much of the costs (see Figure 4.1).

As shown in Figure 4.1, initial expenses of $1.563 million to manage the learning curve of becoming a sustainable company are more than offset by the value creation that occurs in the later years of analysis (approximately $7.1 million, for a net positive income of $5.5 million over ten years). The net present value of the overall investment in time, money, and experience is positive, amounting to $2.1 million, and the internal rate of return (IRR) of 34.5 percent is well above Green, Inc.'s minimum threshold of 15 percent.

3	4	5	6	7	8	9	10	TOTALS
0	0	0	29,000	29,000	29,000	29,000	29,000	145,000
			187,500	187,500	187,500	187,500	187,500	937,500
			50,000	50,000	50,000	50,000	50,000	250,000
			32,475	37,346	38,077	38,187		146,085
0	0	0	450,000	450,000	450,000	450,000	450,000	2,250,000
0	0	0	0	0	0	0	0	0
0	0	0	0	0	0	0	0	0
0	0	0	0	0	0	0	0	0
0	0	0	0	0	0	0	0	0
0	0	0	67,500	10,125	1,519	228		79,372
0	0	0	0	67,500	10,125	1,519		79,144
0	0	0	0	0	67,500	10,125		77,625
							67,500	67,500
0	0	0	150,000	150,000	150,000	150,000	150,000	750,000
0	0	0	0	0	0	0	0	0
0	0	0	0	0	0	0	0	0
	0	0	0	0	0	0	0	0
		0	0	0	0	0	0	0
			22,500	3,375	3,881	4,463		34,220
			0	22,500	3,375	3,881		29,756
			0	0	22,500	3,375		25,875
			0	0	0	22,500		22,500
			1,250,000					1,250,000
				187,500	215,625	247,969	285,164	936,258
0	0	0	2,116,500	1,176,475	1,222,971	1,261,446	1,303,442	7,080,834
								(49,000)
(890,000)	(57,000)							(1,514,000)
(890,000)	(57,000)	0						(1,563,000)
(890,000)	(57,000)	0	2,116,500	1,176,475	1,222,971	1,261,446	1,303,442	
0.751	0.683	0.621	0.564	0.513	0.467	0.424	0.386	
(668,670)	(38,932)	0	1,194,709	603,718	570,525	534,976	502,533	

Years		0	1	2	3
REVENUES					
Shaping values, chapter 1			21,300	21,300	21,300
Creating a culture of green glue, chapter 2			79,000	79,000	79,000
Green from the inside out, chapter 3			3,300	3,300	3,300
Developing expertise in LEED, chapter 4		0	0	0	0
Sale of Green, Inc.					
Total Revenues			103,600	103,600	103,600
EXPENSES					
Shaping values, chapter 1		(166,000)	(114,500)	(91,875)	(97,286)
Creating a culture of green glue, chapter 2		0	(57,600)	(64,631)	(72,717)
Green from the inside out, chapter 3		0	(78,490)	(22,381)	(22,998)
Developing expertise in LEED, chapter 4		(49,000)	(150,000)	(417,000)	(890,000)
Total Expenses		(215,000)	(400,590)	(595,888)	(1,083,002)
Total Cashflow		(215,000)	(296,990)	(492,288)	(979,402)
Discount Factor		1.000	0.909	0.826	0.751
PV Cashflow		(215,000)	(269,991)	(406,849)	(735,839)
NPV		2,300,994			
IRR		28.38%			

Figure 4.2   Amalgamated cash flow analysis for Green, Inc.

But this only tells a part of the overall story. Throughout the early chapters of this book, we have been discussing early year investments in shaping company values, shaping a company's culture, developing an environmental audit, and learning how to develop green. And in each chapter, we have found ways in which these early investments have paid off for the company in ensuing years. But does this picture change when we combine these separate analyses into one ten-year discount cash flow calculation? Is it possible that the heavy investment in learning and discussion and reflection early on in the company's history is simply too much financially, despite the creation of value later on? An amalgamated discount cash flow for Green, Inc., seen in Figure 4.2, tells the story fairly clearly.

Total expenses for shaping values within a company, for creating the green glue that binds a company culture together, and for working through the learning curve of becoming green developers amounts to $4.36 million. Total revenues—through savings and actual incomes—during this same ten-year period conservatively amount to approximately $10.67 million, for positive surplus of $6.3 million. Discounted back into Year 0 dollars, we find a positive net present value of all investments in green values, culture, and experience amounting to $2.3 million. The internal rate of return for all that Green, Inc. has invested in amounts to 28.38 percent, also well above the company's hurdle rate.

4	5	6	7	8	9	10	TOTALS
21,300	21,300	471,300	538,800	548,925	550,444	550,672	2,766,640
79,000	79,000	79,000	79,000	79,000	79,000	79,000	790,000
3,300	3,300	3,300	3,300	3,300	3,300	3,300	33,000
0	0	2,116,500	1,176,475	1,222,971	1,261,446	1,303,442	7,080,834
						0	0
103,600	103,600	2,670,100	1,797,575	1,854,196	1,894,189	1,936,413	10,670,474
(103,490)	(110,605)	(118,765)	(128,125)	(138,866)	(151,190)	(165,335)	(1,386,037)
(82,016)	(92,710)	(105,007)	(119,150)	(135,413)	(154,117)	(175,625)	(1,058,987)
(24,858)	(26,997)	(29,456)	(32,285)	(35,538)	(39,278)	(43,580)	(355,862)
(57,000)	0	0	0	0	0	0	(1,563,000)
(267,365)	(230,311)	(253,228)	(279,560)	(309,817)	(344,585)	(384,541)	(4,363,886)
(163,765)	(126,711)	2,416,872	1,518,015	1,544,380	1,549,604	1,551,873	
0.683	0.621	0.564	0.513	0.467	0.424	0.386	0
(111,853)	(78,678)	1,364,261	778,982	720,464	657,183	598,314	0

Coda: Long-Term and Intangible Revenue Creation

Not factored into our evaluation of the value of going green are those aspects of the business that are longer term in nature, aspects that are more intangible and, as such, harder to evaluate scientifically and objectively. One of those aspects concerns the capacity to access capital funding for our business. At the time of this writing, we are literally inundated by calls from various institutional capital players. They have all been busy cobbling together sustainable real estate funds and are now in the process of competing against one another for developers that can provide the products their investors are seeking. It's a small universe of developers right now that can deliver the goods. We are in the fortunate position of being one of them.

That situation will change as more and more developers become experienced developing a more sustainable product. We hope that's the case. Nevertheless, at the present time, we find ourselves having access to capital that, had we been a conventional developer delivering a conventional product, would simply not have been open to us with the same readiness nor with the same largesse. We have literally held discussions, serious discussions, with several institutional players interested in providing us with enough capital annually that—had we accepted—would have had us developing real estate portfolios every year equal to or greater than the size of the one portfolio we have spent the last seventy years creating! We declined, simply viewing this type of growth as not in keeping with our sustainable, slow-growth ethos. Nevertheless, we do have access to capital that we did not have before. The value of this? Hard to say. We don't have to spin our wheels in time (and money) trying to match a deal with capital. We have the opportunity to choose those projects that best leverage and complement our team's skill set.

Another aspect of longer-term, intangible value creation concerns the enhancement of a company's name and reputation. It's there, I can feel it. But how do you put a value on it? Sometimes, in quiet or down-time moments with the company (moments that don't happen all that frequently by the way), I wonder how much Melaver, Inc. would sell for on the open market. Who knows? The sales of companies with kindred spirits—Ben and Jerry's to Unilever, Stonyfield Farms to Danone, Tom's of Maine to Colgate-Palmolive—are few and far between, the documentation sparse on the premium paid for such companies' brand, reputation, and goodwill.

We could, I suppose, do a back-of-the-envelope analysis, drawing upon some of the revenue/savings calculations in this and previous chapters in our discussion of Green, Inc. In our analysis of that fictionalized company, we felt comfortable with the following projections:

- Additional business development of $450,000 annually because of values orientation (Chapter 1).
- Additional business development of another $450,000 annually due to knowledge of and experience with LEED processes and construction.
- Additional consulting revenue of $150,000 annually.

Let's imagine that at the end of the ten years of analysis we have been discussing for Green, Inc, senior management decides to sell the company, based upon the three projected revenue line items above. These three line items alone, projected out for fifteen years at a discount rate of 10 percent, results in a present value of roughly $20 million (see Figure 4.3).

Year	10	11	12	13	14	15	16
Periods	0	1	2	3	4	5	6
Additional development		900,000	1,035,000	1,190,250	1,368,788	1,574,106	1,810,221
Consulting		150,000	172,500	198,375	228,131	262,351	301,704
Total Cashflow	0	1,050,000	1,207,500	1,388,625	1,596,919	1,836,457	2,111,925
Discount Factor	1.000	0.909	0.826	0.751	0.683	0.621	0.564
PV Cashflow	0	954,545	997,934	1,043,295	1,090,717	1,140,295	1,192,127
NPV	19,906,905						

Figure 4.3   **Evaluating the goodwill of Green, Inc.**

What would be the return for Green, Inc., should it be sold in year ten? What, in short, is the real return on being a sustainable real estate company? Figure 4.4 provides a provisional answer.

As Figure 4.4 shows, the management team of Green, Inc. are looking at a 45 percent return on their various investments in green stuff: values, culture, internal environmental audits, and expertise at green development. The projected goodwill of $20 million ten years from now equates to just over $10 million today, or roughly 40 percent of the hard asset value of the company.

The point of this exercise is not so much to derive an ultimate sales price for Green, Inc. but rather to emphasize the considerable intangible value embedded in a values-centric business. There's an additional intangible value that needs to be considered, one that perhaps may enhance further the overall goodwill of the company. I like to think of this intangible value as a company's "X" factor.

Speaking personally, I have worked throughout the world and have never seen anything like this. This may sound presumptuous. However, I have lived on three continents and worked in over seventy countries at some of the world's largest companies, and there is something going on here that I have not seen or felt elsewhere.

When I moved to Savannah in 2005, I thought I would be moving to a quaint historic town that would allow for a perfect work/life balance. The coast, with all it has to offer, the small-town atmosphere, even the allure of a few golf courses, were all aspects that I looked forward to enjoying upon arrival. Working for Melaver, Inc., a quirky family-owned real estate company, was also something I was going to enjoy in a relaxed way. Were all my expectations met? Yes, Savannah is charming. Yes, the climate, lifestyle, people I have meet are very rewarding. But what has far exceeded my expectations is the quality and passion of my colleagues.

To begin with I simply thought this was a good bunch, experienced and committed to working hard. But then, don't all companies say they have the best employees around? As time went by I realized there was something else, that "X" factor. There was a glue that kept us together, but it is more than a glue. It's something that enables everybody to do more, perform better, take the extra step, and act as a team. People filled in the gaps and picked up loose ends, should colleagues be tied up with

17	18	19	20	21	22	23	24	25
7	8	9	10	11	12	13	14	15
2,081,755	2,394,018	2,753,121	3,166,089	3,641,002	4,187,152	4,815,225	5,537,509	6,368,135
346,959	399,003	458,853	527,681	606,834	697,859	802,538	922,918	1,061,356
2,428,714	2,793,021	3,211,974	3,693,770	4,247,836	4,885,011	5,617,763	6,460,427	7,429,491
0.513	0.467	0.424	0.386	0.350	0.319	0.290	0.263	0.239
1,246,314	1,302,965	1,362,191	1,424,108	1,488,840	1,556,515	1,627,266	1,701,232	1,778,561

Years	0	1	2	3
REVENUES				
Shaping values, chapter 1		21,300	21,300	21,300
Creating a culture of green glue, chapter 2		79,000	79,000	79,000
Green from the inside out, chapter 3		3,300	3,300	3,300
Developing expertise in LEED, chapter 4	0	0	0	0
Sale of Green, Inc.				
Total Revenues		103,600	103,600	103,600
EXPENSES				
Shaping values, chapter 1	(166,000)	(114,500)	(91,875)	(97,286)
Creating a culture of green glue, chapter 2	0	(57,600)	(64,631)	(72,717)
Green from the inside out, chapter 3	0	(78,490)	(22,381)	(22,998)
Developing expertise in LEED, chapter 4	(49,000)	(150,000)	(417,000)	(890,000)
Total Expenses	(215,000)	(400,590)	(595,888)	(1,083,002)
Total Cashflow	(215,000)	(296,990)	(492,288)	(979,402)
Discount Factor	1	1	1	1
PV Cashflow	(215,000)	(269,991)	(406,849)	(735,839)
NPV	10,011,860			
IRR	44.98%			

Figure 4.4   Total return on a green company

urgent matters or important rush projects. I had never seen that type of activity—not in Chicago, not London, not Singapore.

I couldn't understand what was going on here. Why were these employees performing better than all those I had encountered before, even those in companies where I had worked that were among the leaders in their respective industries? It wasn't the pay, nor the workload, nor the perks. I realized it was the sharing of core beliefs, a deep feeling that what we were doing made sense, that something extra each individual could identify with. Call it pride, a sense of accomplishment, a meaning in the work being done. That is the "X" factor that allows everybody to work harder, go the extra mile, and automatically pick up the slack. Everybody knows the ultimate goal. No one needs supervision or a "to do" list.

This "X" factor may actually be the most valuable aspect of developing green. It provides something special for everybody to rally around. This could be the main financial benefit for a sustainable company—this sense of purpose and meaning, this sense that we are truly making a difference—that improves productivity and ratchets up teamwork and enthusiasm, and enables us to attain incredible objectives. Without this "X" factor, we might not be able even to consider the visionary goal we have of becoming and remaining a thought and product leader in sustainable real estate.

4	5	6	7	8	9	10	TOTALS
21,300	21,300	471,300	538,800	548,925	550,444	550,672	2,766,640
79,000	79,000	79,000	79,000	79,000	79,000	79,000	790,000
3,300	3,300	3,300	3,300	3,300	3,300	3,300	33,000
0	0	2,116,500	1,176,475	1,222,971	1,261,446	1,303,442	7,080,834
						20,000,000	20,000,000
103,600	103,600	2,670,100	1,797,575	1,854,196	1,894,189	21,936,413	30,670,474
(103,490)	(110,605)	(118,765)	(128,125)	(138,866)	(151,190)	(165,335)	(1,386,037)
(82,016)	(92,710)	(105,007)	(119,150)	(135,413)	(154,117)	(175,625)	(1,058,987)
(24,858)	(26,997)	(29,456)	(32,285)	(35,538)	(39,278)	(43,580)	(355,862)
(57,000)	0	0	0	0	0	0	(1,563,000)
(267,365)	(230,311)	(253,228)	(279,560)	(309,817)	(344,585)	(384,541)	(4,363,886)
(163,765)	(126,711)	2,416,872	1,518,015	1,544,380	1,549,604	21,551,873	
1	1	1	1	0	0	0	0
(111,853)	(78,678)	1,364,261	778,982	720,464	657,183	8,309,180	0

As the saying goes, even in utopia there's myopia. I have my wish list of things I'd like to see change to make my job and those of my colleagues easier, to make this sustainability movement we are a part of move faster toward mainstream acceptance.

There needs to be a change in how properties are appraised. Many real estate professionals rely on independent appraisers to provide market values on properties. I have yet to see any additional value for green being included in a formal appraisal report. In fact, I have even seen some signs of skepticism when appraisers are asked to value a green building. I gather it is our role to help educate appraisers as to the benefits of green real estate. We need to share with them our metrics for future enhanced operational savings and projected improved discounted cash flows, as well as the potential productivity gains and other intangible benefits that need to be considered. Once appraisers grasp the value of green, their independent validation of the value will be crucial in reaffirming green by the numbers.

The financial sector needs a greater number of green projects to fund, in order to create a green building backed mortgage securities market and pricing. This would create a tiered market that would allow green assets to command a better price than the current commercial-backed mortgage securities. It will take a few years to reach such momentum.

There needs to be a change in the way leases are crafted so that the interests of the landlord and the tenant are better aligned to share in the benefits of green properties.

We need to have real estate brokers who are knowledgeable about the benefits of sustainable development. They should be conversant with the LEED scorecard as well as educated about the technologies that are used in the construction of green buildings. The value brokers can bring to a transaction is dependent on how well they educate themselves, the way they communicate this new green paradigm with buyers and sellers, and how they orchestrate and connect the different parties involved. Like a spider in the web, the broker is often the key person linking places with people. A sustainably oriented broker has a key role to play in orchestrating the interaction between a developer that may not yet have contemplated a green development with a builder that has extensive experience in meeting LEED requirements.

I suppose my wish list is rather extensive. There are days when my colleagues and I are impatient to run down that checklist to see just how far we have come. The financial benefits of developing green are starting to be recognized. As the demand for green increases there will be more developers moving into this arena. With more players, there will be more knowledge and expertise in the market. The cost of green today already is only marginally higher than standard development costs and will continue to decline. Someday soon, people will wonder why a CFO took the time and energy to make the financial case for building green because it will seem so self-evident.

Sometimes, our focus at Melaver, Inc. on such distant goals makes us lose sight of day-to-day blocking and tackling and basic operational processes. We have to slow down, regroup, catch our breath, and make sure we continue at a pace that is challenging but manageable, that stretches our capacity to learn and yet is still fun. We still need to work on the work/life balance thing. And it would be nice if, every once in a while, my notion of the color green was simply the distinct shade of the grass on a golf course somewhere nearby.

NOTES

[1] Jonathan Lash and Fred Wellington, "Competitive Advantage on a Warming Planet," *Harvard Business Review*, March 2007.

[2] Daniel C. Esty and Andrew S. Winston, *Green to Gold: How Smart Companies Use Environmental Strategy to Innovate, Create Value, and Build Competitive Advantage* (New Haven: Yale University Press, 2006), pp. 101–4.

GREEN IS (ALSO) THE COLOR OF INEXPERIENCE: LEARNING FROM A LEED PILOT PROJECT

RANDY PEACOCK

SUMMARY

From 1940 until 1985, the Melaver family ran a grocery business in and around Savannah, Georgia. Abercorn Common (then called Abercorn Plaza) was the company's first shopping center, built during the late 1960s to house its own M&M Supermarket as well as a Western Auto store, a discount general retailer (Sam Solomon's), a small shoe store, and a finance company. In the late 1990s, Melaver, Inc. began to assemble properties adjacent to the original shopping center with the intent of both renovating existing buildings and expanding the development to twice its original size. The site underwent a dramatic transformation to become, in 2006, the first LEED certified retail shopping center in the United States.

There's a lot of green everywhere you look at Abercorn Common. There's more landscaping than you'd expect to see at a suburban strip shopping center on a four-lane thoroughfare, and it's all irrigated with harvested rainwater. Tall sable palm trees provide shade in the parking lot, and a wide brick walkway invites shoppers to stroll from store to store. And there's lots of green you don't see. Its tenants pay lower power bills and use much less water—two factors that directly affect their stores' bottom lines—and they and their customers enjoy improved indoor air quality, among other benefits.

In this chapter, Randy Peacock, Melaver's head of Development and Construction Services, discusses the sustainable features of Abercorn Common, whose tenants include a mix of national chains (Circuit City, Panera Bread, Home Goods, Books-A-Million, and Michaels, among others) and local businesses (F. P. Wortley Jewelers and The Riitz, an Aveda Salon, among others), the first LEED McDonald's restaurant in the United States, and a newer out-parcel addition called Shops 600 that has a green vegetated roof and solar hot water.

The redevelopment of Abercorn Common was a learn-as-you-go experience for the company. Peacock begins with a general history of the center and considers the overarching challenges the development team faced in making this an early LEED project. Section one takes a more specific look at the project itself. Sections two through four look at each of the three particular component parts of the development—the in-line shopping center, the LEED McDonald's out-parcel, and the boutique Shops 600 out-parcel—and the particular green strategies associated with each one. Section five considers the financial implications of being an early entrant into green development. In the concluding section, Peacock considers lessons learned, with an eye toward what the development team might have done differently. Charting new territory has pitfalls as well as benefits. "Do as we say, not as we did," a variation of the old saying, is an apt motto for this chapter, and Peacock's discussion provides a roadmap for other developers.

If buildings could talk, Abercorn Common could tell a lot of stories, stories about values and innovation and long-term commitment to place. Take the M&M Supermarket that was built as part of the original shopping center, Abercorn Plaza. When the store was constructed in 1970, it had things no other grocery store in the region had, such as a deli, a bakery, and a large wine and cheese specialty section. In addition, the store included cutting-edge technology in the form of computer scanners that were used to help manage the inventory through a just-in-time delivery system— and remember, this was almost forty years ago.

About the time the Abercorn Plaza M&M Supermarket readied for its grand opening, owners Norton and Betty Melaver signed an open letter to the city that was published in the *Savannah Morning News*, supporting the then-recent federal decision to integrate the local school system. *"You might agree or disagree with the decision,"* the letter said, *"but it's the law of the land and the right thing to do."* As a result of this letter, the grand opening of the store was hindered by an angry group of picketers trying to stage a boycott of the store—an action that required going to court to enjoin the picketers from continuing and potentially ruining the company. Prominent environmentalist and constitutional lawyer Ogden Doremus, later known as the author of Georgia's visionary Marshland Protections Act, advocated successfully on behalf of the Melaver business.

I am too young to remember those early years of Abercorn Plaza, but growing up on the south side of Savannah, I got caught up in the ongoing story of the shopping center. My part of the story started at age fifteen, on a Saturday. I rode my bicycle to the grocery store, parked it at the bike rack, walked inside the store, and began bagging groceries. I was making good tips, but after a few hours of hard work, the store manager called me over and asked me who I was and what I thought I was doing.

"Bagging groceries and making tips," I replied.

"Son," he said, "come back when you are sixteen, fill out an application, and we'll put you to work." Which I did. It was my first job. Twenty-something years later, I was running the development effort to make Abercorn Common the first LEED retail shopping center in the country.

The long, demanding process of renovating and expanding the center is a collection of stories. Early on in the process of assembling the neighboring tracts of land, a Wal-Mart representative, out of the blue, hired an engineering firm to survey the site and stake it out with the idea of locating one of its Supercenters on the location. We weren't notified of the assignment. We simply discovered one day a field of small stakes with blue flags covering the site. Rumors started to fly of Wal-Mart's intentions, and despite our best efforts to convince neighboring landowners we were not planning on developing a Wal-Mart, folks were not entirely convinced. Land prices started to soar. Assemblage, often a challenging undertaking, became that much more formidable. Another challenge to redevelopment occurred when a tenant, actually a sub-lease, pulled out an attachment to its lease that appeared in no other copies on file, with odd typewritten additional language that stated the sub-tenant had total rights over any and all additions made to the center. This rider to the sub-lease made our redevelopment effort more complex and time-consuming.

But the really challenging pre-development work occurred around the company's decision to make this a LEED project, at a time when only about twenty retailers in the nation had publicly made vague commitments to building out their locations to green specifications. We made our commitment and were proud to do so, but soon the questions started to flow. Where would our tenants come from? We didn't really know. Would they be willing to enter into a cost-sharing green lease with us, since some of our initial investments in energy-efficient technologies would benefit them? (The market simply wasn't ready for that yet, and it still isn't to a large extent, though that is changing.) Who would handle our brokerage representation? We clearly needed a regional or national brokerage firm that had strong connections with a retail clientele. But those firms, by and large, were not interested in pitching a deal that looked different from every other locale. Those firms knew their retail clients' prototype specifications, made their money on doing a volume of same-old deals, and were disinclined to take on an assignment that seemed to be difficult and time consuming. Eventually, we turned to our own brokerage division. They understood the green angle, since they were among the first brokers in the country to become LEED accredited. But even with a product unlike any other on the market and an excellent location, lease-up was slow going.

In addition to the challenges of leasing a green retail shopping center, we found it necessary to share with the city building departments the lessons we were learning about sustainable design. We had an odd list of wishes for the various city departments that would have to provide the approvals and permitting on this project. In an effort to reduce the amount of impervious surfaces and mitigate heat island effects, we were looking for parking ratios at 4.0 to 4.5 cars per thousand square feet—well below the required parking ordinance ratio of 5.5 cars per thousand square feet for a shopping center of our size. It took time to explain why this was important to us and to get over that hurdle. In the end, however, the tenants provided the real challenges we faced during design and construction of this sustainable shopping center.

One early challenge had to do with our landscape plan, which called for significant tree coverage on the site to help reduce the overall heat island effect by providing shading. Trees, some tenants felt, would interfere with sight lines, visibility, and overall customer brand recognition. And that was one of the simpler challenges.

Of much greater significance in our quest to build green was the issue of dealing with each tenant's prototype, which typically called for a very specific look using a formulaic design and predetermined list of materials that could not be changed. Luckily for us, the majority of our scope of work was focused on developing a building envelope and base systems with significantly greater energy efficiency. We intended to have our tenants modify their respective prototypes only to the extent required by our selected LEED program. Even though changes in most cases were minor, lengthy (sometimes extended) negotiations often ensued. There were times when we simply had to say *If you want to be in our center, these are the criteria you will need to abide by.* It's odd when an 800-pound retail giant goes toe-to-toe with a local bantamweight developer that refuses to give ground. None of our tenant prospects seemed to fathom that we meant business. A few walked. Most, however, worked with us. The values-centric orientation of our company helped enormously in enabling us to stand our ground. So, too, did the fact that ours was the premier retail site in the city. If the mantra of a green real estate company is values, values, values, it's still important to remember the old real estate cliché about location, location, location. We were fortunate to have the right locale for this first green retail development. Otherwise, we might not be able to tell this story.

There was also the challenge of creating a group of dedicated professionals around our team who would not only share our excitement of developing a green retail project but would also be at least familiar with, if not knowledgeable about, what such a green development would entail. Our own staff was not a problem. Something like a dozen of my colleagues had been through LEED training and become accredited, as had I. But what about our architect, someone who was expert at delivering conventional retail space but knew nothing of the LEED program? Our general contractor, landscape architect, and civil and mechanical engineers? Everyone seemed excited with the cutting-edge nature of our project, but just how committed they were, no one really knew. As it turned out, over thirty professionals on this project become LEED accredited, on their own time and at their own expense. As it turned out, this factor was a critical piece in the project's overall success.

To sum up, the initial obstacles facing us in delivering a LEED retail project were long and, in retrospect, daunting. They included:

- **Pro-active determination of scope (the "what").** Making the decision to make this a LEED project after initial design work had already been done.
- **Selecting the correct LEED program.** Initially trying to fit this project to what was then the sole LEED product category (LEED for New Construction), which called for control over interior design elements that are typically determined by retail tenants. Eventually, we shifted to becoming a participant in the U.S. Green Building Council's pilot Core and Shell program, which made all the difference in the world.
- **Being clear about the "who."** Ensuring that we built an entire extended team of professionals who were all passionate about delivering a different type of development. Finding brokerage representation that could genuinely support the marketing of a very different product.
- **Working with stakeholders.** Introducing city officials to the LEED program and how some of the program's criteria were improvements upon municipal codes and ordinances.
- **Being clear about the "how."** Learning how to integrate design and development, where all professionals are at the table working collaboratively from the get-go and not in isolated, linear fashion.
- **Clarifying expectations with tenants.** Being a bit inflexible early on about what issues to stand firm on with tenants.
- **Being clear about the message.** Trying to figure out the marketing message of Abercorn Common in ways that would appeal to the various tenant-rep brokers (this issue is discussed at length in Chapter 10). Our initial thought that green should be the primary message was wrong. Ironically, some of our later projects have suffered by not making the green message as strong as it should be.
- **Determining the green details.** Determining which of the many new-to-the-world green products and technologies being proffered to us would actually work as promised.
- **Streamlining administration.** Developing lease proposals, lease forms, and work letters that captured landlord expectations (toward the tenant) of various green compliance issues.
- **Streamlining execution.** Overseeing operations of and material purchased by the general contractor to ensure actual compliance with the LEED program (e.g., making sure that sub-contractors were not smoking inside the premises and confirming that low-VOC paint was being used).
- **Streamlining documentation.** Developing a process for capturing the detailed information for documenting a LEED project.
- **Capturing lessons learned.** Analyzing the actual performance of the project, from both a financial perspective and in terms of resource use.

We are often quick to identify the challenges facing a new endeavor— and those are important considerations, ones we will be pursuing throughout the remainder of this chapter. Before doing so, however, it's worthwhile to pose an inverse question: What challenges do we face by continuing to develop in the same ways our culture has for decades, even centuries?

In the United States today, we lose more than one million acres a year to roads, parking lots, and urban sprawl development.[1] Over the last two hundred years, we've lost 50 percent of our wetlands, 90 percent of our old growth Northwest forests, 99 percent of our Midwest prairies[2] and 98 percent of our longleaf pine ecosystem.[3] Between 1970 and 1990, there were approximately 25,000 shopping centers built in the U.S., or an average of one every seven hours.[4] With approximately five million commercial buildings in this country today, we still add another 170,000 annually. With over 100 million housing units in stock today, we add another two million each year. We also demolish 44,000 commercial buildings and a quarter-million housing units each year, most less than thirty years old and destined for the landfill.[5] And virtually all of this so-called growth occurs using energy strategies (if you can call them that) that are sadly inadequate. And so the question becomes, do we need to make a business case for green development or for the traditional development practices that seem—when you consider the broader context—so unsound? That was the underlying premise of our intent to develop Abercorn Common differently, despite the many challenges we faced.

An Overview of the Project

The phase one portion of Abercorn Common, now built out and leased up, consists of approximately 209,000 square feet (187,00 square feet of leaseable area) spread across three distinct parcels: a main in-line shopping center and two out-parcels—a stand-alone McDonald's and a set of small shops (called Shops 600) fronting the main thoroughfare. Each parcel, developed one after the other, called for distinctive development strategies and constituted separate LEED projects. Phase two, consisting of an adjacent parcel of land contiguous to the main in-line development, has remained undeveloped pending future plans. We will first take a look at the overall site before delving into the specific LEED strategies for the three major components of the project.

If you were to bicycle or drive or ride the bus past Abercorn Common today, you might notice just how normal the development looks, a far cry from the edgy LEED buildings you see in magazine photographs. This design was deliberate, the result of our intention to reassure the market that a new-to-the-world green retail development could be green without looking out of place. The second thing you might notice about Abercorn Common is how literally green it looks, thanks to the large number of trees and shrubs on site. You also might notice it's not a typical flat-facade retail strip with head-in parking spaces along the fronts of the stores. In designing Abercorn Common, we asked our architects to emulate the look of historic downtown Savannah, so each store would have a distinctive appearance and would look as if each building had been built by a different owner at a different point in time. Building heights and exterior finishes are varied. A wide brick walkway that winds along the fronts of the stores is interspersed with fountains, plantings, and places to sit, recalling downtown Savannah's renowned public squares. It's both a shopping destination and a gathering place.

If you were to view Abercorn Common from the air, you'd see lots of white and green—white roof membranes, white concrete parking areas, white "sidewalks" made of crushed seashells on planted islands in the parking lot, extensive tree canopy, and a vegetated garden roof (you'll hear more about that later). All that white and green helps mitigate what's called the heat island effect, a rise in air and surface temperature in an area.

DE-ENERGIZING THE SITE

Heat islands are formed when cities replace natural land cover such as forest, marshland, and pasture with pavement and buildings. These changes contribute to higher urban temperatures in several ways. Displacing trees and vegetation reduces the cooling effects of shade from tree cover and the cooling effect produced when water evaporates from soil and leaves—a process called "evapotranspiration." Dark roofs on buildings and dark asphalt paving absorb and hold heat delivered by sunlight rather than reflect it. Heat from vehicles, industry, and air conditioners can add warmth to the surrounding air, further exacerbating the heat island effect. In warm climates in the summertime, heat islands may contribute to global warming by increasing demand for air conditioning, thus increasing energy use that results in additional power plant

Figure 5.1   Rainwater-fed fountain, Abercorn Common.

emissions of heat-trapping greenhouse gases. Strategies that reduce the heat island effect can reduce the power plant emissions that contribute to global warming.

The south side of Savannah, where Abercorn Common is located, is typically four to five degrees warmer in the summer months than other areas of the city, a difference I could always feel as I drove back and forth from our office in downtown Savannah, with its thick tree canopy, to the south side of Savannah each day during the two years Abercorn Common was under construction. There are three ways to mitigate the heat island effect: through choice of roofing material, with plants, and with reflective paving surfaces. At Abercorn Common, we used all three measures.

As part of our plan to combat the heat island effect, we specified that each roof on our buildings be covered with a high-albedo white thermoplastic polyolefin (TPO) membrane, which reflects heat and allows our roof surfaces to be 20 to 30 degrees cooler in the dead of summer. (Albedo is a measure of a material's solar reflectance, its ability to reflect sunlight.) TPO products aren't more expensive; in fact, they actually cost less than traditional built-up roof membranes. In addition, TPO material can be recycled and reprocessed. The combination of the highly reflective, high-albedo roof surfaces and the tree canopy provided by the oak and palm trees means more than 30 percent of the site is reflective and shaded. See Figure 5.2 for an aerial view.

DE-ENERGIZING THE EXTERIOR LOOK

The exterior facades of Abercorn Common—some brick, some stucco, like old Savannah—look substantial and beefy and are built to last. In our negotiations with tenants

Figure 5.2   Aerial photo of the Abercorn Common site.

during the pre-construction and design phase, we learned it was the policy of most chain store and big box tenants to demolish the existing facade on buildings they rented, regardless of its physical condition, so the tenant could build a new facade to suit its current corporate identity. Since corporate identities seem to change every few years, we saw this practice as wasteful and certainly not sustainable. So we designed our shopping center with a traditional look and substantial facades that we hope future tenants won't want to demolish, but will, instead, be happy to adapt to, just as tenants in buildings in downtown Savannah choose not to demolish two-hundred-year-old facades but instead accept and adapt them.

If you look closely at our buildings, you will see a greater level of architectural detail compared to typical retail shopping center facades. In addition to better quality, we ensured an appropriate level of caulking and painting to provide a better airtight building envelope. To aid our facade performance, the walls are insulated to R-19, and the roofs are insulated to an extreme level, R-30, to save energy.

When considering the design of our storefront facades, we took into account the inevitable heat gain from sunlight through our glazing. As a result, we provided more window glazing in north-facing facades, which receive less direct sunlight, and less window glazing in the west-facing facades, which get much more direct sunlight late in the afternoon, when temperatures in the summer often reach the high 90s at 5 p.m.

One thing I wish we had thought of is the use of dissimilar building-wall thicknesses in appropriate locations. Like most developers, at the time of design we didn't consider that some wall sections battle heat gain, whereas other wall sections, such as north facades, don't. In hindsight, it only makes sense that a building's walls should be thicker on the west-facing facade and, in some cases, the south-facing facade, to accommodate more insulation. Why do we build north-facing walls and west-facing walls the same way? The simple, common sense answer is that we shouldn't. But sometimes the common sense answer isn't the first thing that comes to mind.

Another common sense idea is to build with demolition in mind. After all, no matter how much we don't want to see serviceable buildings torn down, the fact is that any building may eventually be demolished. Therefore, in our design and construction of our steel members we have many mechanical connections instead of welds, so the steel can easily be removed and re-used in future renovations.

In addition to these sustainable design practices, we also included as standard items high-efficiency two-by-four light fixtures that use 28-watt tubes, hot gas re-heat efficient HVAC equipment, and of course, low-emission (low E) window glazing that, in addition to keeping heat inside in winter and out in summer, reflects ultraviolet light, which can damage furnishings and window displays. These energy-saving features, combined with the quality building envelope and the highly reflective white roof membrane, reduce average electricity consumption at Abercorn Common by an estimated 30 percent when compared to a conventional building built during the same time frame.

SAVING WATER

One goal we set for our design team early on was to significantly reduce water use. To that end, the restrooms at Abercorn Common are equipped with low-flow sink faucets with aerators and mechanical timers. The Circuit City store has dual flush toilets (0.8 gallons per flush (gpf) and 1.6 gpf). The remainder of the shopping center has 1 gpf toilets. By selecting such water-efficient plumbing fixtures and installing waterless urinals in higher demand spaces (Circuit City, Michaels, Locos Deli, and McDonald's), we have reduced overall water use by nearly 50 percent (a 46 percent reduction for potable water and a 49 percent reduction for black water). Table 5.1 compares the baseline water use vs. Abercorn Common's water use using water-efficient plumbing fixtures, as calculated by our engineers.

Just how much a tenant can reduce its water use can be seen in Table 5.2, Circuit City Water Usage, which compares a baseline non-LEED design with a LEED design that includes dual flush toilets, waterless urinals, and ultra low-flow faucets and aerators. The comparison is based on a store with 26,600 square feet and 48 people (one person for every 550 square feet). Water savings are more than 40 percent.

The lush landscaping incorporates plants that are native to coastal Georgia because they require less water and maintenance and thrive in Savannah's mild winters and warm summers. Planting beds and islands in the parking lot include numerous trees (palms and oaks), evergreen and flowering shrubs, and perennials. Seasonal planted areas and containers along the brick walkway bloom with colorful annual flowers. All of the landscaping—all this greenery—is irrigated solely by rainwater that is harvested from the roofs of the buildings and stored in an open cistern that's 60 feet wide by 120 feet long and twelve feet deep. It typically has about eight feet of water in it year-round. Using the rainwater for landscape irrigation throughout the property saves about 5.5 million gallons of water each year. We estimate our tenants collectively save about $40,000 a year since they are not paying for water to irrigate landscaping, which lowers the fee they would normally pay for common area maintenance (CAM) elsewhere. Water collected and stored in the cistern also is used to fill the three circulating fountains placed along the brick sidewalk along the fronts of the stores.

TABLE 5.1 ABERCORN COMMON WATER USE	
POTABLE WATER	
Baseline Potable Water Use	227,288 gallons/year
Design Potable Water Use	123,860 gallons/year
Percentage Potable Water Reduction	46%
BLACK WATER	
Baseline Black Water Use	179, 546 gallons/year
Design Black Water Use	91,328 gallons/year
Percentage Black Water Reduction	49%

TABLE 5.2   CIRCUIT CITY WATER USAGE COMPARISON

CIRCUIT CITY BASELINE WATER USAGE (NON-LEED DESIGN)

FIXTURE	USES/DAY	DURATION	FLOW RATE	VOLUME (GAL/USE)	TOTAL (GAL/USE)
Toilets	96	1 flush	1.6 gal	1.6	153.6
Urinals	48	1 flush	1 gal	1	48
Lavatories	144	1 cycle	.25 gal	.25	36

Daily Water Use	238 gal/day
Operating Days per Week	6.5 days
Baseline Potable Water Use	80,309 gal/yr
Baseline Black Water	68,141 gal/yr

CIRCUIT CITY DESIGN CASE WATER USAGE (LEED DESIGN)

FIXTURE	USES/DAY	DURATION	FLOW RATE	VOLUME (GAL/USE)	TOTAL (GAL/USE)
Toilets 1.6 gpf	48	1 flush	1.6 gal	1.6	76.8
Toilets 0.8 gpf	48	1 flush	0.8 gal	0.8	38.4
Urinals	48	1 flush	0 gal	0	0
Lavatories	144	1 cycle	.13 gal	.13	18

Daily Water Use	133 gal/day
Operating Days per Week	6.5 days
Baseline Potable Water Use	45,022 gal/yr
Baseline Black Water	38, 938 gal/yr

Circuit City Water Savings (LEED vs. Non-LEED)	35,297 gal/yr
Circuit City Percentage Potable Water Reduction	44%
Circuit City Percentage Black Water Reduction	43%

As is usual with all new technologies, the rainwater collection cistern comes with an array of maintenance considerations. Algae growth in water is a huge issue in the south, especially during warmer months, and controlling it has proved more challenging than we anticipated. To combat algae proliferation, we specified freshwater plants to be installed in the large water fountain. In the cistern, we installed two pumps that move the water and discourage algae growth. However, we underestimated the aggressiveness of algae.

Because a conventional chemical solution would damage and most likely kill plant life in the fountains and our landscaping, we knew had to attack the algae problem from a slightly different angle. First we installed additional pumps in the cistern, but this time we used de-stratifying pumps that move the water from the bottom to the top instead of the kind of pumps that move the water around in circles. Second, we installed a relatively new technology in the war against algae from Grander Technologies—a metal tubular structure that actually alters the chemical makeup of algae as it passes through the structure's rings. This process kills the algae, and it works without the use of harmful chemicals or electricity. So far, it's doing the job.

STORMWATER MANAGEMENT

At Abercorn Common, we made a great effort to design the site so that stormwater runoff did not negatively impact our neighbors. In early studies of the site and the surrounding area, we learned that during severe rains, street flooding occurred down gradient from the existing shopping center. This concerned us greatly; any part of our design that didn't take people into consideration was deemed unsustainable and revised. Since Abercorn Common is less than one mile from marshlands, we were also concerned about where our stormwater was going and what contaminants it might carry with it to waterways, such as fertilizers, oils, and gasoline. We designed our shopping center to collect and hold rainwater on site and allow it to percolate into the soil, as it would on unpaved land.

To help prevent flooding and contamination of our local waterways, we first used grade inlets and porous concrete to help absorb and hold rainwater. To do this, we limited the number of grade inlets normally seen in a development of this size, forcing the storm water to sheet flow across the parking lot for longer distances and to flow over porous concrete before arriving at a storm grade inlet.

Additionally, we increased the size of our cistern so it would hold more water, which lessened the stormwater impact on our two very small detention ponds. Where we could, we used porous sidewalks made of soft limestone (formed largely from crushed coquina shells) in the parking lot islands. As our out-parcel building, Shops 600, is not connected to the rainwater collection system that feeds the cistern and two detention ponds, we designed and built an infiltration trench (more about this later) to allow water to flow through to the ground, where it is filtered by sands and soils as it is returned to the water table.

Collectively, our efforts have reduced stormwater runoff at Abercorn Common by 30 percent. This is in comparison to years past, when houses in the surrounding neighborhood in the area known as the Fairmont Basin would flood in heavy rains.

Although we can't directly attribute the change to our efforts to reduce stormwater runoff, we do know that flooding in the Fairmount Basin has lessened considerably.

Greening the In-Line Shopping Center

The in-line portion Abercorn Common (187,000 total square feet) posed some interesting challenges, enhanced by having to integrate renovation of the old with development of new sections, while also having to manage the LEED program differently for existing versus new tenants. Not atypically for a renovation project, we were tasked with constructing the site to LEED standards while limiting the daily impacts on existing tenants by working in stages, which caused numerous starts and stops over the two years of construction. For the stores that were tenants at the time—Michaels, Books-A-Million, HomeGoods, and McDonald's—participation in the LEED program was voluntary. Some tenants chose to participate, while others did not. The two that did participate (Michaels and McDonald's) received a new building. Their relocating provided the perfect opportunity for us to work with their building and design professionals to pass along what we had learned about green building practices. We were, in some instances, successful in improving the tenants' base building plans.

The two tenants that opted not to participate at the time of the redevelopment, Books-A-Million and HomeGoods, received cosmetic upgrades to their storefronts so they would complement the new facade design of the shopping center. Their roofs, like those on the rest of the center, were covered with the highly reflective white TPO membrane (which improved their stores' energy performance).

All of the new tenants benefited from the enhanced site improvements. For LEED certification purposes, tenants that could not fully participate in our program were included in the site credits but not for credits where we had no control, such as energy and water modifications in their stores.

For the new tenant spaces (where we were responsible for the design and construction of the building envelope), we provided high-efficiency base lighting, HVAC, and plumbing fixtures. Tenants were responsible for their own interiors. However, our lease provides tenant guidelines that explain the LEED Core and Shell program, offer insight into the LEED Commercial Interiors program, and suggest the use of low-impact building materials.

One tenant, Locos Grill and Bar, considered pursuing LEED Commercial Interiors certification and worked with an architect to incorporate environmentally friendly design elements, including energy-efficient lighting and plumbing fixtures and furnishings made from wheatboard, an emissions-free alternative to particle board that's made from recycled wheat chaff and formaldehyde-free binders. While the tenant probably met LEED standards, it ultimately decided not to complete formal certification.

Other tenants, such as Panera Bread Company and Grand Harbour, worked with us to modify their base lighting plan to include almost all high-efficiency fluorescent light fixtures, which helped our building program meet Abercorn Common's overall environmental goals.

Over the course of the renovation, we demolished over 45,000 square feet of buildings (the unwanted portions of the existing shopping center). This included the front sections of two buildings, three retail buildings, an auto repair shop, two auto showrooms, and a restaurant. Most of this demolished material and the construction debris from redevelopment was either recycled or reused. In the end, we diverted over 80 percent of typical construction waste (weight by volume) from the local landfill. This equates to over six thousand tons of material.

One interesting reuse effort we accomplished was the on-site production of our graded aggregate base (GAB). We brought a machine to the site to convert (crush) the block, brick, cement, and mortar from the demolished buildings into GAB and used the new material as the base for the asphalt parking lot. It would have cost $50 a ton to dispose of the debris plus the wasted use of front-end loaders, dump trucks, and fuel. At the time, we estimated the cost of setting up the on-site crushing station would cost about the same as hauling and dumping the debris and purchasing new materials. It was a wash financially, but environmentally, we saved big.

Prior to demolishing the first building, we compared a proposal to demolish all structures with machinery and haul the debris to the dump with a proposal to hire a local company to disassemble the buildings by hand and salvage the reusable materials. We decided to go with the latter choice and were very pleased with the results. Though it took almost four times as long to demolish all the structures by hand, the cost to us was the same, as we had time in the schedule to allow for a longer demolition period. The crew was able to salvage a variety of materials for reuse, including the wooden roof trusses from the old Western Auto store, three-inch-thick tongue-and-groove cedar roof planks, piles and piles of old pine paneling, stacks of framing lumber, and tons of block. When all was said and done, we saved tons of useful material from the landfill, and we provided employment for several men for over a month. I liked the idea of our money going to pay workers rather than paying a landfill for dumping privileges.

Over 70 percent (by cost) of the more than $2.5 million of new materials and products we used on the project were manufactured within 500 miles of the site, including pipes, structural steel, metal framing, insulation, stucco, concrete, asphalt, glazing, and gypsum board. Using materials and products manufactured close to the site significantly reduces indirect emissions associated with transportation, supports local businesses, and keeps money in local communities. Also, on average, those materials contained 20 percent (by cost) recycled content (both post-consumer and post-industrial).

Circuit City chose to become an Abercorn Common tenant and to participate in the LEED Core and Shell program. However, they chose to use their own contractor, engineer, and architect for the building, and their architect and engineer became LEED accredited. The Circuit City design team followed the Abercorn Common design team's specs and complied with the LEED Core and Shell rating system. They started with a construction budget of $57 per square foot and ended up spending $65 per square foot. With better insulation in the walls and ceiling, the white roof membrane, and a higher-efficiency HVAC unit, they were able to cut twenty tons of HVAC from their prototype. High-efficiency units cost more initially, but less equipment is required and they use less electricity. Circuit City moved into its Abercorn Common store in the summer of 2004.

We believe that the benefits we can provide our tenants accrue directly to their bottom lines, like lower electricity and water bills and lower fees for common area maintenance. Our tenants' lowered cost structure and higher profitability in turn helps make our developments more desirable. Retail is retail, after all, and the bottom line is the priority. A newer tenant, ULTA, part of the cosmetics and skincare retail chain, was able to reduce the HVAC load design for its Abercorn Common store, reducing the load from 55 tons to 47.5 tons, a savings of 15 percent.

There are, however, initial findings that still have us somewhat puzzled. Circuit City, for example, has provided us with figures for their electricity and gas consumption for Abercorn Common as well as for two comparison stores—one in nearby Brunswick, Georgia and another in Atlanta, which has somewhat different climate conditions. The results of this side-by-side comparison indicate that Circuit City's energy costs per square foot are actually higher at Abercorn Common than at the comparison locations (see Table 5.3). Internal debates around this issue have been interesting. Some on staff contend that the energy use of Circuit City would have been higher had a conventional building envelope been constructed. It's a challenging issue, since the ostensible benchmark is a building that was never built in the first place. The local manager of Circuit City has a different theory, believing that the cause of this unexpected utility premium is actually owing to the building's greater thermal insulation. Circuit City runs, among other things, a bank of TV monitors throughout its hours of operation. Circuit City management believes that heat created by the electronic devices builds up within the store, resulting in increased air conditioning use. This is a relatively new finding for us, one that will require further analysis. But this particular story does bear out our belief in the law of unintended consequences, as well as the fact that many aspects of green design and implementation are still not mastered.

Lighting in the parking lots at Abercorn Common was designed to conserve energy and reduce light pollution. Box lights mounted on slightly lower poles shine straight

TABLE 5.3   UTILITY COMPARISON AT THREE CIRCUIT CITY STORES

STORE	SIZE (SQ. FT.)	ELECTRIC/SQ. FT.	GAS/SQ. FT.	TOTAL COST	COST/SQ. FT.
Savannah	34,000	$2.73	$0.02	$93,582.60	$2.75
Brunswick	20,500	$2.24	$0.19	$49,858.85	$2.43
Atlanta	33,700	$2.18	$0.10	$76,994.40	$2.28

down, concentrating the light and allowing the use of five foot-candle illumination, rather than the more conventional ten foot candles. No lights shine beyond the property lines. The strategy reduces light pollution and energy usage, while also reducing the common area maintenance bills to our tenants.

Our buildings also provide benefits that don't directly contribute to our tenants' bottom lines but do increase their customers' comfort and well being. One benefit is the improved indoor air quality that comes from the use of low volatile organic compound (VOC) paints, sealers, and adhesives. Thanks to market demands and greater availability, it doesn't cost more to use low VOC products, you simply have to specify them on the building plans. The well-documented benefits to building occupants from improved indoor air quality include increased worker productivity, less absenteeism, and greater comfort.

Beyond the focus on our tenants is the focus on our general community. Originally, we allocated a space in the shopping center for an education center, to inform shoppers and the general citizenry about sustainable products and technologies. Ultimately, we nixed that idea since it was too expensive to construct, but also because we recognized that the entire center (and not just a stand-alone educational center) would be the

Figure 5.3   Map used for self-guided walking tours.

best resource (see Figure 5.3). As such, we created something of a walking educational trail throughout the project, with signs posted throughout the center that point out sustainable aspects and practices. We also offer guided tours of the center conducted by Melaver, Inc. staff members, and a map handout that offers a self-guided walking tour.

Another piece of our community outreach and education—fortunately a part of the general LEED program—involves making the most of the interactions we have with the general public. Preferred parking spaces, for instance, are provided for shoppers driving hybrid vehicles to Abercorn Common. Those who drive conventional vehicles and park in these spaces are greeted with an educational "ticket," which explains what the hybrid-only parking signs are all about. To encourage bicycling, racks are provided for bicycle parking; showers are provided for bicycling tenants as well.

Retail stores generate an estimated four pounds of trash per employee per day, and much of that can be recycled. To encourage our tenants to recycle, there is a central area behind the stores with large metal receptacles for paper and cardboard, and bins for recycling metals, plastics, and glass. We also encourage shoppers to recycle by providing seven smaller receptacles in the parking lot areas for recycling plastic, metal, and glass. A private contractor collects the recyclables for free each week.

Figure 5.4   Preferred parking spaces are set aside for hybrid vehicles.

A LEED Gold McDonald's

A McDonald's restaurant had been at the old Abercorn Plaza center since 1974. The original location was demolished at the end of the original lease to make room for the Shops 600 out-parcel, and a new out-parcel was created to make room for the new McDonald's (4,800 square feet). The story behind this franchise becoming the first LEED McDonald's in the country could take up a chapter in its own right.

In a nutshell, our company was simply not going to put up a non-LEED building anymore, even for the largest fast-food chain on the planet. McDonald's, for its part, wanted to hold on to this location. We weren't budging, and they weren't going elsewhere. It's a story of location, location, location meeting green, green, green. The local franchisee actually embraced the process, though corporate was less enthusiastic. In fact, at the time we received LEED certification on the facility and wanted to do a major public relations campaign, we were threatened with a lawsuit by corporate if we made any public statements about the LEED certification. Our guess is that the overall profitability of this location, fed in part by reduced operating expenses, will encourage more franchisees to seek LEED certification. That's our hope, in any case.

So what makes the McDonald's at Abercorn Common better? Lots of things. Over 75 percent of the restaurant's floor space receives daylight, allowing McDonald's to reduce its lighting costs. Clerestory windows flood the two-story dining room with light, inviting diners to sit and linger. About 90 percent of the interior space has a view of the outdoors. (Management has noticed more people choose to eat at tables in this McDonald's than they do at most other fast food restaurants. Perhaps that's why.)

When working with the McDonald's design team before construction began, one glaring shortfall we noticed in their plan was the absence of daylight in the kitchen so we suggested a design change. Now windows in the kitchen areas provide workers with desired daylight, reducing the tunnel effect so common in restaurant kitchens. Window canopies on the west side of the building provide shade from the afternoon sun.

A bright yellow metal roof, with overhangs on both gable ends that shade the windows, covers the two-story part of the building. The rest of the roof, like other roofs at Abercorn Common, is covered a bright white membrane. The roof membrane, tight building envelope, and high efficiency HVAC and lighting fixtures all work together to reduced energy consumption 20 percent based on our energy models.

Like the main part of the shopping center, McDonald's has walk-off mats at entrances to enhance indoor air quality and help keep the HVAC system cleaner. Using low-VOC paints, sealants, and adhesives throughout the building helps reduce indoor air pollution. Outside there's high albedo reflective concrete to combat the heat island effect and porous paving to reduce stormwater runoff, racks for bicycles, and preferred parking for hybrid vehicles. In addition, the restaurant operates on 100 percent green power through the purchase of wind power credits.

Other restaurants in the chain have incorporated some of these innovations into their own building designs. When McDonald's constructs a more sustainable building, it strongly demonstrates the mainstream potential of green building.

Figure 5.5 **McDonald's at Abercorn Common (designed by Adams + Associates Architecture, Mooresville, NC)**

Shops 600

Our out-parcel smaller retail building, Shops 600, was built near the end of the first phase of Abercorn Common's redevelopment, on the area previously occupied by the original McDonald's building. The building contains 16,000 square feet of retail space in six storefronts. Because the building facade faces west, the doors and windows on the fronts of the stores have six-foot canopies to shield them from the afternoon sun.

By the time we were ready to design and start constructing Shops 600, we had learned a lot about what worked and what didn't work in creating a LEED certified retail building for coastal Georgia. In our plans, we decided not to incorporate porous concrete, 30 percent fresh air requirements, or MERV 13 air filters because they are more difficult to manage in our climate. (Experience has taught us MERV 11 filters are a better choice, since they allow increased airflow.) Knowledge we gained from the first phase of Abercorn Common's redevelopment paid off dramatically—our design review and value engineering session for this new building lasted about one hour.

Materials used for the project were selected with an eye to high-recycled content. For example, we specified that our drywall come from a plant in Tampa whose product contains 95 percent recycled material. All of the wood used on the project was sustainably harvested and certified by the Forest Stewardship Council. Building Shops 600 has provided us the opportunities to experiment with our first vegetative roof—an intriguing idea we hope to use on future building projects—and using solar energy for water heating. The building's small size makes it an excellent laboratory for experimenting with new (to us) techniques.

The vegetated roof—four inches of engineered soil over a thick rubber mat (thicker than a normal roof membrane)—covers 9,000 square feet of the roof, which is more than half the building. The soil is planted with several types of drought-tolerant plants. (We're still experimenting to determine which ones work best in our climate. We started with four species of sedum; one of them is growing well.) This "garden roof" captures and retains rainwater, eventually returning some of the moisture to the air through evapotranspiration, the combination of the evaporation of water from the soil and transpiration of water to the air through the leaves of plants. The garden roof provides additional insulation to the part of the building it covers, and it helps protect the roof from damage caused by sun and pollution, more than doubling its life expectancy.

The remainder of the roof area is covered by the same type of bright white, high-albedo membrane as the other flat roofs on the property. Refrigerant units, which are CFC-free, were installed on top of the membrane. Both the garden roof and the white membrane help mitigate the heat island effect.

Figure 5.6   **The green roof atop Shops 600.**

The green roof area added an additional $80,000—about $9 additional per square foot—to the cost of the Shops 600 roof. Using the thicker roof membrane (60 mil rather than 45) cost 30 cents more per square foot ($2,700).

A solar panel on the roof provides tenants with hot water that is stored for use in an insulated tank on the ground floor. An auxiliary heating unit ensures a steady temperature, so the water is consistently hot, even on cloudy days. Although this system cost us $8,000 to get operational, we calculate our net additional expense was about $3,000, since we didn't have to buy an individual water heater for each store's bathroom (one per store, for a total of six) and have them installed. It just didn't make sense, from an environmental standpoint, to use electricity to heat water for hand washing (the only thing a retail store uses hot water for), even though the costs of heating the water would be borne by the tenants, not us, as part of their electric bills. Because of the central solar hot water system, water in the building is not individually metered.

The bright white parking lot also helps to mitigate the heat island effect by reflecting rather than absorbing sunlight. It's made of concrete with a high fly ash content. Fly ash is a residue generated in coal combustion. In the past, fly ash was released into the atmosphere from the smokestacks of coal-burning power plants, but new pollution control equipment has made it possible to capture the fly ash from the chimneys before it gets into the air. It is stored on-site at power plants. Naturally white, fly ash can be added to supplement portland cement in concrete production. While not exactly a green product (since it does, of course, come from coal), fly ash that wasn't captured would be a particulate pollutant, and using it in concrete provides a market for an industrial by-product. We were able to obtain high fly ash concrete from a local contractor. It costs about the same as regular concrete.

An infiltration ditch that skirts the parking lot allows stormwater runoff from the parking lot to percolate through vegetated runoff ditches called "bioswales," which are depressions around the perimeter of the pavement with plants and rocks that act as dams, capturing and filtering the water naturally, purifying it in the process, and eventually returning it to the water table. The garden roof, the infiltration ditch, and a strip of porous paving behind the building allow 100 percent of the stormwater from the site to infiltrate and re-charge the water table.

The building's green roof, tight envelope, solar hot water heating, and high-efficiency HVAC units reduce tenants' electricity consumption by over 25 percent when compared to a conventional building, and the building's core energy use operates on 100 percent green power.

The Green Consequences of Green Retail Development

One of the still-persistent mythologies of green development is that it costs a great deal more to do a project to LEED standards as compared to conventional construction criteria. This was true for early LEED projects. Today, however, we can deliver a

LEED building for equivalent costs to conventional construction. Having said that, Abercorn Common was one of those early LEED projects that did cost more. We feel it is important to be open and transparent about that premium cost, even thought we also feel that such a premium is largely—though not entirely—irrelevant today. Let's first address the issues that are still relevant.

One of the main premium expenses for developing Abercorn Common as a LEED project was overcoming the learning curve associated with approaching a project differently. Much of that additional expense we associate with time: the time it took for us to become conversant with the LEED program, the time it took for us to determine which materials and technologies to use, the time it took for us to educate tenant-rep brokers and potential tenants about the nature of the project, the time it took for us to draft and negotiate leases that called for build-out specifications different from those typically called for in our tenants' prototype designs. How much time? It's hard to estimate the delays directly associated with the LEED program, since we lack a counterpart with which to compare it. Our best-guess, conservative estimate is that the overall time it took to become conversant with the LEED program was around six months and probably led to delayed rents of approximately $400,000. Rather than attribute that cost to Abercorn Common specifically, we believe this expense item is one that should be considered a cost to the company as a whole, as the cost of our general effort to attain expertise in green development. As such, that particular cost has been addressed in Chapter 4 as part of our overall learning curve

Another significant cost premium associated with Abercorn Common concerned the additional fees paid for design and engineering. None of the professionals on our team had experience doing a LEED project, and so our architect, landscape architect, and civil and structural engineers all charged their standard fees plus an add-on for addressing the unknowns of this project. These soft fees alone amounted to approximately $300,000, or 2 percent of the overall construction budget for the in-line shopping center. It is telling that our professional fees for McDonald's and Shops 600 fell dramatically, as our designers and engineers felt comfortable with delivering a LEED product within their normal fee-based schedule. All told, the soft-cost premium for Abercorn Common was about $340,000.

Finally, there was the premium we paid for materials and technologies, totaling about $650,000 for all three projects at Abercorn Common. Of this sum, roughly one-third was devoted to a more thermally efficient building envelope, one-third to reflective roofing surfaces, and one-third to pervious paving materials. Overall, we calculate expenditures of approximately $1 million for all LEED items, resulting in an overall premium of 4.6 percent. An overview of the hard and soft costs for all three projects at Abercorn Common can be seen in Table 5.4.

It's worth noting that any company that develops to green standards will face some type of premium as a result of learning the ropes—expressed either in terms of additional time and expense of developing this expertise in-house or the cost of hiring an outside consultant. The good news is that this premium is significantly less than what we experienced, since the marketplace in general is more familiar and comfortable with the LEED program than it was in the early 2000s. The premiums associated with

TABLE 5.4   THE COST OF LEED AT ABERCORN COMMON

ITEM	GENERAL SITE WORK	IN-LINE SHOPPING CENTER	MCDONALD'S OUTPARCEL	SHOPS 600 OUTPARCEL	ENTIRE CENTER
Square Feet		166,000	4,500	16,500	187,000
LEED HARD COSTS					
Pervious Paving	$41,000	$155,730	$16,238	$573	$213,541
Lighting		$47,295	$1,200	$4,800	$53,295
Roof Materials		$151,022	$3,400	$16,000	$170,422
Building Envelope		$180,000	$6,000	$24,000	$210,000
TOTAL HARD COSTS	$41,000	$534,047	$26,838	$45,373	$647,258
LEED SOFT COSTS					
Consulting		$62,000	$5,000		$67,000
Architectural/Structural		$60,000			$60,000
Civil		$6,000	$3,500	$3,500	$13,000
Landscape Architectural		$7,500			$7,500
Mechanical/Plumbing		$66,650			$66,650
Energy Commissioning		$65,000	$8,100	$5,200	$78,300
Energy Modeling		$26,000			$26,000
LEED Registration		$2,500	$2,000	$450	$4,950
LEED Time In-house		$10,000	$5,000	$2,500	$17,500
TOTAL SOFT COSTS		$305,650	$23,600	$11,650	$340,900
Total LEED Items	$41,000	$839,697	$50,438	$57,023	$988,158
Total Cost of Center		$18,293,000	$1,410,000	$1,788,107	$21,491,107
LEED Cost/SF		$5.06	$11.21	$3.46	$5.28
TOTAL Cost/SF		$110.20	$313.33	$108.37	$114.93
Percent Premium		**4.59%**	3.58%	3.19%	4.60%

Abercorn Common are the costs one might typically associate with pioneering efforts that are compensated for by first-mover advantage.

So what of those early benefits? Like some of the more intangible aspects of costing the LEED premium, while we realize that we have benefited by our pioneering efforts at Abercorn Common, it is hard to put a hard dollar number on them. Did the time it took to lease the center at some point become easier as a result of the green brand we were creating? We believe it did, though we lack a benchmark to compare it to. Did we see an increase in the rental rates we were able to negotiate over the time of the development? In fact, we did see an increase, with initial rental rates negotiated in the $15- to $17-per-square foot range, and later rental rates realized in the $35-per-square-foot range. But how much of that increase is simply attributable to the market getting comfortable that a development on paper was becoming a reality, how much to the fact that vacant space was becoming more limited, and how much to the green brand of Abercorn Common? Anecdotally, we feel that the green brand was a small contributing factor to the up-tick in rents—and even a small portion of a 100 percent increase in gross income is significant. However, since we don't feel comfortable assigning a percentage benefit, we will leave it out of our analysis. We did see an overall increase in the valuation of Abercorn as part of our refinancing of an entire green portfolio. That increase is factored into our analysis in Chapter 4.

So what does the overall financial picture of the first all-retail LEED shopping center in the country look like? Not bad. Investment in land and original improvements

Years		0	1	2	3
REVENUES/SAVINGS					
	Cash Flow after debt service		281,800	281,800	281,800
	Sale of Abercorn in year 10				
Total Revenues		0	281,800	281,800	281,800
EXPENSES					
	Equity Contribution	(1,000,000)			
Total Expenses		(1,000,000)	0	0	0
Total Cashflow		(1,000,000)	281,800	281,800	281,800
Discount Factor		1.000	0.909	0.826	0.751
PV Cashflow		(1,000,000)	256,182	232,893	211,721
NPV		731,539			
IRR		25.20%			

Figure 5.7   **Discounted cash flow for Abercorn Common.**

to the center amounted to $7 million. Total cost of construction, including tenant improvements, totaled $24 million, for a total project cost of $31 million. Because of the enhanced value of our green portfolio (discussed in Chapter 4), we were able to place debt of $30 million on Abercorn, leaving $1 million of our own equity in this deal that has been producing just over $280,000 in annual cash after debt service. Internal rate of return on Abercorn Common is just above 25 percent. A discounted cash flow for Abercorn Common can be seen in Figure 5.7.

For seasoned real estate folks, the picture of having only $1 million of equity and $30 million in debt (and positive cash flow!) might raise a few eyebrows. How does one manage to have such little equity in Abercorn Common? As it turns out, shortly after completing the project, we rolled Abercorn Common into a portfolio of green properties that we refinanced (see Chapter 4). Abercorn Common was valued at $34.7 million (as opposed to its actual cost of $31 million). The company was able to extract a value of $3.7 million from Abercorn Common and redeploy that capital for development projects elsewhere.

This does, however, raise the question as to how the project will perform over the long term. Let's imagine that we decide to sell Abercorn Common in Year 10, utilizing a terminal cap rate of 7 percent on net operating income of $2.5 million for the sales price (equates to $36.2 million) and paying off the balance of the long-term loan (roughly $28 million). This analysis (see Figure 5.8) indicates an internal rate of return of approximately 39 percent.

4	5	6	7	8	9	10	TOTALS
							0
281,800	281,800	281,800	281,800	281,800	281,800	281,800	2,818,000
						0	0
281,800	281,800	281,800	281,800	281,800	281,800	281,800	2,818,000
0	0	0	0	0	0	0	(1,000,000)
281,800	281,800	281,800	281,800	281,800	281,800	281,800	
0.683	0.621	0.564	0.513	0.467	0.424	0.386	
192,473	174,976	159,069	144,608	131,462	119,511	108,646	

Years		0	1	2	3
REVENUES/SAVINGS					
	Cash Flow after debt service		281,800	281,800	281,800
	Sale of Abercorn in year 10				
Total Revenues		0	281,800	281,800	281,800
EXPENSES					
	Equity Contribution	(1,000,000)			
	Principal Payoff				
Total Expenses		(1,000,000)	0	0	0
Total Cashflow		(1,000,000)	281,800	281,800	281,800
Discount Factor		1.000	0.909	0.826	0.751
PV Cashflow		(1,000,000)	256,182	232,893	211,721
NPV		3,897,400			
IRR		39.02%			

Figure 5.8   Analysis of Abercorn Common with sale Year 10.

Concluding Remarks

In general, our approach to design and construction is different from that of most developers. We feel there are right and wrong ways to pursue development. It's our policy to use sustainable designs and practices, and we encourage neighborhood involvement. Throughout the design and construction process, we believed we were building something that would be here for a long while, something that would be enjoyed by many. As such, we wanted Abercorn Common not only to have a look of permanence and quality, but also to perform well from an energy and natural resource perspective and be a meeting place where people could come, shop, sit, and stay a while.

We've had great local response to the work we've done at Abercorn Common. The care and quality we've put into the project have set a new standard and raised the bar, and we're proud of that. We know we're making a difference in our community when we get comments like these from fourth-grade science students who visited Abercorn Common on a school field trip.

My field trip to Abercorn Common was great. I learned they use rainwater in their fountain instead of real water. Also they hold water in the back in cisterns. —Caroline

This shopping center is certified by Leadership in Energy and Environmental Design (LEED). You learn so many things at the Abercorn Common. It shows you how much the earth needs your help. —Bailey

4	5	6	7	8	9	10	TOTALS
							0
281,800	281,800	281,800	281,800	281,800	281,800	281,800	2,818,000
						36,211,429	36,211,429
281,800	281,800	281,800	281,800	281,800	281,800	36,493,229	39,029,429
						(28,000,000)	
0	0	0	0	0	0	(28,000,000)	(29,000,000)

4	5	6	7	8	9	10
281,800	281,800	281,800	281,800	281,800	281,800	8,493,229
0.683	0.621	0.564	0.513	0.467	0.424	0.386
192,473	174,976	159,069	144,608	131,462	119,511	3,274,507

The parking spaces at Abercorn Common are different than ordinary parking spaces. These parking spaces are made from porous concrete instead of regular concrete. Porous concrete is special because when it rains the water that hits the concrete dissolves instead of making a puddle. —William

The coolest thing is they built the first environmentally friendly McDonald's in the world! They also have these lanterns that use 27 watts that shine like a 100 watt light bulb. —Jacob

The old building rubble was either recycled or used to build certain parts of the complex. —Sales

When they tore it down they crushed the rock. The used the rock on things so they didn't need to buy them. The rest they gave to people who would recycle the materials. —Alex

They also made the roofs white so the heat would reflect off the roof and not make the building hot. If you have a black roof, the heat would go inside the building. It would cause people who own the building to turn the air conditioner on and waste energy. —Erika

I think that the field trip to Abercorn Common was the most interesting field trip I've been on. The rainwater on the roofs of those buildings gets pumped into a cistern behind the buildings. It is then used to power the fountain instead of using the city water. —Lindsay

Figure 5.9   Randy Peacock and Tommy Linstroth with a group of fourth graders at Abercorn Common.

> *People like the people who built this shopping center are bound to make a difference somewhere.*   —Elizabeth

> *I hope they build more LEED certified shopping centers. I really enjoyed the field trip to Abercorn Common.*   —Preston

In writing this chapter, I've had the chance to take my own field trip to Abercorn Common and reflect upon the decisions we made along the way. That trip back into the past brings up questions about lessons learned: If we had to do Abercorn Common all over again from the start, what would we consider doing differently? While not exhaustive, the following list serves as a litany of things I wish we had done:

- Taking a systems-wide approach to our energy management strategy, specifically by designing a central cooling tower for the various buildings rather than various stand-alone rooftop units for each retail space.
- Working more proactively (and assertively) with our tenants to determine their actual lighting needs from the user's standpoint and then developing custom lighting packages for tenants. Fifty percent of overall energy usage at Abercorn Common is from lighting, and we feel we could have reduced this significantly.

- Being firmer and more knowledgeable about specifying the sizes of HVAC units per tenant. There is significantly greater variability in tonnage per square foot than we feel is warranted.
- Changing how we handle fresh air and dehumidification. At Abercorn Common, the approach has been to supercool tenant spaces and then re-heat them. We believe that an energy recovery system, using exhaust rather than supercooling, would be a preferable strategy.
- Improving energy efficiency in the building envelope by use of spray foam insulation, which removes all unintended air infiltration, rather than fiberglass batts.
- Installing better performing glass with lower solar heat gain coefficient.
- Utilizing pavers in the parking lot rather than porous pavement, which isn't percolating as well as we'd like.
- Working more proactively with the city so as to use our rainwater capture system for sewage conveyance, as well as for landscape irrigation. In 2008, the City of Savannah (working in concert with members of our staff) became one of the first municipalities in the country to formally adopt Appendix C of the International Plumbing Code (IPC) on graywater use. We regret not having made this push earlier on.
- Using a green roof specialist or green roof manufacturer to guide us in the design of our green roof system, rather than having a general architect provide this service.
- Installing 1 pint per flush (ppf) water urinals rather than the waterless urinals in our retail spaces. When a real estate company manages a facility (see the Crestwood Building, Chapter 6), use of waterless urinals is fine since there is direct oversight of the maintenance of the system. However, in a retail environment, where tenants manage the facilities themselves, waterless urinals are not ideal and probably should be replaced with these simpler 1 ppf water urinals
- Inserting language in lease documents that enables us to review energy and water usage for documentation and monitoring.

With this long litany of things we did wrong (or at least inexpertly)—and it's hardly a complete or exhaustive list—the question begs to be asked: Did we do anything right? I think so. We made the decision early on to make Abercorn Common a LEED project, irrespective of the unknowns we were likely to face. Trivial or irresponsible? I don't think so. A quick story will help bring the point home.

There's a scene from the movie, *Shakespeare in Love*, which my colleagues and I sometimes refer to in the context of our green development work. In the film, the upcoming first production of *Romeo and Juliet* is going to hell in a handbasket. The actor playing Romeo has decided not to show up. The actor playing Juliet has contracted laryngitis. The actor who gives the opening speech in the play, a tailor by profession, can't stop stuttering. The Master of the Revels for Queen Elizabeth has decided to close the famous Rose Theater, so the company can't even find a theater to produce the play in. And in the midst of all this mayhem, William Shakespeare is in an absolute panic. He tells Philip Henslowe, the businessman charged with producing the play, that they're lost. And Henslowe, who has been a complete nervous Nelly throughout, is suddenly and absolutely calm, and he tells Shakespeare that everything will turn out well. Shakespeare wants to know how. Henslowe admits he doesn't know.

But Henslowe is right. It all turns out well. It's the magic of theater. This so-called magic has also played out at Abercorn Common. Early on in the project, our staff determined to make it and every other project going forward a green development. We were green ourselves in our experience with LEED, having delivered only one prior project to LEED criteria. It's hard to remember that now, six or seven years later, but we really had no idea what we were getting ourselves into. That may seem foolhardy, but maybe it's the magic of green.

I mean that in both senses of green, which are actually intertwined. There has, historically, been a certain magic associated with our green orientation, of doing well financially by taking the ethical stance of being better stewards of our land and our community. That has never been the overt intent. As a company, we never made the decision to do the right thing *in order to be* successful. It has simply worked out that way. Things have turned out well.

I think a partial explanation is, in fact, that we are often green when it comes to our approach to development challenges. As un-businesslike as it probably sounds, sometimes it pays to approach a development project with fresh eyes, eyes that may not see all the reasons why a new and different approach won't work. And I do think there is something to be said for stepping beyond the bounds of the known, in the interest of furthering a few things that need to go right and to promote smarter, better development practices. Our company finds itself in that same situation today, having signed on to the 2030 Challenge, which calls for dramatic reductions in carbon emissions from all of our development projects from this date forward. We don't know (yet) how we will deliver on that commitment. We simply know it needs to be done.

NOTES

[1] David W. Orr, *The Nature of Design: Ecology, Culture, and Human Intention* (New York: Oxford University Press, 2002), p. 203.
[2] Jason F. McLennan, The Philosophy of Sustainable Design: The Future of Architecture (Kansas City, Mo.: Ecotone Publishing, 2004), p. 80.
[3] Lawrence S. Earley, *Looking for Longleaf: The Fall and Rise of an American Forest* (Chapel Hill, N.C.: The University of North Carolina Press, 2004), p. 2.
[4] Timothy Beatley and Kristy Manning, The Ecology of Place: Planning for Environment, Economy, and Community (Washington, D.C.: Island Press, 1997), p. 157.
[5] Jason F. McLennan, *The Philosophy of Sustainable Design: The Future of Architecture* (Kansas City, Mo.: Ecotone Publishing, 2004), p. 80.

EXISTING BUILDINGS: THE GREEN-HEADED STEPCHILD OF THE SUSTAINABILITY MOVEMENT

SCOTT DOKSANSKY

SUMMARY

The Crestwood, a 94,000-square-foot Melaver-owned and -managed office building in suburban Atlanta with an occupancy rate of over 90 percent, is Georgia's first office building to receive LEED certification for an existing building (LEED for Existing Buildings, formerly "LEED-EB"), and one of only a handful of multi-tenant office buildings in the country to be renovated to LEED specifications. In this chapter, Scott Doksansky, Melaver, Inc.'s Director of Portfolio Management, discusses the work that was done at the Crestwood to achieve LEED for Existing Buildings certification, the costs, and the benefits.

After considering the overall need in this country to be better stewards of our existing buildings, Doksansky focuses on making the most of one particular building—the Crestwood office building. In section two, the discussion drills down into the specifics of renovating this building to LEED standards. Since many of these LEED features may seem abstract when viewed through the lens of a LEED scorecard, section three entails a virtual tour of the facility, highlighting its many subtle distinguishing features. Section four considers the financial implications of this renovation, touching on ongoing issues regarding building maintenance and operations, and plans for additional improvements. Section five addresses

> the more general issue of how an asset manager assesses which existing buildings most lend themselves to a LEED renovation, including a checklist for evaluating an existing building for its green potential.

As a culture, Americans are not known for being kind to buildings. Perhaps that is the legacy of several centuries of having a seemingly infinite frontier, a territory ahead. It wasn't just a go-west phenomenon. In the mid-nineteenth century, American cities in the east began see to low-density development stretch out for miles from city centers to the suburbs along the paths of horse-drawn streetcars, then rail lines, and then highways.

This process of jettisoning the old for the new was given greater impetus starting in 1916 (and then in 1921) with federal subsidies to improve roads, develop state and national highways, and, in general, promote the use of the car, then seen as the most significant influence of the rise of local taxes. While other sectors of the economy during the Depression were flat, highway building saw a boom. The creation of the Federal Housing Administration (FHA) in the 1930s, emphasizing single-family detached homes in the suburbs, synergistically fed off the highway construction. Creation of easy mortgages under the Veterans Administration not only facilitated the demand of returning soldiers from World War II for inexpensive housing but also ratcheted up the development of new subdivisions. So too did a new federal income tax law allowing mortgage interest to be tax deductible. The *coup de grace* was the 1956 passage of the Interstate Highway Act, which called for the federal government to subsidize 90 percent of the construction cost of 41,000 miles of new expressways.[1]

It is not my intent in this chapter to offer up a full-scale critique of urban planning. What I do and am most knowledgeable about is managing the assets of a real estate portfolio. This much, I think I do know: Managing our existing buildings a different way can go a long way toward righting some of the wrongs of our consumptive practices. Learning to treasure what we have, as well as make them more efficient, is a critical aspect of an overall sustainability strategy. One leader in the real estate industry has even gone so far as to say that greening our existing buildings is the "holy grail of the sustainability movement."[2] I think he is right. We in this country have long faced a challenge having to do with not being efficient in our use of materials, of consuming rather than treasuring the things we make a part of us. We build our buildings poorly, for the moment, and then move on down the road to put up stakes elsewhere. An ethic of rapid consumption, particularly of nonrenewable resources, has gone hand-in-hand with a loss of community and social fabric. Re-thinking our approach to existing buildings has far-reaching consequences— for being better trustees of the buildings we construct, for better conserving the resources from which these buildings are made, for being better stewards of our land and our communities.

Stewart Brand, author of *How Buildings Learn*, provides us with as good a jumping point as any from which to re-think our approach to existing buildings. Brand begins with the premise that "no buildings adapt well," that, in fact, "they're *designed* not to adapt," in terms of how they are constructed, the budgets that are created to manage them, how they are regulated and taxed, even how they are remodeled. Having said that, Brand makes a number of cogent points, among them:

- Office buildings, comprising the largest capital asset of developed nations and accommodating over half of their workforces, have to adapt quickly because of intense competitive pressures.[3]
- Over a fifty-year period, changes within a building cost three times more than the original building.[4]
- Changes tend to occur from the inside out, starting with stuff and space plans (five to seven years), then moving to services (fifteen to twenty years), and then finally to the external skin (twenty years) and structure (thirty years and upward).[5]
- Adaptation occurs most easily in cheap buildings that few care about, and occurs in a refined way in long-lasting sustained-purpose buildings.[6]
- A building's longevity is often determined by how well it can absorb new service technologies, such as innovative energy services.[7]
- A building's capacity to adapt over the long term (its *adaptivity*) should be distinguished from a building's graceless turnover from tenant to tenant.[8]
- Age plus adaptivity makes buildings and community humane, and hence more beloved.[9]

Brand's overall assessment of the relative adaptive capacity of buildings is, overall, dead-on. Buildings in our country generally do not adapt well; when they do, they adapt from the inside outward. Adaptations work most readily when there's a capacity to use new technologies, and they occur best at the two ends of the spectrum—with cheap buildings no one loves and with expensive, institutional buildings, buildings that were designed to be loved for an extended period of time. But what about the commercial buildings in the vast middle of the spectrum, buildings that are neither cheap and unloved nor expensive and cherished?

John Lyle, author of *Regenerative Design for Sustainable Development*, in making the argument for our need to understand systems ecologically, relies tellingly enough on the metaphor of a building:

> Ecosystematic order, then, is analogous to the order found in buildings. First, there is the structural order of posts, beams, walls, and roof. Second, there is the functional order of material and energy flows represented by the pipes, valves, wires, switches, circuit breakers, ducts, dampers, and other apparatus. Third, there is the locational order of the floor plan.[10]

Buildings show us how to think in a holistic way, over the long term. We need to think about their structure: what they are made of, everything that goes into them. We need to think about what makes them function: the energy that flows into them, how water is supplied and utilized. And we need to think about a building's location and the fact that how it functions is highly dependent on local conditions like temperature variability, moisture, and available resources.

Lyle extends his metaphor, identifying a dozen design strategies that form the basis of regenerative thinking (letting nature do the work, linking all the parts of a system into an aggregated whole, matching technology to need, managing storage as a key to sustainability). Many of these strategies are relevant to how buildings are designed or re-designed, and thinking about the re-design of existing buildings, particularly that vast category of buildings "in the middle," is training for life in a new paradigm of less consumption (of resources, energy, and water) and less waste.

It's been said that the greenest building is one that's already built. In part that's because older buildings contain what's called "embodied energy"—the energy used to harvest, manufacture, and transport the materials they are made of and the energy used to build them in the first place. When a building is torn down, that embodied energy is wasted (and still more energy is used in the demolition process, which also creates waste and debris). This is the second, more pragmatic reason we cannot simply ignore those "buildings in the middle": a large portion of our building stock over the next thirty years—the places we work and live—will be renovated buildings, not new ones.

A projection that illustrates the opportunities afforded by approaching existing buildings differently can seen in Figure 6.1. We currently have a stock of 300 billion square feet of building in the United States. By 2035, we will have demolished about 50 billion square feet of buildings—between 15 and 20 percent. Of the remaining 250 billion square feet, 150 billion will be renovated. And 150 billion square feet of new construction will be added to overall inventory.

My colleagues and I at Melaver, Inc. hope new building stock that will come on the market will be developed to green standards, and we are very concerned about how existing buildings will be renovated. The U.S. Environmental Protection Agency estimates nearly half of America's greenhouse gas emissions comes from the construction and operation of buildings. Preserving the embodied energy of still-useful buildings and improving their energy consumption and environmental impact present both a significant challenge and an enormous opportunity.

Making the Most of an Existing Building

Three of Melaver, Inc.'s LEED-certified projects are renovations of existing buildings, and all three have multiple tenants. (Most LEED office buildings are occupied by a single tenant, typically the building's owner.) Two of them—the Telfair Building (LEED for Commercial Interiors, where the company's main office is located) and the Whitaker Building (LEED for New Construction Silver, 2004, the first LEED certified building in Savannah and home to the company's brokerage division, Melaver | Mouchet) —are mixed-use (retail and office) commercial properties in the historic district of

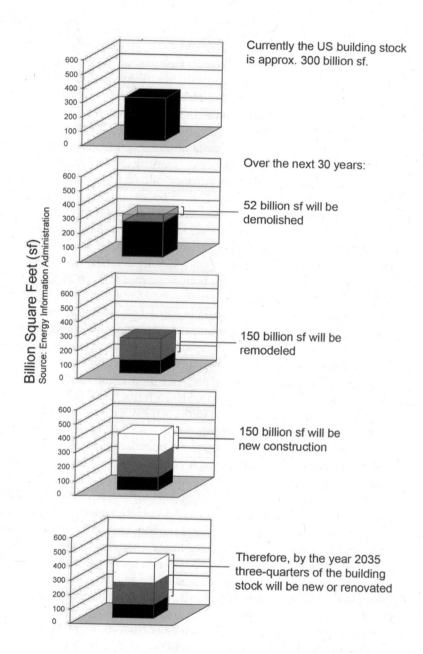

Figure 6.1 **Projected construction over the next thirty years.**[11]

© 2007–2008 2030, Inc./Architecture 2030; used by permission.

downtown Savannah, Georgia. The third, which received LEED for Existing Buildings certification in 2007, is the Crestwood Building, a twenty-two-year-old, multi-tenant office building in Gwinnett County, Georgia, an area of suburban Atlanta that has experienced tremendous growth in the past thirty years. The Crestwood Building is located in a commercial area that includes office complexes, office parks, hotels and restaurants, shopping malls, and strip shopping centers.

Built in 1986, the Crestwood Building has five stories and contains 93,554 square feet of rentable office space. Competition for tenants in the immediate area includes existing office buildings and newly constructed office buildings. The Crestwood Building boasts a 92 percent occupancy rate (at 2007 year end) in a submarket with an occupancy rate of 81 percent. The 5.57-acre property includes the building, wooded areas of mixed pine and hardwood trees (19 percent of the site is shaded), areas landscaped with small trees and ornamental plants, and paved parking for 271 cars.

Melaver, Inc. purchased the Crestwood in 1998 for $9.4 million, at a time when the company was looking for an opportunity to diversify its portfolio. Crestwood presented an opportunity to acquire both a sizeable office building and expand our holdings in the Atlanta market. At purchase, $884,000 was allocated for improvements, including a renovation of the existing first floor lobby, the addition of a conference room, break room and fitness room, and furnishing and equipping the aforementioned. An additional $452,000 was allocated to reconfigure Melaver, Inc.'s office suites operation on the fourth floor. Improvements included upgrades to the common areas to facilitate a small break room, conference rooms, and management offices. Total capital investment in the building was $10.75 million. Table 6.1 provides an overview of the initial capital costs involved in the acquisition and upgrades to the building.

There are some additional things worth noting in Table 6.1. For one thing, the analysis reveals that at about $910,000 in net operating income, the building was showing a return of about 8.47 percent, which conservatively is about one hundred basis points higher than comparable office buildings nationwide. That's a double-edged piece of information. On the one hand, this gap between the returns on the building (8.4 percent) and what the market would expect (7.5 percent conservatively) indicates that there's additional value-add of about $1.394 million. On the other hand, this additional value doesn't allow for a lot of room for major upgrades to the building. Any strategy for achieving LEED certification would have to be shaped with due fiscal restraint. Every $100,000 of green improvements would translate into an eight basis point reduction in our returns. As you might expect in a company built around shared leadership, debates over how much of an investment to make in green and how much of an immediate hit we can afford to take on our returns are hot contests. Our CEO wouldn't blink at spending the full $1.4 million, while our CFO is, well, more financially disciplined.

In our thought process, we needed to take into account:

1 a soft market with potentially greater vacancies and/or downward pressures on rents, and

2 immediate cash flow needs for the property specifically and for the company as a whole.

TABLE 6.1 INITIAL CAPITAL INVESTMENT IN CRESTWOOD

IMPROVEMENTS	COST
First floor construction	$ 665,000
First floor furniture and equipment	$ 185,000
Artwork	$ 44,000
Fourth floor office suites improvements	$ 452,000
Total improvements	$ 1,346,000
Acquisition price	$ 9,400,000
Total capital investment	$10,746,000
Gross square feet	93,544
Common area factor	18%
Leaseable square feet	76,706
Market vacancy factor	8%
Rental rate per square foot	$17
Gross potential revenue with vacancy factor	$ 1,199,683
Expense ratio	24.1%
Net operating income	$ 910,559
Cap rate on total capital investment	8.47%
Typical cap rate for comparable office building	7.5%
Value of Crestwood at market cap rate	$12,140,793
Additional expenditures available for green upgrades	$ 1,394,793
Every additional $100,00 results in cap rate change	–0.08%

Nevertheless, this overview provides a flavor for the type of calculation that needs to be factored in more when it comes to considering the greening of an existing building. Delivering a LEED for Existing Buildings project is challenging but it is considerably easier when the focus is less on maximizing a return on a building and more on optimizing its triple bottom line performance within market constraints.

MAKING A DIFFERENCE

This shift in approach toward long-term fiscal management of an asset is just the first, albeit important, step in greening an existing facility. A second step was our overall approach to property management.

When our company committed to constructing all of its developments to LEED standards in 2002, there was uncertainty regarding what to do with our existing portfolio, which ranged from office to retail to industrial warehousing. We decided to revamp our approach to managing these properties by developing a set of sustainable property management guidelines, known collectively as the Mark of a Difference (MOAD) program. This program, a precursor to going LEED, focuses on improving the health and well being of building occupants, as well as on reducing financial overhead by decreasing electricity and water costs. MOAD covers six areas of property management (interior care, exterior care, tenant relations, mechanical systems, buildings and systems, and tenant build-out and common area improvement). Each topic section includes a goal. Each section's goal is the "what," the result we want, the bigger sustainability issue we're working to address. The goal for "Interior Care: Entryway Systems," for example, is to reduce building occupants' and maintenance personnel's exposure to potentially hazardous contaminants. According to the Environmental Protection Agency, indoor air quality is estimated to be five times worse than outside air, and Americans spend approximately 90 percent of their time indoors. The specific property management challenge entails lessening the amounts of dirt (dust, mud, gravel), debris (paper, leaves, chewing gum), and pollutants (oil, grease, pesticides) that people accumulate on their shoes that get tracked into our buildings. To attain the goal, we've developed requirements and strategies that will achieve the desired result, which in this case is a more passive variation of something your parents probably told you, over and over: Wipe your feet before you come in the house. Our requirements cover the "where," "who," and "when" of the story—specific actions we take to get the results we want. For entryway systems (to continue this example), we use grilles, grates, or mats to trap dirt, dust, pollen, and other particulates at the door. As people walk over them, the soles of their shoes get cleaner, and they bring less of what they were carrying on their shoes into our building. Requirements also designate who maintains the system (in this case, it's the janitorial staff) and how often the system requires maintenance (daily, weekly, monthly, quarterly).

The approach gets even more detailed in the strategies section for each topic, which includes additional actions we can take that will help us meet our goal. The specific strategies for "Interior Care: Entryway Systems" include design considerations for renovation or construction that affect entranceways (like providing a water spigot and electrical outlet near or accessible to entranceways for maintenance and cleaning activities) and guidelines for landscaping practices at building entrances.

Some goals and requirements of the MOAD program are more straightforward, incorporating both sustainability issues and financial ones. For example, the goal for "Exterior Care: Parking Lot Lighting" covers reducing electricity used for external lighting and reducing levels of light pollution. Three things are at work here: reducing electricity consumption to mitigate environmental impacts of electricity generation, reducing electricity consumption to spend less money on utility bills, and reducing the amount of light a property contributes to light pollution. The requirements for reaching that goal are simple and direct. First, install photo-sensors on exterior parking lights (so they aren't on when they don't need to be, thus reducing consumption utility bills and creating a solution that works automatically). Second, when retrofitting, shield ballasts so that light is reflected toward the ground instead of out into the sky (shielding ballasts means the light falls where it's needed [on the ground], not where it isn't [in the sky]). By focusing light on the task at hand, hopefully less can be used, ultimately conserving energy and saving money.

Other goals and requirements of our MOAD program relate to basic and routine practices that are a standard part of how we manage any facility—primarily having to do with how we source materials and how we manage our waste. The goal for "Buildings and Systems: Procurement Practices," for example, is a dual one—to decrease demand for virgin raw materials and increase demand for recycled and environmentally friendly products. The requirements section for "Procurement Practices" recognizes that to increase demand for recycled products means we need to be buying them ourselves, and so lists categories of paper products (kitchen paper towels, bathroom paper towels, toilet paper, facial tissue, napkins, and paper plates), specifies 100 percent recycled content, and provides the names of brands and sources that meet the recycled content requirements.

The Mark of a Difference program was a group effort by Melaver team members in property management, construction, and sustainability. Many of the program's property management processes, which are intended for company-wide implementation, were developed and tested at the Crestwood Building.

Mark of a Difference Program Table of Contents

1. Interior Care
   Sustainable Cleaning Products
   Storage and Collection of Recyclables
   Daytime Janitorial Cleaning
   Optimize Use of IEQ Compliant Products
   Entryway Systems
   Integrated Pest Management
   Interior Plants for IAQ

2. Exterior Care
   Bicycle Storage and Changing Room
   Preferred Parking for Carpooling/Hybrid Vehicles
   Exterior Landscaping with Drought-Resistant or Native Species
   Parking Lot Lighting
   Irrigation
   Exterior Window Cleaning
   Roof Repair, Maintenance and Replacement
   Asphalt/Cement Repair/Replacement

Mark of a Difference Program Table of Contents (*continued*)

3. Tenant Relations
   Education and Outreach—Tenant and Visitor Engagement
   Tenant Feedback Policy
   Training Provision for Tenants and Managers
   Building Operation and Management Staff Education
   Travel Information Kiosk

4. Mechanical Systems
   Building Systems Maintenance

5. Buildings and Systems
   Document Sustainable Building Costs and Benefits
   Best Management Practices
   Waste Stream Audit
   Building and Systems Audit
   Procurement Practices

6. Tenant Build-out and Common Area Improvement
   Best Management Practices

Obtaining LEED for Existing Buildings Certification

As part of our company-wide commitment to LEED, Melaver began pursuing LEED certification for the Crestwood in 2004. We began our analysis of whether LEED certification was feasible by conducting an Energy Star assessment. With an Energy Star rating, we felt LEED certification would be obtainable. Failing to qualify for Energy Star would mean LEED certification was impossible. Energy Star, a joint program of the U.S. Environmental Protection Agency and the U.S. Department of Energy, began rating commercial buildings for energy efficiency in 1999. An Energy Star certification indicates a facility rates in the top 25 percent of buildings in the country for energy efficiency. There are approximately 4.6 million commercial buildings in the U.S., with 170,000 more built each year. As of this writing, only 4,398 commercial buildings have the Energy Star certification, and of that total, approximately 37 percent (around 1,645) are office buildings. There are thirty-six Energy Star office buildings in Georgia. Crestwood is one of them (see Table 6.2).

TABLE 6.2   ENERGY STAR BUILDING INVENTORY

CATEGORY	NUMBER	% OF TOTAL
Energy Star buildings in U.S.	4,398	100%
Energy Star office buildings	1,645	37.4%
Energy Star office buildings in Georgia	36	2.19%

Obtaining an Energy Star Rating

Obtaining an Energy Star rating for a building is no slam-dunk. You've cut your electricity bills by 10 percent. The chiller is operating smoothly, lights are off after hours, the building automation system is working on schedule—overall, you have a pretty efficient building. Compared to what? While your utility bills may be lower this year, how do you know what your target should be? How are other buildings of your size, occupancy, and location operating?

One way to find out is to call facility managers in your area, compare the scope of your buildings, and analyze your utility data together. But this is burdensome and time-consuming, even if the other facility managers have this information and are willing to share it. Fortunately, a mechanism exists for facility managers to compare their buildings with other buildings of the same size, occupancy type, and location. This tool is the Energy Star Portfolio Manager program.

Portfolio Manger is an online tool that allows users to track and assess building energy consumption. Portfolio Manager can help you set investment priorities, identify under-performing buildings, verify efficiency improvements, and receive EPA recognition for superior energy performance. It's free and very easy to use. You can track one building or an entire portfolio. After you register with Portfolio Manger, you create a profile for the building you want to analyze. All that is needed is basic information, such as square footage, occupancy counts, estimated number of computers, operating hours, and, of course, energy consumption and cost data. For qualifying facilities,[12] Portfolio Manager rates your building on a score of 1 to100. This ranking is relative to similar buildings nationwide, based on an energy consumption survey conducted by the Department of Energy, the Commercial Building Energy Consumption Survey (CBECS). CBECS is conducted every four years and gathers data on buildings and energy use from thousands of properties across the United States, giving you the ability to measure your building's performance against that of thousands of similar buildings.

> ### Obtaining an Energy Star Rating (continued)
>
> A rating of 50 from Portfolio Manager means a building performs better than 50 percent of similar buildings nationwide, while a score of 90 means a facility is performing in the top 10 percent of buildings across the country. If your building earns a score of 75 or higher, it qualifies for the Energy Star label, indicating it uses, on average, 35 percent less energy than typical similar buildings and generates one-third less carbon dioxide. For a building to achieve Energy Star certification, an independent professional engineer must verify the data and ensure that the building meets code requirements in a number of areas, such as proper ventilation rates, humidity, indoor pollutant controls, and adequate lighting. Even if your building does not qualify for the Energy Star label, Portfolio Manager provides a host of benefits for the property or portfolio. You can use Portfolio Manager, for example, to set investment priorities and verify/track progress of improvement projects. To set investment priorities, Portfolio Manager provides a financial calculator to estimate returns based on energy savings in your building. This helps you focus on the projects that offer the fastest return on investment to quickly and efficiently reduce energy consumption. Once the baseline energy consumption is established, Portfolio Manager provides a reduction analysis so you can quickly and easily see how well improvements you make are actually working. It also allows you to efficiently communicate energy performance with owners and tenants.

Overall, we spent approximately $63,000 on the Energy Star program for film coatings on windows, lighting upgrades, and improvements to our HVAC system—just under half of the total amount we devoted to achieving LEED for Existing Buildings certification. Once the Energy Star certification was in hand, we were well on our way to LEED certification for Crestwood.

The LEED for Existing Buildings Scorecard (see Figure 6.2) lists five categories for building improvements and one category for innovation and design. We started the certification process by examining the categories and making a yes/no/maybe list to determine our plan. We knew, for example, that we could increase our water efficiency and reduce the amount of water for landscaping purposes. But we knew, owing to county ordinances, that it wouldn't be feasible to treat wastewater on site, or cost-effective to re-plumb the building to use graywater for flushing toilets. We knew we could implement green cleaning practices and use sustainable cleaning products and materials—in fact, we were already doing so as part of our MOAD sustainable property management practices—but we knew it wouldn't be practical to re-roof a building that was still under a roof warranty.

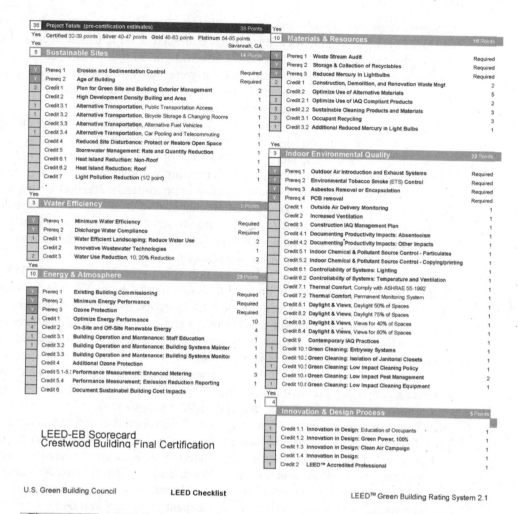

Figure 6.2   LEED for existing buildings scorecard for Crestwood.

We explored non-roof heat island reduction (Credit 6.1 under the Sustainable Sites category), thinking that since our site had lots of trees already, we might be able to pretty easily meet the 30 percent shaded requirement for heat island reduction. We hired an environmental design firm, Ecos Environmental Design, to calculate the amount of shade on our site according to the LEED handbook guidelines, and found we were 19 percent shaded with our current tree cover. Only 11 percent to go. Ecos created two plans for adding trees and shade to the parking areas. One option proposed adding fifty-two canopy trees and six ornamental trees and removing seven parking spaces; the other option proposed adding forty-eight canopy trees and removing eleven parking spaces. Ultimately, a simple cost-benefit analysis indicated that additional canopy coverage would be a low priority. We felt a better time for such enhancements would be when the parking lot needed to be resurfaced or we had made a commitment to a stormwater management project. We also recognized that we needed to push for shorter-term payback opportunities, and we were concerned about losing parking spaces since our parking lot contains fewer spaces than the county technically requires. (There's more about parking later in the chapter.)

Whether we are building or renovating, we make it a practice to review the potential to generate energy on site. Since Crestwood has over 15,000 square feet of exposed roof surface, we explored the possibility of solar power. But after we evaluated the cost and potential impact of implementing photovoltaics (PV), our preliminary calculations indicated we could generate approximately 260,000 kWh (only 20 percent of the building's need) at an installation cost of $1.4 million. If our calculations were correct, the payback from PV would have been approximately thirty years, not considering net present value (NPV) and assuming all the generated power would be sold to the grid. Generating solar power was not a feasible option.

When all was said and done, we were able to earn 35 points on the LEED Scorecard—enough to meet the minimum of 32 points (out of a possible 85) for Existing Buildings certification. In the Sustainable Sites category, five points were earned: two points for the plan for green site and building exterior management, and three points for alternative transportation (public transportation access, bicycle storage and changing rooms, and carpooling and telecommuting).

Three points were awarded for Water Efficiency, one point for reducing water use for landscaping, and two points for a 20 percent reduction in water use. In the Energy & Atmosphere category, four points were awarded for optimizing energy performance, four for renewable energy use, and two for building operation and maintenance (one point for staff education and one point for systems maintenance), for a total of ten.

Ten points were also awarded in the Materials & Resources category—two for waste management in construction, demolition, and renovation; two for optimizing use of IAQ (indoor air quality) compliant products; three for the use of sustainable cleaning products and materials; two for occupant recycling; and one for using light bulbs with additional reduced mercury content. The three points awarded in the Indoor Environmental Quality category are all in the area of green cleaning—one point for entryway systems, one for a low-impact cleaning policy, and one for using low-impact cleaning equipment.

In the Innovation & Design Process category, three points came from innovation in design—one each for educating occupants, for 100 percent green power (which is purchased from renewable energy sources), and for being a partner with Georgia's Clean Air Campaign. An additional point came from having at least one LEED accredited professional as a member of the project team. (Actually there were six: Trey Everett, Crestwood's building engineer; Rebecca Riggins, assistant property manager for Melaver's Atlanta portfolio; Angela Walden, the general manager of Melaver Executive Suites, which occupies the 4th floor of the Crestwood Building; Rhoda Brown, head property manager for the company; Tommy Linstroth, Melaver, Inc.'s head of Sustainable Initiatives; and I.). An overview of points earned can be seen in Table 6.3.

TABLE 6.3 OVERVIEW OF LEED FOR EXISTING BUILDINGS POINTS FOR CRESTWOOD	
CATEGORY	**POINTS**
Sustainable Sites	5
Water Efficiency	3
Energy & Atmosphere	10
Materials & Resources	10
Indoor Air Quality	3
AInnovation & Design	4
TOTAL	35

The work was completed in February 2007. Certification was awarded in October 2007 and was celebrated with a reception in January 2008, when the LEED medallion in the lobby was ceremoniously unveiled.

What Does Green Look Like?

Perhaps the most extraordinary thing about the Crestwood is that although it performs much better in terms of energy use than a typical office building of its age and size and contributes in meaningful ways to the comfort and health of its tenant occupants, the differences are subtle—perhaps even unnoticeable—to the casual observer. Let's take a tour of the building to see how it looks and works.

EXTERIOR IMPROVEMENTS

The numerous tall trees on the site provide shade for the window glass on two sides of the building. The window glass on the remaining two sides was tinted to reduce heat gain and glare, at a cost of $33,986. The heat gain reduction accomplished by the window tinting improved the temperature balance between the exterior and interior office spaces and significantly reduced the number of tenant complaints about comfort —being too hot (in offices with windows where solar heat gain raised the room temperature) or too cold (in interior offices without windows).

Our maintenance contractor uses a non-toxic window cleaner (which is actually a dishwashing liquid) to clean the windows on the outside. Window cleaning thus does not contribute volatile organic compounds or fumes to the atmosphere, harm the people who are doing the cleaning by exposing them to toxins, or pollute the soil around the building.

Landscaping includes plants native to Georgia, perennials (which don't require re-planting every year), and small trees like Japanese maples, which provide seasonal color and interest. Seasonal beds that had previously been planted in flowering annuals have been reduced in size. Planting expenses have decreased $3,800 annually, from almost $5,000 a year to $1,200 a year, and the plants we have require less water. In 2007–08, because of Georgia's continuing drought, which first prompted outdoor watering restrictions and then a ban on outdoor watering, no seasonal plantings were installed. (This watering ban has also reduced the amount of water the building uses and the amount of money spent for water, both discussed in detail later in this chapter.) Abundant mulch in planting beds conserves water and currently covers areas designated for seasonal plants.

Because the paved parking area of the property does not on its own meet the county requirement for number of parking spaces, we have a reciprocal agreement with an adjacent Marriott hotel for shared parking. Crestwood tenants and guests have use of the hotel's parking lot for overflow parking during the week, when the hotel's parking need is less. On the weekends, when the hotel's needs are greater, its guests can use the Crestwood lot, which is less full at these times. This arrangement allows both parties to devote less site area to paving. Green space is preserved and stormwater runoff is not increased. As our county recently began assessing fees for stormwater runoff based on the amount of impervious surface on a site, having not covered more of the site with impervious paving will also save money.

Designated preferred parking for tenants who carpool is located close to the entrance to the building. At this time, preferential parking for hybrid vehicles is not designated. (This was deemed too self-serving, since several hybrid vehicle drivers are members of our staff.)

THE PUBLIC SPACES

Walk-over metal grates at the building entrance, one on the outside and one on the inside, act as permanent, passive doormats—as people walk over them, dirt, debris, and contaminants from their shoes fall through a removable grate and into a trough that is periodically cleaned. The system reduces indoor pollution and wear and tear on the carpeting, and the grates won't wear out like doormats, which require more frequent cleaning and periodic replacement, and are often made of petroleum. It also means we don't need an outside vendor for mat service, so we're eliminating an expense to us and the pollution caused by someone else's driving to our location.

The polished red granite panels that cover some of the walls in the two-story lobby were in place when the building was purchased and were left in place. The remaining wall surfaces were originally covered with fabric panels that puckered in damp weather and couldn't be cleaned. The decision was made to remove them in favor of creating a surface that could be cleaned and would look better with the red granite. Removing the panels caused minor damage to the wall surfaces, which were repaired in large part by skim coating, though surface imperfections remained. A multi-color, multi-coat Zolatone finish in colors that complemented the red granite panels was applied. The multiple colors provide visual texture and superior coverage, hiding the minor surface imperfections that had been visible when the bright afternoon sun floods the lobby area, and are LEED compliant. (Each exceeds Green Seal GS-11 and complies with California law.) Since Zolatone finishes are thicker and more durable than wall paint; they require less maintenance, are easy to repair if damaged, and don't have to be replaced as often as paint. The finish has been in place for three years and so far no repairs have been necessary. It was, however, more expensive than paint. Zolatone cost $2,300 for materials and labor, while painting would have cost $700.

Carpet squares from Interface cover the floors of the main lobby and the elevator lobbies, and are used in the hallways. Carpet squares have many advantages—they provide the coverage and sound insulation of wall-to-wall carpet, but can be replaced in sections or by the piece, as needed, when they become worn or if they are damaged or stained. These carpet squares are manufactured in Interface's LaGrange, Georgia plant that is ninety-five miles from the building. They contain recycled content and are fully recyclable.

TENANT SPACES

Four floors of the building are office spaces that are configured to suit the tenant. Generally, our tenants sign five-year leases and pay rates that are competitive for the area. To minimize adverse impacts and ensure compliance with building environmental standards, prospective tenants meet with a space planner and reach an agreement about the design of the space before a lease is signed. Building standards included in the lease spell out the types of carpeting, cove base, and paints that may be used. There are five color choices for carpet and a selection of paint colors. The carpeting is Green Label wall-to-wall carpet with low VOCs (volatile organic compounds) and a high-recycled content, the molded cove base has recycled content, and paints must be low VOC. Tenants work with our LEED accredited contractor, who specifies materials and checks them as they come into the building. Tenants are encouraged to use glass in interior walls to allow more daylight into offices and hallways. Carpet, drywall, metal wall studs, porcelain, and copper wiring removed during renovations are recycled by the contractor and/or our building engineer.

The fourth floor of the building offers office suites with short-term leases. The suites serve small and start-up businesses and offer a telecommuting option. Several businesses have incubated from Melaver Executive Suites to become long-term tenants in the main building. It's also where our Atlanta offices are located. Amenities for tenants in the office suites include a shared reception area, a person to answer the telephone, and shared copiers, fax machine, break room, and two conference rooms. Tenant improvements are limited to carpet and paint, and MOAD standards apply.

HVAC IMPROVEMENTS

The HVAC system, now twenty years old, needed a thorough servicing. The goals were to maximize the efficiency of the system and to add features that would improve the indoor air quality of the building. To accomplish this, the chiller was rebuilt, a new fan motor and CO_2 sensors were installed, and electronic fresh air dampers were added. The largest investment was rebuilding the chiller, a project that needed to be done on weekends to avoid a negative impact on our tenant partners. The chiller rebuild totaled $15,420. The new fan motor, CO_2 sensors, and fresh air dampers were an additional $5,147.

The building's heating and air conditioning system was retrocommissioned—a separate process done by a commissioning agent that guarantees a system operates as the manufacturer specified—at a cost of $7,128. Although it's not possible to directly calculate if retrocommissioning has saved money, we do know that maintaining the system properly helps it work more efficiently and affords better control of air flows, increasing tenant comfort and reducing complaints.

WATER EFFICIENCY

All the restrooms were retrofitted with low-flow toilets (1.6 gallons per flush), touchless faucets with aerators, and (in men's restrooms) waterless urinals in 2005. Break room sinks have faucets with low-flow aerators. Water use dropped by over 20 percent after we upgraded the restroom fixtures in the building (by changing the toilets and urinals) and made changes to the landscaping that reduced the need for outdoor irrigation. Per-day water usage dropped from an average of nearly 3,000 gallons per day to 1,600 gallons per day in 2007, a reduction of almost 50 percent. Dollar savings during that same six-year period ranged from $240 to over $4,700 per year. (See Table 6.4.) That drop in consumption is in part due to the installation of the reduced-flow faucets with aerators in the restrooms and break rooms. Also, because of Georgia's continuing rainfall deficits and drought, we made changes to our landscaping plans to reduce annual plantings and reduce water used for irrigation. In the summer of 2007, when the ban on outdoor watering was enacted, we stopped all landscape irrigation, which further reduced our annual water usage.

The building contains six sets of restrooms (one set per floor, plus men's and women's facilities in the fitness room). The public restrooms were original to the building and in need of updating. The restrooms were fully deconstructed, floor tiles and countertops were recycled, mirrors and partitions were reused, and new floor and wall tiles and countertops were installed. Plumbing improvements were a portion of the renovation project and totaled $20,567 of the $100,000 spent for restroom renovations.

Water rates increased by only a moderate 7 percent annually from the base year to 2007, with savings of $10,985 during the sample period. For our accounting purposes, the payback is calculated from the period it was expensed, indicating an average savings of $1,465 per year and a payback of fourteen years. Water usage and savings can be seen in Table 6.4.

We do still use paper towels for hand drying in restrooms. Because of the location of the restrooms, electric hand dryers are simply too noisy, and their operation would result in sound transference to adjacent offices. The paper towels and toilet tissue we use are 100 percent recycled content, of which 30 percent is post-consumer.

TABLE 6.4 WATER USAGE AT CRESTWOOD

YEAR	WATER USAGE (IN THOUSANDS OF GALLONS)	% DECREASE FROM BASE YEAR	ANNUAL WATER BILL	SAVINGS FROM BASE YEAR	% DECREASE FROM BASE YEAR
2002	3100		$15,913.64		
2003	2800	9.68%	$15,670.06	$ 243.58	1.53%
2004	2800	9.68%	$15,351.04	$ 562.60	3.54%
2005	2300	25.81%	$14,139.35	$1,774.29	11.15%
2006	2200	29.03%	$12,215.91	$3,697.73	23.24%
2007	1600	48.39%	$11,207.90	$4,705.74	29.57%

ELECTRICITY CONSUMPTION

The Crestwood Building is rated by Energy Star to be in the top 25 percent of energy efficient buildings in the country. After damage in the wake of Hurricane Katrina in August 2005 caused increases in the cost of utilities of 25 to 40 percent, the rates Georgia Power Company, our electric utility, was able to charge its customers increased, but our electric bills in 2005—in real dollars—stayed nearly the same as the previous two years ($86,425 in 2003, $86,932 in 2004, $89,703 in 2005), thanks to additional energy conserving measures implemented in the building that year, which resulted in an additional 10 percent reduction in our average kilowatt hour (kWh) per-day use. Had rates remained the same per kilowatt hour since 2005, we calculate that we would have saved around $26,000 over the most recent three-year period. The figures are shown in Table 6.5.

Average per-day energy use continued to decline in 2006 and 2007, though our costs have continued to increase by about $10,000 per year because of rate increases from our electric utility. However, we (and, in turn, our tenants, since increases in utility

TABLE 6.5 ENERGY USAGE AT CRESTWOOD

YEAR	TOTAL KWH USED	COST	SAVINGS FROM BASE YEAR (KWH)	SAVINGS IN % OF KWH USAGE	SAVINGS FROM BASE YEAR ($)
2003	1,374,600	$ 86,425			
2004	1,384,400	$ 86,932			
2005	1,263,200	$ 89,703	(121,200)	–8.75%	$7,611
2006	1,229,000	$ 97,665	(155,400)	–11.23%	$9,758
2007	1,235,400	$106,111	(149,000)	–10.76%	$9,356

costs are passed on to tenants) would have paid much higher prices in real dollars. Control of costs for our tenants is a factor in our being able to retain tenants and reduce turnover, is part of fiduciary responsibility and, ultimately, saves everyone money. Though the numbers vary depending on lease rates, commissions, tenant improvements, downtime, etc., the incremental cost of losing a tenant is estimated to be six times the cost of retaining that tenant.

One sure thing about energy costs is that they are unlikely to decrease. Looking for ways to increase our conservation of energy remains the best strategy for avoiding higher expenses. One obvious way we have reduced electricity consumption is through our lighting choices. Low-mercury, high-efficiency fluorescent tubes (25 watts instead of 32 watts) were installed in all light fixtures, replacing less efficient fluorescent tubes at a cost of $7,060. The high-efficiency tubes use less electricity and produce less heat, thus reducing the amount of energy required for indoor cooling in warm weather and providing additional controllability for occupants. Photocell light sensors automatically turn off the exterior and lobby lights during daylight hours. Motion detectors turn on the lights in halls, common areas, and restrooms, further conserving electricity when no one is present. All light bulbs are recycled. Energy-efficient vending machines were installed in the break room. These vending machines don't have the promotional lighting of standard drink machines and operate similarly to a refrigerator, keeping snacks and drinks cold but using less electricity than conventional vending machines.

RECYCLING

Building occupants are encouraged to recycle, and many do. Our comprehensive recycling program includes metals, plastics, glass, batteries, paper, cell phones, cardboard, newspapers, toner cartridges, and light bulbs. The money earned by selling the recyclables is put toward an ice cream social and holiday events for our tenant partners. Recycling towers, with separate bin compartments for glass, plastic, and aluminum, are provided to each tenant and in the common break rooms. These bins are emptied daily by the cleaning crew. Recyclables are collected and stored on-site. When the volume we've collected justifies the trip, they are recycled at a local facility a few miles away by our building engineer. Office paper for recycling is collected on each floor and gathered in a central location. An outside vendor picks up the paper when called. The vendor sells the paper and doesn't charge us for the service.

The storage bins cost approximately $900. The recycling towers provided to the tenants were an additional $1,500 for a total initial investment of $2,400. Each week, on average, the building engineer spends and hour and a half collecting and transporting recyclables. Time allocation and mileage reimbursement are estimated at $2,860 per year.

CLEANING AND MAINTENANCE

We have a low-impact cleaning policy that includes the use of low-impact cleaning equipment. An outside vendor, our cleaning contractor, handles all of the cleaning for the building. Our contract with the cleaning firm outlines what gets done on a daily basis (five days a week), what gets done periodically, and specifies that all cleaning agents supplied by the contractor must be Green Seal approved. Green cleaning greatly contributes to good indoor air quality in the building and means we use fewer products. Our cleaning contractor, for example, uses one glass cleaner and one surface cleaner when cleaning the bathrooms. We meet with our cleaning contractor every two weeks to ensure that things are running smoothly and check the cleaning supplies closet regularly to insure compliance with our policies.

COMMUNITY BUILDING IN AN OFFICE BUILDING

Our building amenities—the conference rooms, the break room, and the fitness area—have helped us attract tenants and retain them. Having community areas in a building are a trade-off—they make the building more appealing to tenants and can foster community, but they reduce the overall amount of space available for rent. (About 18 percent of the square footage of the Crestwood Building is in the common areas; we use a 15 percent common area allocation when calculating the rent per square foot. The BOMA [Building Owners and Managers Association International] standard is 12 percent.)

The three conference rooms are available to tenants free for eight hours per month on a first-come, first-served basis. Tenants who have only an occasional need for a large meeting space appreciate having these spaces available.

An attractive, comfortable break room that's always accessible provides space for socializing during the workday, and has tables and chairs where tenants can eat lunch (many brown bag) without having to drive to a restaurant, saving them time, energy, and money. The break room also provides tenants space for informal meetings and the opportunity to get away from their desks without having to leave the building. An outside vendor provides vending machines, keeps the machines stocked with drinks and snacks, and services the machines as needed.

The fitness area was added to the building as part of the improvements made shortly after acquisition of the building. More recently, the lighting fixtures were changed to ones that accommodate high-efficiency fluorescent tubes. Equipped with treadmills, stationary bikes, a rowing machine, stair-climbing machine, elliptical trainer, free weights, and weight machines, the fitness area is available to all tenants 24/7 via pass key access. Changing rooms with showers are provided. The fitness area is cleaned by the cleaning crew, and repairs to the equipment are handled on an as-needed basis through an on-call maintenance contract with an outside vendor.

A touch screen monitor in the lobby serves as an interactive directory and provides information to tenants and visitors about networking opportunities, working in a sustainable building, and area amenities. We also communicate with tenants via an electronic newsletter (see Figure 6.3) that informs them about upcoming events in the building (like the chocolate fountain we bring in on Valentine's Day), and welcomes and introduces new tenants.

We also use the newsletter to disseminate information about transportation options, such as Georgia's Clean Air Campaign, a not-for-profit organization that educates

 melaver, inc.
^DIFFERENCE

CRESTWOOD NEWS

Feb | 2008

Have you checked out the new interactive touchscreen monitor in the lobby? Learn about networking opportunities and working in a sustainable building. Please review your company information in the directory and let us know if changes need to be made. *Coming Soon*: Area amenities, hotels, restaurants. Questions or news items? Drop us a line: reception@melaver.com

Celebrate a Difference
Thanks to all who were able to attend our event on January 24th here in the Crestwood lobby. We celebrated Georgia's First Office Building to Receive LEED Certification for an Existing Building (LEED-EB). Learn more about Leadership in Energy & Environmental Design here: www.usgbc.org

The Chocolate Fountain is Back!
Join us on Valentine's Day for the overflowing chocolate fountain in the 4th Floor Breakroom from 1:30 - 3:00 pm, and bring your sweet tooth on February 14th.

Welcome to our New Tenants - Suite 400
ExcelBlaze Consulting
FedExpress Passports & Immigration Services
The Cauble Center for Skin Care
TRANE, A Business of American Standard, Inc.
Visionary Productions, LLC

The Clean Air Campaign
Thursday, February 14th 11:30am - 1:30pm
Come by and see us in the lobby!

Interested in carpooling, walking, biking, or riding transit? Do you already carpool or ride transit? Visit CleanAir to register and find out what program(s) are right for you!

FREE GIVEAWAYS - Enter drawing for a $25 Target Gift Card
It's Easy - To enter into the drawing:
1) Register for our incentive programs on the clean air website before the 14th
www.logyourcommute.com/melaver
2) Stop by our table on the 14th and sign up

Figure 6.3   Information and outreach, distributed electronically.

people about air pollution and traffic congestion issues, and encourages carpooling by linking interested commuters. We inform tenants about the service and host presentations about carpooling for interested tenants. The program offers financial incentives and guarantees rides home in the form of taxi vouchers to those who carpool but find themselves unexpectedly unable to participate in their carpool arrangement due to unforeseen events (e.g., a child's illness or unscheduled overtime).

Assessing the Financial Aspects of LEED Certification

Record-keeping and documentation can be difficult and tedious, but it's necessary so we can monitor and document our progress over time, determine the financial benefits of the choices we make, and maintain records for re-certification, which, to maintain LEED for Existing Buildings, is required every five years. For tracking purposes, we know it's important to establish procedures that allow us to control the materials that come into the building. Doing this yields better, more involved relationships with vendors and suppliers. We know we need to check everything. Our Mark of a Difference program aids in these efforts, providing written standards and methods and procedures, and keeping us focused on particular goals.

Much of the work required for obtaining LEED for Existing Buildings involves choosing appropriate materials for improvements and working with suppliers, contractors, and vendors to be sure the specified materials are properly installed and properly maintained. Some changes, such as water-conserving plumbing fixtures and higher-efficiency lighting are investments that save money over time. Depending on the structure of the lease, it can (and will) be argued that the cost savings accrues to the benefit of the tenant. It can also be argued that a full-service office building lease is based on the premise of total occupancy cost (rent + common area overage + RE tax overage + insurance overage). A consistent increase in the overage payments by a tenant will lead to a rollback on rental rates (assuming an annual increase in base rent) and/or a dissatisfied tenant. Limited or no overage should allow for an increase in rental rates. Other changes, like using sustainable cleaning products, don't save any money (although they don't cost more, either), but they do contribute to everyone's health and well being.

If we refer back to Table 6.1, we see that the Crestwood Building was acquired for $9.4 million and that an additional $1.346 million was invested early on for various upgrades, for a total initial investment of $10,746,000. Total gross income amounted to $1.2 million, net operating income annually was approximately $910,000, and the cap rate for this building was about 8.47 percent. We now need to update these numbers with data that reflect upgrades to the building.

Before entering the LEED for Existing Buildings program, we pursued an Energy Star rating for zthe building, entailing $62,664 of investment. We followed that

Energy Star certification with additional upgrades tailored to the LEED for Existing Buildings program, amounting to another $74,457. Total improvements to the building amounted to $137,121, for a total investment over the years of $10.88 million. The initial cap rate of 8.47 percent has been reduced ten basis points to 8.37 percent. An overview of these calculations can be seen in Table 6.6.

But what is the benefit of taking even a ten basis point reduction in the cap rate for this building? Our quick response is simply this: As long as our overall cap rate is within the range of market comparables, we feel we have been true to our mission of

TABLE 6.6   REVISED CAPITAL INVESTMENT IN CRESTWOOD

IMPROVEMENTS		COST
Acquisition price		$9,400,000
Improvements after acquisition		$1,346,000
LEED-EB and Energy Star improvements		
Window film	$33,986	
Zolatone upgrade	$ 1,600	
Restroom retrofit	$20,567	
Lighting retrofit	$24,824	
HVAC retrocommissioning	$ 7,128	
HVAC enhancements/repairs	$37,840	
Recycling program	$ 5,260	
Entry grates	$ 3,916	
LEED documentation/submittal	$ 2,000	
Total LEED-EB and Energy Star		$ 137,121
Total capital investment		$10,883,121
Gross square feet		93,544
Common area factor		18.00%
Leaseable square feet		76,706
Market vacancy factor		8.00%
Rental rate per square foot		$17
Gross potential revenue with vacancy factor		$1,199,683
Expense ratio		24.10%
Net operating income		$ 910,559
Cap rate on initial capital investment		8.47%
Cap rate on total capital investment		8.37%
Change in cap rate		0.11%

optimizing the triple bottom line performance of this building. (The general real estate market, however, is unlikely to find that response wholly convincing.)

Let's consider a bit more closely the benefits associated with this particular LEED for Existing Buildings project, utilizing Esty and Winston's four-category schema, employed throughout much of this book. That schema, as you may recall, looks at the benefits of a green program through four lenses: short-term and tangible revenue creation, short-term and tangible cost reduction, longer-term and more intangible benefit creation, and longer-term and more intangible risk and liability reduction. I'll consider each briefly in turn.

Short-term cost reductions through the LEED program come primarily through lower operating expenses (for water, energy, landscaping, etc.). Water savings amount to $4,700 annually, while energy savings are calculated to be approximately $9,400 each year. We have also seen a reduction in the cost of landscaping as a result of migrating to drought-resistant native plants. Insurance is a less promising story. So far, LEED certification has not resulted in a lower insurance premium for Crestwood per se, although our underwriter is working on it. And, as pointed out in Chapter 4, our overall portfolio has seen insurance savings of around $29,000 annually, a savings we attribute to the overall green orientation of the company. One insurance company that promotes its green program quoted us its rates for LEED buildings. Unfortunately, the rates were double what we're currently paying. In short, savings in operations annually amount to $14,100, for a total savings over a ten-year period of $140,000.

Short-term revenue creation is a bit trickier to calculate. We know that our 2007 year-end occupancy at Crestwood was 92 percent, compared with an occupancy for other office buildings in this sub-market of 81 percent, for an 11 percent differential. With total rentable square footage at 76,000 square feet, that differential means that Crestwood has approximately 8,400 square feet more than other office buildings in the general vicinity, which translates into additional gross rent of $143,000, or net income of $109,000. The big questions, of course, are how much of this revenue is owing to the building's green orientation, how much is due to our strong emphasis on service, and how much is related to various building amenities that enable it to compete favorably against other properties in the area. It's hard to say for certain.

At least initially, during our early ownership of Crestwood, we knew that our occupancy rates were more or less comparable to competitive properties in our sub-market. This is no longer the case, so something positive has occurred that has enabled us to differentiate our building from others. Special amenities (the fitness area, conference rooms, and break rooms)? Certainly. Service emphasis? Probably. Green orientation? Probably, as well. In fact, we would argue that our service orientation and our green orientation are inextricable parts of an overall emphasis on our tenants. For the sake of our financial analysis here, we will assume that 75 percent of our over-market occupancy is owing to the building's amenities, while the remaining 25 percent ($27,000) is related to our service offering.

The longer-term benefits are, if anything, trickier to calculate. We have certainly seen a significant amount of positive publicity as a result of delivering the first multi-tenant LEED office building in Georgia. However, this benefit of enhanced reputation is too abstract to calculate with any degree of confidence. Moreover, since the overall intent of this book is to make the business case for green buildings generally, the intangible benefit we might see by delivering a first-to-the-state product is not replicable as more LEED for Existing Buildings projects are developed. As such, we don't include it in our cost/benefit analysis here.

The same logic applies to the longer-term intangible benefits that occur as a result of reducing our overall liability and risk. With conservation measures in place, are we less prone to be cited for violating outdoor water use regulations? Yes. Can we place a value on that reduced exposure? Probably not.

With these various costs and benefits of delivering a LEED for Existing Buildings project in mind, we can now look at a basic discounted cash flow analysis for Crestwood. As can be seen in Figure 6.4, total costs of the LEED program amounted to $163,000 ($137,000 initially plus the on-going cost of the recycling program of $26,000). Total benefits are conservatively estimated at $450,000 (with a primary contributor being our over-market occupancy). The net present value (NPV) of the LEED investment comes to $135,000, with an internal rate of return on our green investment of 42.3 percent.

Years		0	1	2	3
REVENUES/SAVINGS					
	Native landscaping		3,800	3,800	3,800
	No external mat vendor		425	425	425
	Water savings		1,700	3,700	4,700
	Energy savings		7,600	9,800	9,400
	Over-market occupancy owing to service/green		27,254	27,254	27,254
Total Revenues			40,779	44,979	45,579
EXPENSES					
	Window film		(33,986)		
	Zolatone upgrade		(1,600)		
	Restroom retrofit (water conservation)		(20,567)		
	Lighting retrofit		(24,824)		
	HVAC retrocommissioning		(7,128)		
	Cost of recycling program		(5,260)	(2,860)	(2,860)
	HVAC enhancements/repairs		(37,840)		
	Entry grates		(3,916)		
	LEED documentation/submittal		(2,000)		
Total Expenses			(137,121)	(2,860)	(2,860)
Total Cashflow		0	(96,342)	42,119	42,719
Discount Factor		1.000	0.909	0.826	0.751
PV Cashflow		0	(87,584)	34,809	32,095
NPV		135,573			
IRR		42.30%			

Figure 6.4   Discounted cash flow of LEED for Crestwood.

LOOKING TO THE FUTURE

The Crestwood Building's LEED for Existing Buildings certification will be up for renewal in 2012. In addition to maintaining the standards we have set and recording information about the materials usage, processes, and procedures we have in place, we are always on the lookout for small improvements and adjustments to our day-to-day routines that can save energy, reduce waste, and improve tenant comfort, satisfaction, and retention. We constantly ask ourselves, "What can we do better?" That's our policy.

We have a five-year plan for projected improvements that includes some big-ticket items, like the overhaul of our elevator system (we've hired an outside consultant to help us with this) and adding greater shade canopy in our parking lot. Our first consideration is energy efficiency. At some point, we will need to replace our aging HVAC system, which is nearing the end of its useful life. At this time, our efforts at conscientious maintenance have allowed us to defer replacement. Being resource-efficient means replacing things when they need replacing, rather than simply replacing them to have something different.

4	5	6	7	8	9	10	TOTALS
							0
							0
3,800	3,800	3,800	3,800	3,800	3,800	3,800	38,000
425	425	425	425	425	425	425	4,250
4,700	4,700	4,700	4,700	4,700	4,700	4,700	43,000
9,400	9,400	9,400	9,400	9,400	9,400	9,400	92,600
27,254	27,254	27,254	27,254	27,254	27,254	27,254	272,536
45,579	45,579	45,579	45,579	45,579	45,579	45,579	450,386
							(33,986)
							(1,600)
							(20,567)
							(24,824)
							(7,128)
(2,860)	(2,860)	(2,860)	(2,860)	(2,860)	(2,860)	(2,860)	(31,000)
							(37,840)
							(3,916)
							(2,000)
(2,860)	(2,860)	(2,860)	(2,860)	(2,860)	(2,860)	(2,860)	(162,861)
42,719	42,719	42,719	42,719	42,719	42,719	42,719	
0.683	0.621	0.564	0.513	0.467	0.424	0.386	
29,177	26,525	24,114	21,921	19,929	18,117	16,470	

Considerations When Choosing a Property

Our company's goals in the twenty-first century have been centered around refreshing an aging real estate portfolio, diversifying geographically and by asset type, and positioning ourselves to be opportunistic. The foundation for meeting these objectives was in place when I joined the company in late 2003. We sold holdings that the company had held for a number of years and reinvested the equity in stabilized office properties—historic mixed-use as well as development properties—in Huntsville and Birmingham, Alabama and in Augusta, Savannah, and Atlanta, Georgia.

As a company, we can't devote all of our resources solely to development work. That would be too future-oriented—we would run out of money needed to run day-to-day activities long before our development projects came to fruition. It also doesn't make much sense to devote all of our resources to solely managing stabilized properties. We need the enhanced returns that development work brings in. It's more complicated when you add acquisition work to the mix, particularly for a green development company. On the one hand, the potential acquisition needs to be a good fit with a stabilized portfolio, a portfolio with a relatively smooth, relatively low, non-volatile income stream that helps pay the bills and supports the development activity. Newer buildings tend to fit the bill well, but virtually all of the ones on the market aren't built to green standards, and it's hard to justify replacing their relatively new systems. Try telling your CFO that you've just bought a building that is three years old and you've decided to replace the standard roof (which has a twenty-year life expectancy) with a green roof and you get the idea.

On the other hand, older buildings, ones with aging systems, might seem to be perfect acquisition targets, except that they are often unstabilized, with poor occupancy rates, perhaps located in challenging sub-markets, and are facing lots of deferred maintenance. Or maybe the owner improved the property before placing it on the market, thus inhibiting our opportunity to meet our sustainable objectives. Ideally, we identify opportunities before they go on the market and buy them at prices that discount the cost of the improvements. Faced with satisfying a company expectation to green all acquisitions (environmental bottom line) while also satisfying company expectations that acquisitions provide a stable income stream (financial bottom line), what would you do? Welcome to my world.

A GREEN ACQUISITION CHECKLIST

In general, we have found the following considerations helpful when thinking about acquiring properties that will be absorbed into a LEED for Existing Buildings program:

- **Pace.** Converting an existing building to LEED does not necessarily have to happen overnight. Proactively planning the renovations is a critical part of an overall acquisition strategy, since you will want to factor into a financial pro forma the point in time you are likely to make the move to LEED and account for the management time it will take to implement the process.

- **Discipline.** Stick to the plan you made as part of your acquisition. Set and meet deadlines. Evaluate how you're doing. Much easier to say than do, however. Our ability to execute a disciplined approach is challenged by the rapid advancement of technology, new government standards (building codes), and evolving incentives for innovations. You may plan to install a new chiller in three years, but you cannot predict where the technologies may be at that point in time.

- **Building age.** It's probably helpful to consider potential acquisitions that are more than ten years old. At that point, roofing materials are beginning to break down, HVAC systems are ripe for improvements, parking surfaces are calling for more than just a simple cosmetic slurry coat, and common areas will be showing their age.

- **Location: Site orientation.** This is not something you have much control over, obviously, nor should it be a make-or-break criterion if a building that you find to be a compelling purchase is unfortunately oriented north to south. Still, all things being equal, as with a LEED for New Construction project, it pays to consider how well a building is situated to take advantage of daylighting while being insulated from heat gain.

- **Location: Access to wind and solar power.** As a company, you may wish to consider focusing acquisition work in areas where either or both of these energy-generation alternatives are in abundant supply.

- **Location: Support for sustainable strategies.** Acquisitions in areas where there is strong support for sustainable strategies, either in the form of incentives (e.g., for alternative energy) or in the form of progressive regulations (e.g., ordinances that support graywater treatment systems).

- **Leveraging weak points.** Some buildings simply perform poorly in terms of their energy efficiency. That can be a good thing from an acquisition standpoint. Look for current areas of under-performance that, with a deft sustainable touch, can be opportunities for value-add. Chances are these areas are getting a look during due diligence, but not with a sustainable eye.

- **Tenant mix.** Requirements for financing vary for each asset type. In an office building, for example, there should be diversity among the tenants that is consistent with the area. Beyond the risks of a dependence on one market segment (as recently exemplified by the sub-prime lending sector) there is a social element to the tenant mix. Sustainable development must consider the community where the asset is located.

- **Building size.** As our culture generally and the real estate industry particularly begin to focus greater attention on decentralized (i.e., localized) forms of energy delivery, having a certain critical mass to justify on-site generation of energy is likely to be an important criterion of acquisition work in the not-too-distant future.

- **Building skin.** This issue may be already covered as part of an energy audit on the building, but you would certainly want to consider how much glass is part of the exterior envelope, the degree to which there are operable windows, how thermally efficient the shell is, how much daylight is provided, etc.

A green acquisition checklist should also include a list of things to be on the lookout for, in terms of the outside of the building, such as:

- **Complementary users and/or tenants.** Ideal is a set-up whereby the waste by-products of one tenant are a critical need of another. So, for instance, a small manufacturing company has on-site generation of electricity by capping the methane from a municipal landfill nearby. The excess heat from the electric generation is captured and used by a neighboring restaurant for its kitchen ovens. It sounds a bit farfetched today, though there are already numerous examples of leveraging the potential symbiosis among users in a development. By the way, it's worth noting that such symbiosis does not have to involve esoteric technologies—it could be as simple as having a tenant mix whose work hours somehow prove complementary. The shared parking arrangement between Crestwood and the neighboring Marriott hotel is a good example, since it reduces the amount of impervious parking surface these two developments would require if viewed independently.
- **Tree cover.** Does the building site have trees? Trees can reduce heat gain and glare from the sun and reduce the heat island effect of paved areas. They can also provide shade for outdoor gathering areas and parking lots.
- **Landscaping.** Is the landscaping replete with exotic plantings that are water hogs as well as costly to maintain? Or is it characterized by native plantings that manage well enough on their own? Could the cost of landscaping and the amount of water required be reduced by the use of native plants or by planting perennials? Evergreen trees typically require less maintenance than deciduous trees, which drop their leaves in the fall. How much sod is on the site? Grass must be mowed and maintained; in many areas, it must be irrigated. Would it be possible to eliminate the lawn or at least reduce the amount of sod? Could stormwater be captured and stored for irrigation?
- **Parking.** What's the life expectancy of the parking lot? Are there opportunities to cost-effectively move to a semi-permeable surface? How many spaces are required by code, and are there opportunities for getting creative about reducing that number?

Beyond just the physical issues associated with the building, both on the inside and in the outside environs are more general questions to consider: Is the building well located for the use of mass transit or other modes of transportation (walking, bicycling, car-sharing program access, etc.)? Is the project a good fit with key company objectives (e.g., revitalizing a downtown urban core area, addressing a critical community need such as affordable housing)? This part of the checklist might seem to extend well beyond issues of LEED for Existing Buildings adaptivity, but the connection between social and environmental justice issues (as we will see in the next chapter) is a critical one. The important thing to keep in mind is that, fundamentally, a strategy for LEED for Existing Buildings certification should extend well beyond the bricks and mortar of a building.

Concluding Remarks

While playing a lead role in greening the existing buildings we acquire is something I feel passionate about, there is a high degree of common sense to our efforts. The conversion of older, less-efficient buildings to a green program is a critical part of an over-all need to reduce our overall environmental footprint. The work requires time, reflection, a patient eye toward finding the right time to embark on such a program, a willingness to think creatively, and fiscal discipline. It's a balancing act. The greening of Crestwood was fundamentally a team effort. I feel fortunate to work alongside a team of people who pitched in to make it come together.

Tenants in a building are another important part of the team because renovating an existing multi-tenant facility to LEED standards is likely to occur while the tenants are occupying their offices and trying to conduct business as usual. Managing an office building is much more challenging when a building is undergoing a renovation, but managing an office facility during a green rehab provides an unparalleled opportunity to educate tenants about the benefits of going green and explain why it's important. It's also an opportunity to enlist their help in the process, perhaps even to make the building perform better than initially planned, through education and outreach (ideally in non-threatening and fun ways).

There's a lot of work to be done. The nation has 1.5 billion square feet of existing buildings that need to be sensibly renovated over the next thirty years. We have somewhere between 500,000 and a million brownfield sites that need to be restored. We have 6.3 million kilometers of roads, and, despite the fact that 75 percent of transportation funds ($60 billion) is devoted to maintenance, we need over twenty times that amount ($1.3 trillion) to adequately redress aging highways. The infrastructure ostensibly distributing clean water to us all is, on average, over fifty years old, with most of the older cities in the United States relying on systems that date back one hundred and fifty years. Just restoring and protecting old steel bridges with toxic lead-based anticorrosion compounds costs us $350 billion a year.[13] The list goes on and on.

The long list reminds me of the computer game Sim City, where happy-go-lucky building sprees in the early rounds yield to a growing weight of expenditures later to manage aging and failing infrastructure. The point of the game, I think, is two-fold: Manage your growth at a deliberate pace, and devote sufficient resources to managing what you have already built rather than blithely building more projects.

Working with projects such as Crestwood, and devoting time and money and other resources to making existing buildings function more effectively, comprise the first steps in what my colleagues and I hope will be ever-greater attention to restorative or regenerative building practices. It's not the easiest journey in the world. Of all the LEED projects our company has been involved with over the years—including new construction projects at the Platinum level—the greening of Crestwood was our most difficult and challenging undertaking. There's no lack of green consultants out there willing, for a fee, to advise on this process. But there is most certainly a lack of actual LEED for Existing Buildings product—only around sixty as of this writing, many of which are owner-occupied and/or single-occupancy.

Storm Cunningham, author of The Restoration Economy and a leading proponent of devoting our resources to restoring what we have already built (as opposed to building new), notes this difference between conservation and sustainable development:

> Conservation is, in concept, relatively simple (not easy): protecting irreplaceable natural assets, entities, and functions. Sustainable development, on the other hand, is a complicated dream, an intoxicating brew of enlightened policies, unpopular lifestyle changes, expensive (in the short term) process improvements, innovative designs/products/materials, hazy goals, and splendid values.[14]

Cunningham concludes that restorative development, by focusing on the already-built environment, serves both to conserve and sustain. I find the concept compelling.

Although my colleagues and I often refer to our company's aspiration to be a sustainable developer, we also aspire to be a restorative developer; focusing on urban-core locations, avoiding greenfields, staking out a preference for in-fill and rehab work, looking for opportunities to link our bricks-and-mortar activities to enhancing community. While there is tremendous opportunity for green development that accrues to a company's bottom line, there is even greater opportunity to realize value and values by green redevelopment.

NOTES

[1] James Howard Kunstler, *The Geography of Nowhere: The Rise and Decline of America's Man-Made Landscape* (New York: Simon & Schuster, 1993), pp. 55–56, 90, 97, 103–107.

[2] From Martin Melaver's notes from an address by Tom Bisacquino, President of the National Association of Industrial and Office Properties (NAIOP), at the opening session of the DevelopGreen Conference, Glendale, Arizona, March 12, 2008.

[3] Stewart Brand, *How Buildings Learn: What Happens After They're Built* (New York: Penguin Books, 1994), pp. 5, 7.

[4] Ibid., p. 12.

[5] Ibid., p. 13.

[6] Ibid., p. 11.

[7] Ibid., p. 19.

[8] Ibid., p. 23.

[9] Ibid., pp. 21–23.

[10] John Tillman Lyle, *Regenerative Design for Sustainable Development* (New York: John Wiley & Sons, Inc., 1994), p. 23.

[11] From the Architecture 2030 website. See http://www.architecture2030.org/current_situation/hist_opportunity.html.

[12] Due to the nature of the statistical model and data available, Portfolio Manager only works for certain facility types. These include banks/financial institutions, courthouses, hospitals, hotels and motels, K–12 schools, medical offices, offices, residence halls, retail stores, supermarkets, warehouses, and wastewater treatment plants. For more information on Energy Star and Portfolio Manager, go to www.energystar.gov.

[13] Storm Cunningham, *The Restoration Economy: The Greatest New Growth Frontier; Immediate and Emerging Opportunities for Businesses, Communities, and Investors* (San Francisco: Berrett-Koehler Publishers, Inc., 2002), pp. 133, 156, 166, 257.

[14] Ibid., p. 253.

DOLLARS AND (COMMON) SENSE: REALIZING THE VALUE OF GREEN FOR KEY USERS

COLIN M. COYNE

SUMMARY

In this chapter, Colin Coyne, COO of Melaver, Inc., focuses attention on a complex redevelopment project in Birmingham, Alabama, highlighting the need to address the value of green to key constituencies: the tenants who occupy a green building, the financial institutions that provide funding, and the community within which a project is located. In the first section, Coyne focuses on prospective tenants for the office portion of the Birmingham Federal Reserve & Tower, making the business case for how rent in a green building can be viewed as an investment, not simply an expense. In the second section, Coyne turns his attention to providers of financial capital—both debt and equity—and how a green building makes dollars and sense for this stakeholder group. The third section addresses the community and how sustainable development offers social as well as financial and environmental benefits to a locale.

Having considered the three main constituencies of any development project—direct users, financial investors, and citizens of a community—section four focuses on how all of this makes dollars and sense for the green developer charged with integrating and addressing all the needs of various stakeholder groups. Of particular interest here are critical tools that enable the developer to work in an integrated fashion: stating one's philosophy clearly, making that ethos public, being inviting of third-party verification of one's efforts, and opening up to public scrutiny. In the final section, Coyne concludes that sustainable development should not only make dollars and sense for the present generation but should, at its roots, be driven by a sense of stewardship for future generations. Beyond the numerical analysis presented, readers should gain insight from the common sense elements that define sustainability.

The Fabric of Community

A few years ago, I was sitting in a movie theater in Birmingham, Alabama watching the Disney/Pixar animated feature, *Cars*. On my lap was my ten-year-old daughter, Ariana; to my left was her twin brother, Dallas; and to my right was my stepdaughter, Zoe. For those who've never seen it, at one level *Cars* is a movie about (not surprisingly) cars. But at a far more complex level, *Cars* is really a movie about a town that time forgot. More accurately, *Cars* is about a town that let time forget it, and about that town's quest to recapture its soul. It's a movie about the search for character. At its most fundamental level, *Cars* is a movie about re-establishing the fabric of community.

This chapter discusses a pragmatic and holistic approach to sustainability, from the very practical application of sustainable business principles to broader philosophical considerations. We thus begin with the specifics of making dollars and sense for tenants, drawing upon a real-world presentation that we make to prospects wherein we focus on the financial benefits we feel are realized by occupying a green building. At another level, though, this chapter is about the benefit of green to the larger community, and the catalytic role that projects such as the Birmingham Federal Reserve & Tower can play in fostering a sustainable society. A project of this scope has distinct challenges. I will try to be forthright about the difficulties as well as the opportunities this project has afforded us, as there is a great deal to be learned from our difficulties and mistakes. In fact, I am willing to bet the struggles represented by this project are emblematic of similar challenges in your town. The green bottom line, after all, is more than just financial value creation for the developer. As I will note throughout, it includes the financial benefit to specific users—tenants and residents—and extends beyond them to include financial, social, and environmental benefits for the community at large.

Overview of the Birmingham Federal Reserve & Tower

Early in 2006, Melaver, Inc., purchased two vacant buildings in downtown Birmingham—an historic 1920s-era Federal Reserve Bank and an adjoining annex built in the 1950s—as well as a parking lot adjacent to the two structures. Some of our objectives for this property have remained unchanged since the acquisition, dovetailing with our overall development objectives for all our development projects:

- Help revitalize the downtown area both economically and socially;
- Ensure that the project addresses a diverse mix of needs and users;
- Create a high-performance project (minimum of LEED gold certification) that serves as an exemplary model for other green projects in the region;
- Make the project a teaching tool about sustainable practices; and
- Engage a broad collection of city-wide stakeholders in the development to educate, advocate, and facilitate change.

Some objectives for this development have changed over time as our vision of the project has adjusted to market conditions. Initially, we envisioned a complex of office and hotel, with a major law firm occupying much of the office space. When this major tenant prospect ultimately decided to remain in the premises it had occupied for the previous fifteen years, we considered creating a complex that would include a cooking school (with classrooms), restaurants, a performing arts complex, and an office/hotel. The resultant iteration, the one used to approach institutional investors, is described this way:

> The Birmingham Federal Reserve & Tower is a $150 million, sustainable mixed-use development featuring a 235-room four-star hotel, 150,000 square feet of Class A office space, an executive conference center, luxury residences, remarkable restaurants and bars, street-level retail shopping, and a first-class spa. The project will feature cutting-edge technology to minimize its impact on the environment while increasing the comfort and productivity of its occupants. The unique service model of our hotel and the manner in which we will care for our employees and our community will help redefine the notion of gracious service. All the while, the project will be managed with a keen eye toward maximizing the financial results of our investors, proving that environmental responsibility, social accountability, and financial opportunity are, indeed, mutually reinforcing principles.

This project's complexity (see Figure 7.1) has proven to be a challenging one for our company. It is our first major development in a market that is in the nascent stages of a downtown renaissance. Equally, Birmingham is a community that seems to require outside eyes to validate its potential since, early in the process, some longtime residents were skeptical as to whether our project would occur. It is larger than any project we have attempted to date, and is our first foray into hotel development. It is a more complex financial structure than anything we have ever attempted, involving historic tax cred-

Figure 7.1   **A digital rendering of the Birmingham Federal Reserve & Tower project.**

its, New Markets Tax Credits, an incentive package from the City of Birmingham, and outside equity partners. While we need to stay true to our ethos of under-promising and over-delivering, we somewhat paradoxically need to generate enough excitement from key constituencies to make it a reality without overstating what we can deliver. More than anything else however, this project has forced us, as has no other project yet, has to make a compelling case for sustainable development to three distinct but critical groups:

1 **The potential end-users** we are seeking as tenants;
2 **The institutional investors** that will provide the necessary equity infusion; and
3 **The city**, which in being persuaded by the social and financial benefits of our project, is providing an economic assistance package in support of the project.

In short, we need to prove to these diverse groups the financial, social, and environmental benefits of becoming stakeholders in our project. We are speaking the same language of benefits to each constituency. However, the basket of benefits to each looks somewhat different.

Making Dollars and Sense
for Our Tenants

When comparing the Birmingham Federal Reserve & Tower to conventional developments, it is natural to focus on key differentiators. The project doesn't lack distinct elements to highlight, including many of significant importance: energy efficiency, water conservation, waste management, renewable energy sources, nontoxic materials, recycled/reused materials, reduced environmental impact on the building site, wise use of space that reduces total building size, the balance of economic and environmental and social considerations, consideration for cultural and historical factors of the site, and proximity to transportation. The list is extensive. But when we prepare for a presentation of a green project to a potential end-user, we seem to stand ready to confront any and all resistance to green with a plethora of facts and statistics supporting our case. I wonder sometimes if we are mis-reading our audience.

A significant number of technological innovations like those in the Birmingham Federal Reserve & Tower give exciting previews of what buildings will look like and how they will perform in the not too distant future. It's easy to become lost in the technical elements and be wowed by the coolness of it all. It is also fairly easy—and typical—for those familiar with the LEED program to rattle off basic attributes of the program: self-selective point system, five major categories, third-party verification of performance, how many LEED buildings have been built to date, etc. But in framing a green project in this way, we ignore the cardinal rule of business: focus on the client.

Too often, developers of green projects—and we are as guilty of this as any—ignore the characteristic that sets a LEED building apart from all others. Plainly stated, a green or LEED building is a structure designed foremost from the tenant's and community's point of view, not the developer's. So basic and simple and obvious. And yet it is so easy to lose the forest for all the cutting-edge trees, to forget about the basic purposes of a structure: to provide for basic human needs for shelter and protection and comfort.

A green building provides the occupants with greater comfort and control over the temperature of individual workspaces. Design that maximizes light yet reduces heat gain also minimizes eyestrain by reducing glare on computer screens. The improved air quality means reduced airborne allergens and irritation from fumes. From the community standpoint, a green building fits into context, respecting its surroundings and those who live and work in the area. It consumes less land and water, common resources to all. It encourages vibrant activity in the surrounding environs. A sustainable building is a place where its occupants are proud to come to work, are more productive throughout the day, and leave with more to give their families when they return home.

There's one other element: money. A green structure improves upon the occupant's work experience, enhances a tenant's reputation, and delivers better financial performance. Viewed this way, rent might more appropriately be viewed as an investment rather than an expense for a tenant, one that demonstrates a specific rate of return to the tenant. That's a fairly radical, paradigm-shifting notion, one that merits attention in any discussion of a green bottom line.

Let's consider this last part in more detail. Based on our analysis of the Birmingham Federal Reserve & Tower, rental payments could conservatively be reduced by 50 percent owing to specific green features of the building. These savings, which may be on the low end, are contingent upon the particular tenant, its business model, etc. Much is made of the fact (in literature on green buildings generally and in several chapters of this book, specifically) that green pays. But financial analysis is almost always viewed from the perspective of the developer or the owner. However, despite paying no premium for the green building it inhabits, a tenant receives a host of benefits.

When a company calculates the value it receives for the rental rate it pays—and it absolutely ought to—it should view its evaluation through four lenses. The company should consider the direct and indirect savings that a certain building may provide in operating expenses. It should also consider the direct and indirect revenue-enhancing aspects this building could provide. Let's consider each lens in greater detail, in the context of the Birmingham Federal Reserve & Tower project. As with all other chapters in this book, we will identify the financial values of green whenever it meshes with our own direct experience, affix numerical valuations where we have been able to validate these assessments, and make a concerted effort to be conservative in our estimates and projections. In the calculations that follow, we will be using 170,000 square feet as the rentable area of Birmingham Federal Reserve & Tower's office component, and $28 per square foot as the average base rent for all office tenants, resulting in gross rental income of $4,760,000 annually. It is estimated that the total tenant workforce will be approximately five per thousand square feet, or 850 people.

DIRECT COST SAVINGS

Relative to current alternatives, The Birmingham Federal Reserve & Tower project proposal contemplates these direct cost savings for tenants: energy savings, water savings, lower common area costs over time, and lower cost for reconfiguring office space. Let's examine each one.

Energy Savings

We anticipate energy savings of 40 to 50 percent. In our proposal to prospective office tenants, our company structures common area cost savings below a pre-determine expense stop on an equal basis with the tenant.[1] As such, energy savings accrue directly to the benefit of the tenant, even before considering the cost of operations outside normal business hours (when a tenant absorbs the full costs of utilities). Energy savings could effectively reduce the tenant's occupancy costs by approximately $0.65 per square foot annually.[2]

Water Savings

We anticipate water savings of 30 to 40 percent. Our proposal shares common area cost savings below a pre-determined expense stop on an equal basis with the tenant.[3] As with energy, water savings accrue directly to the benefit of the tenant. Such reduced water consumption is typical of the green building industry in general and is borne out by our own projects, primarily through use of low-flow water fixtures.

Lower Common Area Costs Over Time

We believe long-term deterioration of the building—mechanical and electrical systems particularly—is reduced through sustainable design and materials. As a result, costs passed on to the tenant for maintaining the building will be less from initial occupancy, and savings will grow over time. A 5 percent improvement in maintenance costs—a conservative estimate based on our own experience—would reduce expenses by $0.33 per square foot annually. It's worth noting that, over time, the value of these savings over a conventional building become greater due to compounding.

Lower Reconfiguration Cost

The cost of reconfiguring office space may be substantially lower. As an option, we offer the incentive of an additional tenant improvement allowance to encourage a tenant to construct its office build-out to LEED Commercial Interiors standards. We estimate that in doing so, the cost of changing office configurations in the future could be reduced from a national average of $2,500 per employee to as little as $200 per employee. Use of more expensive carpet tile in the initial build-out, for instance, enables the landlord and tenant to replace only worn-out areas rather than full runs of flooring. The inclusion of motion sensors for lighting extends the life of bulbs and fixtures. Raised flooring reduces air conditioning and wiring costs. It is our belief that additional cost savings will be realized through the integrated design of interior and exterior elements if both landlord and tenant are pursuing similarly guided design principles. In the analysis of the financial benefits of tenanting green, we have not factored in these particular cost savings since the option contains too many variables. But it is another area of potential cost savings.

INDIRECT COST SAVINGS

Relative to current alternatives, the Birmingham Federal Reserve & Tower Project contemplates indirect cost savings for tenants from reduced absenteeism, reduced health claims, and reduced turnover. A move to a green building is a real and tangible investment in the well being of a tenant's employees.

Reduced Absenteeism

Improved daylight conditions, better thermal comfort, and lower airborne allergens: These and other features built into the project do translate into hard cost savings for tenants. For example, at the Internationale Nederlanden (ING) Bank in Amsterdam, worker absenteeism fell 15 percent after a move to a sustainable office. The importance of reduced absenteeism is magnified when a tenant's business model is based on the leveraging of administrative staff. If, for example, one support staff member for a group of three lawyers is absent, three professionals' work product is negatively impacted. As a conservative estimate, we assume absenteeism is reduced by one day a year for every employee in a green building.

Reduced Health Claims

On average, Americans spend more than 90 percent of their time indoors, and according to the U.S. Environmental Protection Agency, quality of indoor air is, on average, five times worse than outside air. With more than 17 million Americans suffering from asthma, with $15 to 40 billion in health-care costs estimated to be directly related to "sick" buildings owing to mold spores and other allergens, the Birmingham Federal Reserve & Tower's emphasis on superior indoor air quality translates to real cost savings in today's dollars, as well as reduced liability exposure to the tenant. We have not put a dollar figure on this savings, but feel that this benefit should be highlighted in a presentation to potential tenants in a green building.

Reduced Turnover

Believed to be the highest hidden operating cost, employee turnover remains one of the least attended cost centers of American business. In professional sectors, this cost conservatively exceeds 100 percent of the departed employee's annual salary.[4] Direct costs blend into indirect costs: separation costs (administrative expense, separation/severance pay, unemployment compensation, litigation costs); vacancy (overtime for existing staff, temporary employees); replacement costs (recruiting expenses, interviewing, skills testing, medical exams, travel/moving expenses, administrative expenses); lost productivity (loss of expertise, loss of customers loyal to the departed staff member, mistakes made by the new employee); and morale issues (turnover creates stress, resentment from existing employees as they assume an increased workload).

Tenant Savings from Reduced Turnover

In our analysis, we take as our starting point the fact that businesses today see 100 percent turnover roughly every four years. Further, we assume 10 percent of a company's staff members are key people whom the business would prefer to retain. In his book *The Sustainability Advantage*, Bob Willard reports it costs about $25,000 to replace just any staff member and somewhere between two to three times the annual salary of a key employee to replace one. Moreover, it costs about $7,000 to hire a new staff member. With these assumptions in place, we calculate the following for the 850 people employed by tenants at Birmingham Federal Reserve & Tower:

- **Regular staff turnover.** 191 regular staff members (one quarter of 90 percent of all those working in the building) are likely to leave for other jobs each year, reduced by 3 percent for an annual savings of $143,250, or $0.84 per square foot (191 × $25,000 × 3% = $143,250, which, divided by the 170,000 square feet in the building, comes to $0.84 per square foot).
- **Key employee turnover.** 21 key staff members (a quarter of 10 percent of all those working in the building), averaging annual wages of $50,000, are likely to leave for other jobs each year, reduced by 3 percent for an annual savings of $78,750, or $0.46 per square foot (21 × $50,000 × 2.5 × 3% = $78,750 which, divided by 170,000, comes to $0.46 per square foot).
- **Total expenses for turnover.** 206 staff members total (97% of 212 because our retention rate is higher than the national norm) who leave each year require expenditures on the part of the company ($7,000 per hire). It has been our own experience that having a green orientation makes the process of recruitment smoother,

TABLE 7.1   INDIRECT SAVINGS FOR A TENANT IN A GREEN BUILDING

Building square footage	170,000
Rent per square foot (annual)	$28
Rent savings per square foot	$1.72
% savings on rent per square foot	6.14%

CATEGORY	# OF PEOPLE ANNUALLY	COST PER PERSON	% REDUCTION FROM NORM	CALCULATION OF SAVINGS	SAVINGS PER SQUARE FOOT)
Turnover of regular staff	191	$ 25,000	3%	$143,250	$0.84
Turnover of key staff	21	$125,000	3%	$ 78,750	$0.46
Reduced cost of hiring	206	$ 7,000	5%	$ 72,100	$0.42
TOTALS				**$294,100**	**$1.72**

less time consuming. How much so? Hard to say. Willard suggests a 5 percent reduction in hiring costs as a result of one's sustainable orientation is appropriate, a reasonably conservative estimate. The result is a savings annually of $72,100, or $0.42 per square foot.

In short, we conservatively estimate the total indirect savings to a tenant in a green building to be just over $294,000 a year, or $1.72 per square foot, 6.14 percent of the total annual rent, as shown in Table 7.1.

EXPLICIT REVENUE ENHANCEMENT

Relative to current alternatives, the proposal for the Birmingham Federal Reserve & Tower project contemplates explicit revenue enhancement from improved productivity and naming rights.

Improved Productivity

Labor costs are by far the largest expense for most companies. In fact, they account for 92 percent of the life-cycle cost of a building, more than 72 times the cost of energy. So small increases in productivity result in a tremendous improvement to revenue and/or reduction in expenses. Sustainable buildings increase worker productivity. In a well known-example, the West Bend Mutual Insurance Company documented a 16 percent productivity gain in the early 1990s following a move to a new 150,000-square-foot green building.[5] When multiplied by the tenants' annual billings, such an increase when factored on a professional's time alone yields significant production. If, for example, we assume the average salary at Birmingham Federal Reserve & Tower to be $50,000 and gross production to be 2.5 times salary, total gross revenues from all tenants would amount to $106 million. A 16 percent increase in productivity would add over $17 million to tenants' gross revenue.

In environments where building occupants are regularly tested (e.g., schools), simply adding daylighting has resulted in significant increases in productivity. In Colorado and Washington, for example, end-of-year test scores from students in classrooms with the most daylighting were found to be 7 to 18 percent higher than those with the least. In California, students in classrooms with the greatest amount of daylighting were found to be 20 percent more advanced in math skills and 25 percent faster on reading tests in one year when compared to those students in classrooms with the least daylighting.[6]

The tricky part, of course, is assuming analogous productivity gains when moving from the school environment, where quantitative notions of success are codified and measured, to an office environment, where the data is challenging to acquire and the metrics are highly subjective. Despite this, we do feel that some increase in productivity is found in green buildings. How much? Bob Willard, citing studies from Joseph Romm's *Cool Companies* and Hawken, Lovins, and Lovins' *Natural Capitalism*, notes that increased productivity in green buildings has been documented to range from 6 to 16 percent. Based on these studies, Willard feels comfortable calculating productivity gains as a 7 percent increase for 50 percent of a company's workforce. Our own analysis is more conservative. It assumes that employee productivity is 2.5 times employee cost ($50,000 per staff member) and that tenants will see a 6 percent increase in productivity for 20 percent of its workforce. The result is increased productivity of approximately $1.275 million annually. This increase amounts to slightly more than a 1 percent hike in gross revenues but covers over 30 percent of the company's rent, or $7.50 per square foot in savings. (The calculation is $50,000 × 2.5 × 850 total employees × 20% of this workforce × 6% productivity increase.)

How realistic is this? If we consider the analysis of Green, Inc. in previous chapters, we'll recall that this small (20 employees) fictionalized company is recording additional net income of approximately $1 million through additional development work and consulting, boosting its annual profits from $3,720,000 to $4,720,000. How much of that increased profitability is owing to the increased productivity that comes from locating in a green office? Again, it's difficult to say precisely. But even if a tiny percentage (say 3 percent) of Green, Inc.'s $1 million increased net revenue is linked to the company's environment, the numbers it is posting will bear out the assumptions regarding location in a green building. The analysis is obviously more art than science, particularly since we are using a fictionalized company's performance to underscore projected benefits for a generic tenant elsewhere. For many readers, this will seem just too farfetched. Perhaps. But consider the alternative: A company could occupy conventional space for the same rent as office space in a green building, but productivity could simply be expected to be the same it historically has been. The object of our discussion is the degree of upside value creation in a green building versus a conventional alternative, where no additional value creation can be expected.

Naming Rights

In our presentation to an anchor tenant for Birmingham Federal Reserve & Tower, we proposed as an option the opportunity to place the tenant's name on the building. The opportunity can objectively be quantified by market comparables for billboard signage (estimated conservatively to be worth $35,000 annually), with the added cachet that the tenant would link its brand to the first green high-rise complex in the city. Although we have not factored this into our analysis, it is worthy of mention simply because there is potential value capture.

INDIRECT REVENUE ENHANCEMENT: BRANDING VALUE

Relative to current alternatives, the Birmingham Federal Reserve & Tower project contemplates the following indirect revenue enhancing opportunities through branding value. Companies (such as Melaver, Inc.) that are early entrants into the sustainability movement are integrally associated with green. So too are early signature single-tenant owner-occupied buildings such as the Chesapeake Bay Foundation's Philip Merrill Environmental Center in Annapolis, Maryland or Interface, Inc.'s showroom in Atlanta, Georgia, or the Adam Joseph Lewis Center for Environmental Studies at Oberlin College in Ohio. Various first-generation books touting the financial merits of sustainability inevitably include branding and enhanced reputation as aspects of value creation that a business should factor into its decision to embracing a triple bottom line. As green becomes more mainstream, however, the issue of branding and reputation becomes more complicated.

While the speed at which green is becoming more mainstream is rapid, there still, unfortunately, are only a fraction of actual green buildings out there. In many markets—Birmingham among them—the green product is limited or non-existent. These are markets that offer the promise of first or early-mover advantage. As such, there is an added value to a company's reputation by making a move, literally, into a LEED building in a market where such buildings are few and far between. However, two caveats are in order:

1 Solely occupying space in a green building is not much of value-add to a company's reputation. Such occupancy needs to go hand in hand with a company's underlying values and practices. Otherwise, the lack of alignment between where a company's offices are located and how it conducts business are not only readily apparent but could constitute a possible liability, potentially exposing the company to charges of greenwash (more about this in Chapter 10). As Bama Athreya of the U.S.-based International Labor Rights Forum has noted regarding brand campaigns, "Let's face it, hypocrites are far more interesting than mere wrongdoers."[7] A business should shy well away from promoting its green digs unless its offices are a visible manifestation of a much deeper commitment to green practices. In short, while there are still branding opportunities out there, proceed cautiously and let your practices do the talking.

2 As discussed in Chapter 4, putting an actual dollar figure on branding value is notoriously difficult. It's best to focus on more tangible values, such as reduced operating expenses and enhanced productivity and, at best, simply know that in various unpredictably surprising ways, your company's reputation is enhanced by taking a more sustainable orientation.

My colleagues and I at Melaver, Inc. believe business in general is moving to embrace sustainable tenets, to the point where a company's reputation can be negatively affected by not embracing a sustainable ethos. With this mindset, the question becomes the extent to which a company can afford not to get on the bandwagon. If, in an s-shaped diffusion curve, the early innovators have the marketing advantage that comes with being a first-mover, then the fast-followers need to hasten their activities toward sustainability less they face a potential negative backlash. Some may be aware of the huge wave of negative publicity Shell Oil faced over its initial (and then rescinded) decision to scuttle its Brent Spar oil platform in the North Sea. Or the attention Nike has received over its contract labor practices in the developing world. Or the obesity lawsuit McDonald's faced in the United Kingdom. Granted, these are well-

Years	0	1	2	3	4
REVENUES/SAVINGS					
Direct Savings					
Energy & Water		110,500	110,500	110,500	110,500
Common area maintenance		56,100	56,100	56,100	56,100
Indirect Savings					
Reduced absenteeism		170,000	170,000	170,000	170,000
Reduced turnover: staff		143,250	143,250	143,250	143,250
Reduced turnover: good worker		78,750	78,750	78,750	78,750
Hiring of new personnel		72,100	72,100	72,100	72,100
Direct revenue enhancement					
Improved productivity		1,275,000	1,275,000	1,275,000	1,275,000
Indirect revenue enhancement					
Branding value					
Subtotal Revenues					
COSTS					
Rent		(4,760,000)	(4,760,000)	(4,760,000)	(4,760,000)
Total Cashflow	0	(2,854,300)	(2,854,300)	(2,854,300)	(2,854,300)
Discount Factor	1.000	0.909	0.826	0.751	0.683
PV Cashflow	0	(2,594,818)	(2,358,926)	(2,144,478)	(1,949,525)
NPV	(17,538,438)				
Savings as % of Annual Rent	36.57%				

Figure 7.2   Dollars and sense from the tenant's perspective.

known examples of transnational corporations being held accountable for the brands they have carefully crafted. But I don't think it's too much of an overstatement to suggest that this growing political movement is likely coming to a town near you.

Even to the harshest cynic, the question any company should be asking itself these days is what the cost is in terms of time, money, and brand reputation for resisting a more socially and environmentally responsible approach to conducting business. Our analysis of the benefits of going green does not include a line item for brand enhancement—although for the Birmingham Federal Reserve & Tower project we might be able to justify it since the building provides first-mover cachet. Nor will our analysis include a line item for the liability of not locating in a green building: the market is not yet mature enough. But that time is coming.

ASSEMBLING THE DATA OF SAVINGS AND REVENUES

We have devoted considerable space to discussing the various benefits to a tenant locating in a green building. It's now time to assemble all of the data into one spreadsheet (Figure 7.2).

5	6	7	8	9	10	TOTALS
110,500	110,500	110,500	110,500	110,500	110,500	1,105,000
56,100	56,100	56,100	56,100	56,100	56,100	280,500
170,000	170,000	170,000	170,000	170,000	170,000	1,700,000
143,250	143,250	143,250	143,250	143,250	143,250	1,432,500
78,750	78,750	78,750	78,750	78,750	78,750	787,500
72,100	72,100	72,100	72,100	72,100	72,100	721,000
1,275,000	1,275,000	1,275,000	1,275,000	1,275,000	1,275,000	12,750,000
						18,776,500
(4,760,000)	(4,760,000)	(4,760,000)	(4,760,000)	(4,760,000)	(4,760,000)	(47,600,000)
(2,854,300)	(2,854,300)	(2,854,300)	(2,854,300)	(2,854,300)	(2,854,300)	
0.621	0.564	0.513	0.467	0.424	0.386	
(1,772,296)	(1,611,178)	(1,464,707)	(1,331,552)	(1,210,502)	(1,100,456)	

There are a number of things worth pointing out in this analysis. First, unlike all of the cash flow projections we have seen in previous chapters, this one does not compute an internal rate of return. This is simply because the net present value of our analysis is negative. The rent our collective tenants would pay in the Birmingham Federal Reserve & Tower project ($28 per square foot for an annual amount of $4.76 million and a ten-year total of $47.6 million) is greater than the accumulated savings and revenues we have calculated (ten-year total of nearly $18.8 million). That's OK. In fact, it's a lot better than OK. Essentially, this analysis indicates a tenant gets back over a third, or 36.57 percent, of its rental expense in the way of savings and revenue enhancements, paying an effective rate of $17.76 per square foot ($28 minus $10.24 in savings and revenues). Rather than paying rent over ten years of $47.6 million (which, when discounted back to Year 0 is $29.2 million), our collective tenants effectively pay rent of just over $30 million (which, when discounted back to Year 0, is $18.5 million). Compare that result to a tenant who moves into a conventional office space of the same size, paying exactly the same rental rate of $28 per square foot but getting no added bangs for the buck.

To put it another way: If our tenants paid us $1 (not a square foot, just one dollar) for the benefit of residing in a green building, they would realize an infinitesimally high return for that initial investment over ten years (so high that my spreadsheet software can't calculate it). If for argument's sake, our tenants decided to pay us $1 million up front for the benefit of occupying a green building, they would see a whopping 190 percent internal rate of return for that investment over ten years. In fact, if our tenants all conducted businesses that made an annual 30 percent return on their activities, they would be "willing" to pay $8 million in Year 0 for the benefits they would receive over the ensuing ten years.

It might be easy to assume that the presentation above is too analytical for tenants to grasp. On the contrary, we've been able to use this presentation effectively with prospects. What matters most, however, is that Melaver is not the one inputting the assumptions, but, instead, the tenant. We simply provide a spreadsheet for the tenant's use at their leisure. In one recent tenant presentation, we completed the spreadsheet together, calculating in real time that the tenant's return of rental investment exceeded 200 percent. We did not make a single assumption; the managing partner of the firm made all of them.

Given such benefits, the question becomes not what are the financial justifications for going green, but rather what are the financial justifications for not doing so?

Making Dollars and Sense for Capital Providers

Making the business case to prospective tenants for a green building is only part of "selling" sustainability. Other key stakeholders need to be brought on board. The

Birmingham Federal Reserve & Tower project is challenging in several respects: it was Melaver, Inc.'s first project that was too large to manage with solely its own capital, and it entailed several layers of funding involving multiple parties, each of which needed to be convinced that this project made sense. Somewhat unexpectedly, each party focused on different aspects of the project, and each made sense of the project in different ways.

The project included Operation New Birmingham, a city agency charged with the economic revitalization of the downtown area; the mayor and City Council, which would ultimately have to sign off on a package of financial incentives critical to the deal; banking institutions that would provide both standard construction debt as well as New Markets Tax Credits; and institutional investors who would provide the equity above that already invested by our company. The first two parties—Operation New Birmingham and the City of Birmingham—are discussed in the next section, which is focused on making dollars and sense to the community. Here, we focus on the financial entities, the capital providers for both debt and equity.

Timing is everything. In 2006, Melaver, Inc. had developed a small portfolio of sustainable properties and was seeking to sell a significant portion of it to an outside investment group, ideally one aligned with our sustainable ethos. At the time, the institutional investment community knew next to nothing about green real estate, and we determined that a more viable route for us was simply to refinance these properties and retain 100 percent ownership, which we did. (Details of this transaction can be found in Chapter 4.) Ironically, about the same time we were searching for outside, green-oriented capital, I attended a very small conference in Arizona mostly comprised of socially responsible investment (SRI) groups looking to remedy the fact that their respective funds had no weighting in real estate because there were no green properties to acquire. Fast forward two years, and the situation has changed dramatically, with numerous funds formed or in the process of being formed to invest in green real estate. Two years ago, there was nowhere to go for sustainably oriented capital. Today, the list is long and deep, and we have discussed with numerous institutions various projects—among them, the Birmingham Federal Reserve & Tower project.

The question is: How do you talk dollars and (common) sense to the financial community about sustainable real estate? It's a complex question, since it involves matching up the needs of a developer with the needs of a financial institution, something easier said than done. As it turns out, there are a number of moving parts to this overall challenge. Let's consider some of the major issues.

THE NEEDS OF THE PROJECT

As the largest private development in the City's history, the Birmingham Federal Reserve & Tower project needs money, lots of it. As is often typical with projects focused on revitalizing downtown core areas that have undergone years—even decades—of neglect, this project has risks as a new market trying to recover. As such,

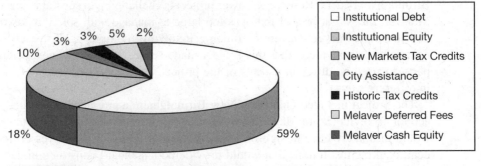

Legend:
- ☐ Institutional Debt
- ▨ Institutional Equity
- ▨ New Markets Tax Credits
- ▨ City Assistance
- ■ Historic Tax Credits
- ☐ Melaver Deferred Fees
- ▨ Melaver Cash Equity

3% 3% 5% 2%
10%
18%
59%

Figure 7.3   **Tranches of funding for Birmingham Federal Reserve & Tower.**

the cash needs of the project are typically spread through numerous tranches, so as to reduce the financial exposure of any one capital provider. A simplified image of the sources of funding can be seen in Figure 7.3.

Each source of funding has its own needs and expectations. Melaver, Inc., for instance, focuses on urban core infill, mixed-use projects that address environmental and social justice issues. Instead of looking to maximize our financial returns, we look to optimize triple bottom line returns, which means that we are prepared to increase our overall investment in a project for social and environmental benefits until our minimum return expectations are reached. We look for institutional entities that are comfortable with this approach.

The entities involved in historic tax credits—local, state, and federal agencies such as the Department of the Interior—are focused on the preservation of key historical aspects of a building. The City of Birmingham is looking to provide incentives that will make this project successful (and thus a good provider of tax revenues in the future), serve as a catalyst for additional investment in the downtown urban core area (providing additional tax revenues), and address social and economic problems facing the municipality. Those institutions providing New Markets Tax Credits are looking for developers with a track record of successful projects in designated impoverished areas. Institutional equity players (in the case of this project) are looking for developments that will meet their investors' expectations for sustainable investment. And institutional debt entities are looking for projects that demonstrate a strong likelihood of financial success. In short, while the interests of each source of capital overlap, they each have particular needs that must be addressed.

THE NEEDS OF FINANCIAL STAKEHOLDERS

Melaver, Inc. is providing a relatively small piece of equity in the Birmingham Federal Reserve & Tower project. The vast bulk of the equity is provided by outside investment from an institutional fund. Not surprisingly, different funds have different needs. They have different exit strategies, with some looking to monetize (i.e., exit) their investment in a relatively short three-to-five-year time frame. Because sustainable development is typically oriented toward a longer-term hold for ideological reasons (belief in investing in communities for an extended period of time) and/or financial reasons (belief in maximizing returns by optimizing the life-cycle capital investment), short-term funds are not typically ideal candidates. Other institutional investors such as CalPERS, which looks to match the long-term outlook of pension investors with long-term investments, offer better alignment for sustainable projects such as Birmingham Federal Reserve & Tower.

Despite the different orientations of every investment fund, all have one thing in common: the returns on a sustainable development have to meet their financial performance objectives. Those objectives differ with every institutional investor, but most expect returns in the mid to high teens. Anything less is simply not going to cut it. For Birmingham Federal Reserve & Tower to make dollars and sense to institutional capital, it has to meet expected market rates of return. In short, a triple bottom line focused development has to deliver on the financial metric first and foremost. Otherwise, the other two set of metrics, social and environmental, simply are of little interest. But, since building a LEED building (as we have seen) really doesn't cost more than a conventional project, the Birmingham Federal Reserve & Tower development projects a return to its institutional investor in the high teens—attractive enough for numerous institutional funds, and certainly not a deterrent. Financial stakeholders do not have to accept sub-par returns in order to invest in green.

A green project like Birmingham Federal Reserve & Tower does address needs of certain financial stakeholders that extend beyond financial returns. There is a growing trend in the market for specific funds that target investment in green projects. Our own experience indicates approximately $2 billion seeking green projects which, when leveraged at a three-to-one debt-to-equity ratio, indicates (at the time of this writing) approximately $8 billion focused on green projects. (In 2007, Wells Fargo alone passed $1 billion in loans for LEED certified projects.) Demand in the market for green development well exceeds market supply of green product. Competition is keen among institutional funds to partner with qualified, knowledgeable, reliable developers who can meet basic hurdle rates and deliver on a green project. At the present time, the number of developers that do both is small—good news for companies such as ours.

Another key need among financial stakeholders that merits discussion is more problematic for companies such as ours: Institutional capital is not interested in partnering on one deal with a green developer. It is looking for partnerships that provide for multiple deal flow. The time it takes for institutional capital to become conversant with green practices and the effort invested in doing due diligence on smaller green developers is significant. That up-front investment in time and effort is obviously more richly rewarded if a developer and its institutional investor commit to multiple projects over an extended period of time. Melaver, Inc.'s capacity for doing multiple projects the size of Birmingham Federal Reserve & Tower is limited. We could certainly scale up to meet this demand. But such scaling up, particularly if it occurs rapidly, tends to move us away from the care and attention we like to give to each project we develop. The very things that should make green developers attractive to institutional capital—sensitivity to place, attention to the needs of land and community—mean that those developers are less likely to have a pipeline of cookie-cutter projects to replicate from locale to locale. This is certainly our own company's orientation, something we wrestle with all the time.

The financial stakeholders in Birmingham Federal Reserve & Tower are not, however, limited to our major institutional equity provider. There are also financial institutions that have been allocated a portion of federal tax credits focused on "new markets" that are strong candidates for revitalization. While the qualifications for being a new market are somewhat complex, the basic criteria for investment include social as well as financial components. Whereas institutional capital that targets green investment typically vets a project on financial performance and green components, entities that provide New Markets Tax Credits vet a project based on financial performance and social criteria. In short, being green in not enough.

For entities providing New Markets Tax Credits, the green component of a development project would seem to be a nice add-on, but one that doesn't make such a project any more attractive from a financing standpoint. As it turns out, however, competition for providing New Markets Tax Credits on Birmingham Federal Reserve & Tower has been intense. In retrospect, this makes perfect sense. A company with a triple bottom line orientation is focused on the social as well as the environmental consequences of its activities. A triple bottom line development company is much more likely to be developing in urban core areas, particularly urban core areas where development can be a catalyst for revitalizing a community. In fact, a triple bottom line developer could be seen as perfectly situated to be a key recipient of the New Markets Tax Credits program. It was never conceived that way, but that is the result. It's worth noting that our projected performance on the Birmingham Federal Reserve & Tower project moves from rather tepid low-teen returns without the credits to solid high-teen returns when those credits are factored in.

Making Dollars and Sense for the Community

While the various financial stakeholders in a project like Birmingham Federal Reserve & Tower are primarily focused on the economics of a deal (but also see the added advantages of going green), another group of stakeholders is focused primarily on the social aspects of a development. In this case, those stakeholders are Operation New Birmingham and the City of Birmingham itself. These key stakeholders have different needs that a triple bottom line company is well positioned to address.

THE BIG PICTURE

We cannot ignore the tremendous social impact that development has on a community—one of the three legs of the triple bottom line of sustainability. What we've discovered,

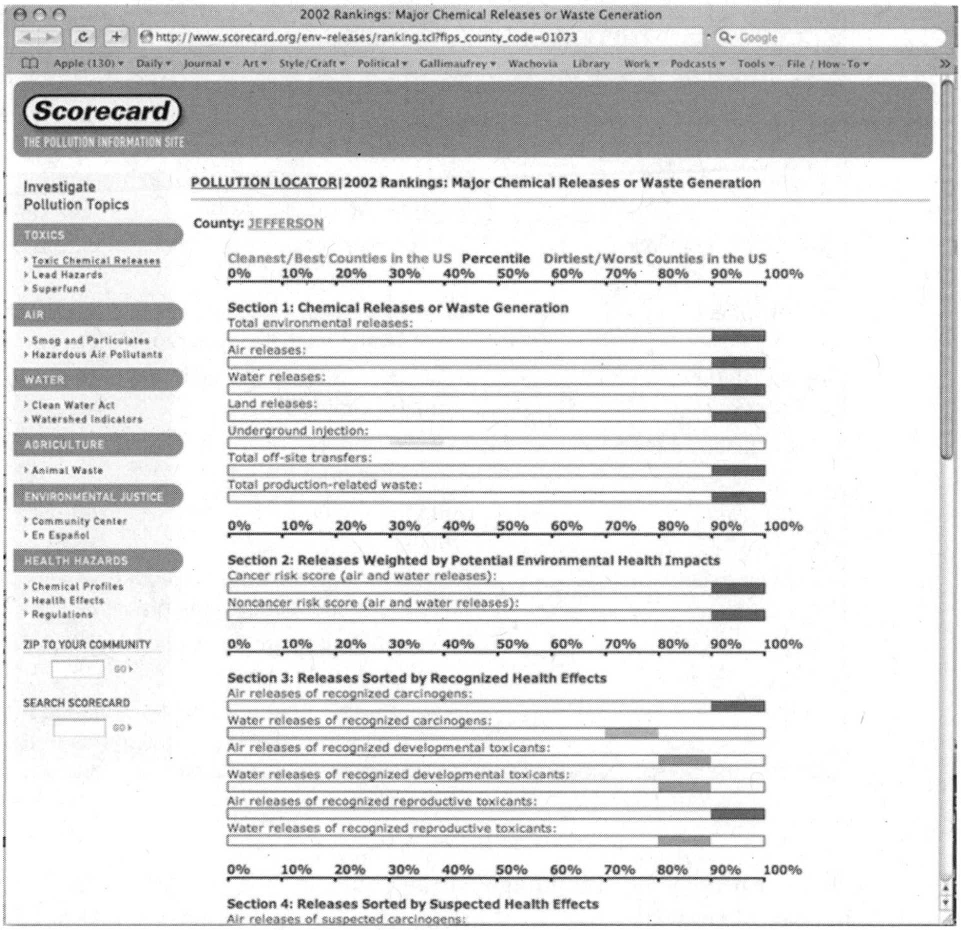

Figure 7.4   Environmental Scorecard for Jefferson County, Alabama.
Source: www.scorecard.org.

however, is that few understand the direct links among the legs of sustainability. Environmental irresponsibility is connected to social inequity. Social inequity, in turn, impacts financial instability. Financial instability, clearly, can be very bad for business.

An examination of the environmental report cards of Jefferson County, Alabama (home to Birmingham) and Hillsborough County, Florida (Tampa), indicates these areas rank among the worst 10 percent of all counties in the United States in terms of major chemical releases and waste generation. Figure 7.4 shows a small portion of Jefferson County's report card.

Why does this matter? When we overlay the environmental scorecard with social metrics our company uses (see Table 7.2), certain inequities become quite apparent. One example: in Jefferson County, Alabama, people of color are 5.75 times more likely to live near a toxic air-pollutant discharger than whites. Stated differently, people of

TABLE 7.2   MELAVER, INC.'S SOCIAL METRICS FOR VETTING POTENTIAL DEVELOPMENTS

QUALITY	QUESTION
Alignment	Are our partners, local decision makers, etc. aligned with our values?
Extending the circle	To what extent does the project engage a broader stakeholder base?
Ripple effect	Does the project facilitate catalytic development that would revitalize an area?
Addressing real community needs	To what extent is this a "pull" (as opposed to a "push") project?
Social justice	To what extent does the project attend to social inequities in a community?
Diversity	Does the project lend itself to a mix of uses and users?
Educational outreach	To what extent does the project serve as a teaching tool about sustainable practices?
Aesthetics	To what extent might the project facilitate the creation of lasting beauty?
Context	To what extent does the project enhance the unique historical and social aspects of a place?

color live near facilities emitting toxic air pollutants at a rate that is 575 percent of that in areas where whites live. In Hillsborough County, Florida, the rate is 152 percent. Here's another example: The rate for low-income families living nearing toxic facilities is 159 percent; for those below the poverty line, it is 169 percent. For children in poverty, it's 188 percent.[8] Environmental irresponsibility exacerbates social inequality. This is true from Birmingham to Boston.

SOCIAL MODELING IN ACTION

A triple bottom line development company by and large understands the connection among social, environmental, and financial issues. Civic leaders also get this connection and are proving more inclined than ever to link incentive packages for developers to projects that address social and environmental justice issues.

But the triple bottom line developer can't just be spouting rhetoric. Careful thought must be given to the various communities or constituencies a development company serves. Three fundamental communities come to mind: the citizens of the surrounding area; those who physically occupy or visit a building; and those who are in one's employ, either directly or through supplier or agency relationships.

Serving the Community in Our Vendor Relationships

We have broadly defined a set of operating parameters for selecting our upstream and downstream vendor relationships, focusing on:

- Businesses indigenous to the area;
- Businesses that are cost-competitive;
- Businesses with a reputation for quality;
- Businesses that employ or help the disadvantaged;
- Businesses with a stated policy of social responsibility; and
- Businesses owned or operated by an under-represented class.

In practical terms, this means we will work with groups like Workshops, Inc., a non-profit organization that hires the mentally impaired to perform piecemeal work, such as preparing stationery or amenities packages for our hotel rooms. Our sourcing for hotel supplies looks to favor woman- and minority-owned businesses. That our vendors utilize fair labor practices is a prerequisite, one that will be monitored closely throughout our relationship. We work with some of Birmingham's most established and respected businesses (Alabama Power, Alagasco, McWane, Inc.), largely as a means to leverage their social capital to seed additional change in the community. Partnering is a critical aspect of our overall focus on advocating, educating, and doing outreach for conducting business differently.

Serving the Community within Our Business

Simply stated, we want the employees of our hotel to be among the best cared for hospitality workers anywhere. We believe if we care for our employees in a personal way, they will care for our guests in a like manner. If they remain with us over an extended period of time, that retention enhances the capacity to deliver a level of guest service beyond that of competing hotels. Examples of the type of care we envision include: coordination with the University of Alabama at Birmingham's Nursing School to provide outreach to staff in the areas of preventive medicine, prenatal care, breastfeeding, exercise and nutrition; working with our food and beverage department and providers to provide one free meal per shift and weekly family meals; providing a health plan that pays 100 percent of the employee's premium; and providing for transportation—the single greatest cost facing low-income wage earners.

Pie in the sky? Maybe. It's too early to say for sure. On a granular level, if we treat our employees with unique dignity they are less likely to leave, thus reducing turnover costs. But at a higher level, we have discovered in the Birmingham Federal Reserve & Tower project and in others, an astonishing willingness among civic organizations throughout the community to assist us in areas outside our core competence. We are, after all, developers first and foremost—or envelopers, as we prefer to call ourselves. But part of the job of development is reaching out into the community and asking others to help us do our job better. Our particular industry seems so narrowly focused on putting up a building that it forgets about the quality of life that a building engenders. Others in the community are thrilled when we ask them what they can do to make that quality of life better—and they have the knowledge and experience and passion to help make it happen. You just need to ask.

Serving the Community that Surrounds Our Business

On the construction barricade surrounding our Birmingham project, we posed a series of reflective questions that are addressed to the community, but which are also ones we ask ourselves. Among the questions we ask:[9]

- What does this building stand for?
- Can one building really make a difference in our community?
- Can the influence of a building extend beyond its footprint?
- Can a building serve as a catalyst for change?

These questions begin the journey of social responsibility in our development, by using the project as a gathering point for dialogue and collective thinking. If we can begin a different type of dialogue, one that departs from the tired model of developers squaring off against stakeholders in a community and replaces that model with something built around a congruence of values, then positive change becomes a greater possibility. All too often, it seems a development project is plunked down as an unidentified (and unwelcome) *fait accompli*, lacking any sense of ownership by the community. We feel that a community should take spiritual ownership of its locale. If a project is to be in Birmingham, we believe that to the greatest extent possible, it should be an indigenous, home-grown product, for everyday use. Likewise, we believe that an infusion of new ideas and possibilities can enliven a community. The challenge, of course, is finding the appropriate balance between the concepts we try to introduce as developers and the vision community residents have.

With Birmingham Federal Reserve & Tower, we are looking to restore a historic treasure while creating a technologically advanced building of broad significance. In this way, we pay respect to the community's rich history while pointing to its potential in the future. We are working with local architectural firms to create the interior design and character of our structure, and have hired BNIM of Kansas City—well known for its sustainable orientation—to design, model, and engineer the core and shell. While showcasing indigenous expertise, we also bring world-class talent to Birmingham.

But building a beautiful, iconic structure, even one with a reduced environmental impact, isn't enough. The community that immediately surrounds our property has a profound impact on the ultimate success of our property. That community deserves special attention. We are fortunate to have organizations such as Harvest Birmingham help us plan carefully to see that food prepared on site doesn't go to waste and that fewer in our community go hungry. We look to work with local shelters such as Pathways (for abused women) and the Firehouse Shelter (Birmingham's only shelter that provides housing for indigent men with children), to see that the various resources utilized in Birmingham Federal Reserve & Tower's ongoing operations (sheets, towels, and blankets, for example) have cradle-to-cradle utility.

We envision our project as serving as a catalyst for positive growth in the community. Each piece of the Birmingham Federal Reserve & Tower project should contribute to a richer community fabric: our hotel to a better employment model; our office tower to a greater regard for healthy working environments; our food and beverage service to the value of gathering as a community; the historic preservation of the Federal Reserve to a respect for our past, and the hope for our future. These goals may sound lofty, yet they are grounded in a basic practicality: the business that neglects its community is not sustainable. In serving our community, we serve our shareholders, investors, and employees. If we subject ourselves to the scrutiny of a higher standard, others will hopefully be compelled to do the same.

What is the business good in this? Vagrancy declines, vandalism declines, security costs decline. Our guests and our tenants enjoy an increased sense of relaxation and comfort. The downtown area, which has for almost three decades had a nine to five o'clock existence, should begin to expand its hours of operation, bringing people into the core city area during the evening for cultural events and nightlife. Street life becomes more vibrant.

Time and again, we find that those who take care of the community get taken care of by the community, a virtuous circle that extends well beyond the footprint of any building. This virtuous circle, in which social, environmental, and financial aspects of a project coalesce, is captured in Figure 7.5.

The bottom line? The City of Birmingham has weighed in with substantial financial incentives for our project. That assistance is a down payment on a new way of doing business. It is up to us to deliver.

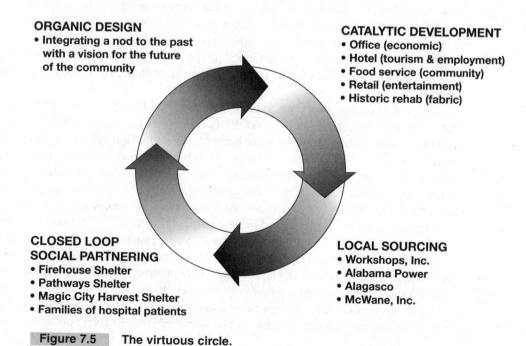

ORGANIC DESIGN
- Integrating a nod to the past with a vision for the future of the community

CATALYTIC DEVELOPMENT
- Office (economic)
- Hotel (tourism & employment)
- Food service (community)
- Retail (entertainment)
- Historic rehab (fabric)

CLOSED LOOP SOCIAL PARTNERING
- Firehouse Shelter
- Pathways Shelter
- Magic City Harvest Shelter
- Families of hospital patients

LOCAL SOURCING
- Workshops, Inc.
- Alabama Power
- Alagasco
- McWane, Inc.

Figure 7.5   The virtuous circle.

Making Dollars and Sense for the Green Developer

Our company has pursued a holistic approach to measuring return on our own development projects, balancing the long-term needs of the community with the more immediate financial concerns of tenants and financial institutions. This synergy and balance among environmental stewardship, social obligation, and financial opportunity goes by many names today—sustainability, the triple bottom line, socially responsible property investing.

Regardless of the name, a business strategy that embraces and enhances the community fabric is the one to which we aspire. To achieve this, we've attempted to define a set of performance metrics that captures the simultaneously competing and reinforcing forces of place, people and profit. We work on refining and redefining our metrics on an ongoing basis, with the pursuit of a community-enveloping methodology at the forefront.

That is what this chapter is about—a pragmatic, holistic business approach to sustainability, from very broad notions of being stewards of land and community, to the very specific ways in which that values-centric framework is implemented, from the theoretical to very practical application of sustainable business principles. Our approach tends to be one that lays broad foundations first, and then drills down in to specifics.

Without viewing a specific project like Birmingham Federal Reserve & Tower in a broader context, the individual application of dollars makes little sense. It would be easy to follow the numbers and end up in the wrong place entirely. Much of the past resistance to sustainability was embedded in a deeper mindset that held business and communities hostage, believing our choices must be framed in terms of "or" rather than "and," that we can have environmental responsibility or social accountability or financial opportunity. That is just not true. These are reinforcing concepts, not preclusive ones.

As business professionals, we can talk about the world being flat; we can discuss tipping points, and we can even freak out about economics. In each instance, what we are talking about is seeing the daily tasks before us as part of a much larger landscape. Is it not reasonable, then, to address change in our business in an integrated fashion rather than a siloed one? Environmental educator David Orr has noted that quantum leaps of change on a global scale are most efficaciously addressed community by community, arguing rightfully that the environmental movement has "grown out of the efforts of courageous people to preserve and protect particular places."[10] If this is true, then we need to—no, we must—begin constructing our thought processes around "and," not "or."

Applied to sustainable communities, the "or" mindset holds us back in any number of areas: we can have affordable housing or innovative, sustainable design; we can push economic growth for a region or retrench ourselves in an older, slower, more cosseting sense of community from a generation or two ago; we can revitalize our urban core areas in ways that enrich them, or abandon them to the urban poor and go live in the suburbs; we can have a big city or a tight community. But what if we change the paradigm? What if development and community planning and sustainable thinking ceased operating as if our choices were binary opposites? What if we considered instead the spectrum of opportunities lying in between?

Thinking in an integrated fashion: Linking the value for tenants, for financial institutions, and for the community. It's unquestionably a challenge, particularly since there is no easily designated "promised land" that indicates you have "arrived" at integration junction. There are specific tools we have found useful in guiding our process toward thinking and implementing in more integrated ways, all having to do with aspects of transparency and accountability:

- State your position clearly;
- Make it public;
- Seek out independent verification; and
- Allow for public scrutiny.

Let's consider each element briefly.

STATE YOUR POSITION CLEARLY

I am constantly amazed by the extent to which businesses want to hedge their bets when it comes to stating intent. Whether it is fear of lawyers and stock analysts or an unwillingness to commit to a clear and unqualified vision, the preponderance of businesses that fail to state a policy or course of action is staggering. Yes, most businesses have broad mission statements, but when asked for a specific set of guidelines by which employees and stakeholders can govern their actions, few firms will commit to what they are willing to do. Even more rare is a statement of what a firm is *not* willing to do. Drawing clear lines in the sand in terms of business practices is a key step.

Social and environmental concerns are a huge part of our development equation. But what does that look like? Working collaboratively over a period of about a year, our entire company articulated a set of real estate guidelines that guide all of our development work. The process was long and involved, with individual words and phrases becoming fodder for intense arguments. At the end of the day, we arrived at a statement of principles that had individual buy-in from all staff members. Equally, we had a clear statement of what we would and would not do as a sustainable real estate development firm. See "Melaver, Inc. Real Estate Investment Guidelines."

MAKE IT PUBLIC

When a firm creates governing policy, the guidelines should be made as public as possible to ensure clear understanding by all stakeholders and to provide a constant

Melaver, Inc. Real Estate Investment Guidelines

I. We will give preference to urban, infill projects that are catalytic to positive redevelopment of the surrounding area in a manner that re-establishes/reinforces the community fabric.

II. We will consider a project outside of the urban landscape under the following conditions:

    a. If it is catalytic to preservation, conservation, or responsible development in the face of unrestrained growth;

    b. If there is a compelling social benefit associated with the project that supports the community fabric, in which case we will still seek to develop in as sustainable manner as possible;

    c. If it promotes equal access to quality of life.

III. We will not develop greenfield sites. In the context of being a viable, sustainable developer, "greenfield" means:

    a. Virgin forest;

    b. An area that would require substantial land disturbance of previously undisturbed areas;

    c. An area that encompasses endangered flora or fauna (rural or urban);

    d. An area where development would change the historic character or communal fabric of the area;

    e. Land where development would negatively impact migration routes, productive farmland, productive forest land, critical habitat, or unique ecological features.

Where we have the ability and resources available, in instances where we are presented with a project in a greenfield location to which we are opposed, we will attempt to influence local zoning and legislative action so as to minimize the negative impact of developing the property.

Melaver will pursue third-party verification of all projects. Today, this means LEED, but the company will also consider other pre-vetted standards such as EarthCraft, etc. To prevent situational application of a standard, and to discourage the use of standards that have become outdated, diluted, or irrelevant, we will review our standards annually and identify which standards will apply in the coming year.

reminder to employees of the standards that are expected. Melaver's real estate guidelines are a part of every investor package we distribute. We want to ensure a clear alignment of expectation: our investment partners know what to expect of us; we know what our partners demand. If the institutional capital funding one of our projects has minimum return expectations, we obviously need to know this. Conversely, our investment partners need to expect that we are very willing to expend additional funds and devote additional design time in order to reduce the overall carbon emissions of our developments. Each of us needs to be very clear about the other's expectations and to understand how those expectations dovetail with one another. Clear articulation of expectations and alignment of those expectations is important. So too are guidelines that are not so abstract that they lack clarity of meaning and implementation.

For example, according to our guidelines, Melaver does not develop greenfield locations. And we are rather public about that statement. As a policy, this would appear clear enough. But what about a site that has already been disturbed, creating a significant stormwater runoff situation? Such a site exists near Savannah, where the state had cleared land to attract an automotive manufacturer to Georgia. Ultimately, the auto plant located elsewhere and Savannah is left with a wide scar on the landscape. Some within our company argue we should reclaim the land and develop it as a livable, transit-oriented neighborhood development. The land area is large enough to create a vibrant community. Others at our company argue that the land never should have been disturbed to begin with, that if we were to come in now, we'd send a clear message: "Melaver doesn't do greenfield unless you cut down the trees before you call us." Suddenly, the choice seems more difficult. Our internal debates over the application of general principles (such as saying "no" to greenfield) to specific projects is ongoing. It sharpens our sense of how much reflective thinking is required as part of an ethos of integration.

SEEK INDEPENDENT VERIFICATION

It's not enough to have a set of published standards, and self-monitor (and evaluate) the standards you put in place. For credibility, a business must have clearly articulated, independently quantifiable, independently verifiable standards. An excellent application of this methodology is the U. S. Green Building Council's LEED program. While the LEED program has proponents and detractors, it has created a standard by which success is defined, it has quantifiable components, and its components must be reviewed and verified by a third party.

Why is this an important hallmark of a sustainable business? The logic runs like this: because if we cannot articulate it, we cannot quantify it; if we cannot quantify it, we cannot measure it; if we cannot measure it, we cannot determine our ultimate success or failure, let alone articulate that success to others; and, without the ability to articulate our success, capital sourcing disappears and our business is out of business. This is hardly a sustainable outcome.

ALLOW PUBLIC SCRUTINY

Allowing public scrutiny doesn't mean getting the PR department in high gear and putting a shiny gloss on everything. And it doesn't mean the only way we can allow scrutiny is if we always have a good story to tell. So we'd better make sure we don't make too many mistakes, right? Wrong. A truly sustainable business is always breaking new ground and making mistakes. If we are not making mistakes, we are not extending ourselves into new paradigms of business conduct that may possibly redress some of the major crises facing our planet. We are not reaching for the kinds of accomplishments that allow a business to not only rise above the competition, but to realize its own highest potential. When we allow for public scrutiny, we must showcase our shortcomings as well as our achievements. This means openly discussing our failures.

For one thing, the sustainable movement needs an open airing and sharing not only about what has worked but what has not worked. We simply don't have the time to reverse the serious consequences of global warming if we are reticent about sharing what we are doing that just isn't measuring up. For another, a business that intends to sustain itself over the long term needs this critical scrutiny.

When people sometimes tell us, "Man, you guys have it right," I wonder if it's perhaps the reverse, that conventionally speaking, we have it all wrong. There is very little we do that would be considered typical of a successful businesses. We are opportunistic to a fault, using our small size to remain highly agile. We frequently jump from one property segment to another, developing cutting edge properties but seldom repeating a property model. While our executive management team has titles for external efficiency, actual roles shift on a regular basis within the CEO/COO/CFO triumvirate. We attend board and management meetings and openly disagree with one another. Where we see this as the only way to pursue the truth in any given situation, our board tells us that we don't appear to be on the same page. Consistently, by the norm, we do things wrong. Yet our financial performance is commensurate with other real estate developers and strong enough to source well-established capital partners.

It's our shared belief structure that makes us comfortable with public scrutiny. Though we're not afraid to admit mistakes and be judged for them, we hope we'll be judged for intent and authenticity as much as for the result. As a practical matter, we have to be right a whole lot more often than we are wrong, but cannot let fear of being wrong dissuade us from what be believe is the correct path. Nor can we be afraid of public scrutiny when we mess up. I think of it this way: if I am afraid to defend my decision, I shouldn't have made the decision to begin. If I am afraid to answer to public scrutiny, I didn't think things through well enough. Fault me for the outcome, but I should never be afraid to defend my intention.

Concluding Remarks

We have been discussing throughout this chapter how one makes a compelling case for green buildings, focusing attention on key users such as tenants, communities, and financial institutions. In doing so, we have touched upon the financial, social, and environmental aspects of sustainable practices. But what about the people within an organization who are charged with making that pitch, such as myself? Does it matter that we, too, realize the value of green?

I'd like to conclude this question by considering the value of green in a personal context. After all, the best businesses understand that there has to be a larger relevance to each person's job if we want the business to achieve truly great things. For me to do my job well, to make of my job what it truly is, a vocation, it is important to connect the values I believe my company offers to others with values I hold dear.

On February 1, 1968, I was a six-year-old boy living in Sydney, Australia, very happy with a wonderful family. That night, I went to bed a fundamentally different person, because during the course of the day, I'd found my four-year-old sister drowned in our swimming pool. We took Kathy to the hospital and by all accounts, she fought valiantly. Still, it was her time, and God called her home.

All of us have experienced loss. Some in the abrupt, dramatic way I lost my sister, some in more gradual, time-softened ways. It's striking to me the degree to which the powerful, significant events in our personal lives are typically excised from our professional lives, as if those critical events are somehow not meant to intrude upon our workaday existence. Think for a minute about your immediate reaction to the paragraph preceding this one: Did it seem jarring, a bit misplaced, perhaps too personal for a book on business practices? I think most people would say, "Yes, it's too personal." And that is precisely my point: It shouldn't be.

Sustainable practices, above all, call for the integration of all things: economic, social, and environmental aspects of business activities, a professional sense of providing added value to the various stakeholders in a business melded to a personal sense of finding meaning in what one does. It literally makes no sense to be one person at work and another person at home. Just as it makes no sense for a business to focus solely on its economic bottom line without integrating social and environmental performance into the mix.

Losing Kathy was a life-defining moment for me. We all have such moments. What do we do with them? How do we integrate them into all that we do in ways that are deeply authentic, particularly in our culture, which seems structured to maintain silos among the disparate parts of our lives?

For me, such integration starts with a few basic beliefs:

1 Everything we love can be taken away in a heartbeat.
2 We don't have time to be lost in anger and bitterness and blame. We only have time to think, to feel, to do.
3 In those rare circumstances when we have the chance to get back that which we love and was taken, we must fight like hell to get it back.

These three personal touchstones imbue every aspect of my work, day-in and day-out. I connect them to the presentations I make to tenant prospects and to civic governments like the City of Birmingham, and to potential institutional investors. I connect them to the sense I have of a planet deeply degraded, and of time running out in which to reverse that degradation. I suspect that if we enable ourselves to let down our guard and jettison our ill-conceived notions of professional decorum, most of us would have a similar sense of things larger than ourselves that are worth struggling for.

Sustainability is more than just a "green" thing. It's really about understanding our role as stewards of the future. Seeing the world this way allows us to grasp the idea that our environmental vision, our social goals, and our financial aspirations really are intertwined. Seeing the world this way speaks to our values, and it's in our congruent values that we find the fabric of a sustainable community. Everything we love can be taken away—it is already happening, every day. We don't have time to be lost in anger and blame—we need a positive, practical vision for how to correct the situation. And we need to fight like hell to do so. It is from this standpoint that I personally feel sustainability has a quiet fourth bottom line, a bottom line having to do with spirituality. Sustainability, deeply felt and practiced, is about locating oneself in a larger, largely selfless endeavor having to do with re-establishing, reinforcing, nurturing, and protecting the fabric of community and the habitat that gives that community its sustenance. Sustainability is about the deep practice of respect for all things. That is the legacy I want to leave my kids, as we leave the movie theater in Birmingham after watching *Cars*.

Chief Seattle of the Suquamish Indians reportedly wrote to the United States government in the 1800s: "This we know: the earth does not belong to man, man belongs to the earth." This view is as revolutionary today as the Copernican reworking of planetary movements was in 16th-century Europe, placing humankind not at the center of its natural home, but at its service. It's an ethos that may indeed provide for dollars, but of necessity it makes common—and communal—sense.

NOTES

[1] The base used to establish the expense stop was that of the most directly comparable property, in both location and size.

[2] Assumes a similar usage pattern to other office buildings in this submarket.

[3] The base used to establish the expense stop was that of the most directly comparable property, in both location and size.

[4] Bob Willard, *The Sustainability Advantage: Seven Business Case Studies of a Triple Bottom Line* (Gabriola Island, B.C.: New Society Publishers, 2002), p. 34. The calculation for losing and replacing a good employee is actually estimated at an astounding two to three times the person's salary.

[5] Joseph J. Romm and William D. Browning, "Greening the Building and the Bottom Line: Increasing Productivity Through Energy-Efficient Design" Snowmass, Colo.: Rocky Mountain Institute, December 1994, Revised 1998), p. 10.

[6] These statistics are cited in a number of studies, with some indicating improvements as high as 33 percent. Specifically, "Greening America's Schools—Costs and Benefits," by Gregory Kats, *Capital E*, October 2006, is an excellent source of detailed information regarding productivity and the health benefits of green building.

[7] Bama Athreya, Executive Director, International Labor Rights Forum. Used with permission.

[8] Source: www.scorecard.org. Visiting this website allows you to examine the environmental scorecard for any county in the United States and overlay similar social measures.

[9] To view how we attempt to answer these questions, visit www.rethinkrebuildrenew.com.

[10] David W. Orr, *Earth in Mind: On Education, Environment, and the Human Prospect* (Washington, D.C.: Island Press, 2004) pp. 160–161.

SUSTAINABLE BROKERAGE:

DIFFUSING GREEN PRACTICES TO

GAIN BROAD MARKET ACCEPTANCE

RHETT MOUCHET AND CLARA FISHEL

SUMMARY

The issue of sustainable brokerage moves the whole discussion of a green building bottom line beyond the confines of one company doing sustainable work to the general marketplace. Melaver | Mouchet, the brokerage division of Melaver, Inc., devotes 95 percent of its activity to third-party commercial representation, which means most of the work it does today involves brokering deals that are not green. The market is still too young to enable brokers to make a livelihood solely brokering green developments. As such, this chapter is as much aspirational—an indication of where green brokerage is headed in the future—as it is a recounting of actual case studies, which are still few in number.

However, Rhett Mouchet, Broker in Charge at Melaver | Mouchet, and Clara Fishel, Sustainability Associate for this division, contend that brokers can play a pivotal role in ramping up the diffusion of green practices. Sustainable brokerage embodies the awareness that a broker's work affects the place where he or she lives and does business, and acknowledges that brokerage can't help but be IMBY (in my back yard). The close professional and personal relationships that are often forged between brokers and clients in the course of a real estate transaction provide brokers with unique opportunities to educate their clients, promote green building, and encourage focus on a triple bottom line.

Mouchet begins the chapter with a personal story and then discusses the brokerage group's evolution. In section two, Mouchet and Fishel compare and contrast "traditional brokerage"—brokerage as it has been practiced and is widely practiced today, emphasizing productivity above all and a single financial bottom line—with what the company calls "sustainable brokerage"—brokerage that facilitates real estate choices that are good for business, good for the community in which a broker plies his or her trade, and good for the environment. Sustainable brokerage, no different from sustainability in general, is shaped by a triple bottom line that acknowledges the broker's appropriate involvement in community planning and environmental stewardship. In section three, Mouchet and Fishel delve into the business case for sustainable brokerage, a process that involves re-framing the notion of brokerage, empowering brokers to play an active role in shaping the types of development that occur within a community, and learning to serve as change agents. Section four entails a brief consideration of competitive advantage, since sustainable brokerage is something of a new-to-the-world business proposition. Section five looks at the actual business case for sustainable brokerage by considering the additional resources it costs to be sustainable as well as the benefits that accrue. The chapter concludes as it began, with a personal story from Mouchet.

The Backdrop: A Personal Story of Sustainable Brokerage

I won't lie to you. Despite the fact that I'm head of Melaver, Inc.'s brokerage division, it has taken me a long time to embrace this whole sustainability thing. It's not that I don't think we need to be better stewards of our land—I do. In fact, for years, I chaired the board of a local land trust organization devoted to creating permanent easements on critical habitats in and around Savannah. I'm an outdoorsy guy in general. I kayak in the coastal estuaries near my home, and I like to hike and backpack. But it's taken me a long time to really understand and implement the business case for sustainable brokerage. I've been a broker for almost my entire adult life, nearly forty years now, and the business model all those years has pretty much remained the same.

I got my start in real estate by taking a real estate course at Savannah Technical College while teaching in the local public school system. I was so enamored with the course that I resigned my teaching job, discontinued my graduate studies, and went into real estate full time. During my early years, I specialized in the sale of historic buildings, including the Manger Hotel, one of the first major office renovations in Savannah's historic district. In the late 1980s I brokered my first large scale land assemblage deal—a 250-acre office park. The asset manager on the deal, who had a proven track record in developing successful office parks, believed that by preserving green space and creating an attractive and pleasant workplace, the value of the asset would gain more than the typical sprawling developments of the day. No one was calling it sustainable development back then. No one thought of it as what scientist and author E. O. Wilson later termed biophilia, humankind's natural affinity for the natural world.[1]

But that asset manager was right, and his message stuck with me over the years. The message was one I grew up with, really. My grandfather, who was a farmer in the small community of Starr, South Carolina, preserved old growth forests, used land conservation practices, and was an active outdoor adventurer. He and my grandmother, who took me on long walks and taught me to fish, raised eight children and still found time to be active community players, participating in their church and on a number of civic boards and organizations. My grandfather was self-educated and determined that his children would all get a college education—and they all did. As a school board member, my grandfather was so dissatisfied with the quality of the minority schools that he funded a better school for African-American kids, who weren't allowed into the white schools in the late 1940s. I feel I inherited my sense of what's right from my grandparents and my Uncle John Rhett, who carried on my grandfather's community service traditions and love for the outdoors.

I have operated as a professional commercial real estate broker in and around the Savannah area for over three decades. And I'm as typical a broker as you will ever see. I'm friendly and outgoing. I love the independent nature of brokerage work, love the fact that you set your own hours and live or die financially and professionally not by coming in and punching a time clock but by making things happen. I'm definitely most at home when I'm out on my own, setting my own course of action, determining my own destiny.

But I'm also deeply plugged into the community, into deals and potential deals and deals that will carry over from initial assignments. I've run my own brokerage company for a number of years, overseen a local office of a big regional brokerage operation, and been involved in a number of large land assemblages. I do civic involvement not just for the business contacts, but because I believe in giving back. I am the past president and have been a Million Dollar Club Member of the Savannah Board of Realtors, have served as a director and local president of the National Association of Realtors, and am often asked to provide an overview of the real estate market at the Chamber of Commerce's annual meeting.

Maybe it seems a bit corny and old-fashioned these days, but I've always been troubled by the notion of a broker being closely associated with a slick salesmanship, get-in-get-out-make-the-deal-move-on mentality. Laugh, but I view my profession as a noble one. I have always tried my best to incorporate the highest ethical standards, environmental stewardship, and community service into my business practices and professional service. I also take my faith in God very seriously and believe He tells us to be good stewards of the land. I believe that God commands us to respect others and treat everyone honestly and fairly. Personally, I have found the community service aspects of my business to be as rewarding, if not more rewarding, than the money I've made.

In 2003, my friend Martin Melaver and I formed Mouchet & Associates as a subsidiary of Melaver, Inc. The company name was changed to Melaver | Mouchet in 2005. We formed the brokerage division of Melaver, Inc. to handle third-party brokerage opportunities and properties within the Melaver, Inc. real estate portfolio. Melaver, Inc.'s tradition of community service fit with my ideas about how real estate in general, and brokerage in particular, ought to be done. Two aspects of Melaver, Inc.'s triple bottom line—the financial and the social—fit well with my own sense of how brokerage should operate. The rub, however, was that third aspect, the environmental one.

For a number of years, our brokerage group was the laggard division in the company, resisting learning about LEED and green practices even as the rest of the company was taking leadership roles in this regard. We just couldn't get our hands around marketing a product that hardly existed in our area. We could see, if not feel viscerally, the connection between giving back to the community and being better stewards of the environment. But it was more an academic, intellectual connection than something we could connect with personally and emotionally. And how about the connection between the environmental bottom line and the financial one? How were brokers going to make money promoting sustainable development? The vast majority of our business was not, and still is not, brokering sustainable projects. My colleagues and I would starve if it were. Nevertheless, over the years, my brokerage colleagues and I have come to recognize a few key things:

- **The need to re-frame the notion of brokerage.** The currency of value held by brokers is information and our wealth of business and social relationships. The information we share with our social network of clients and how we educate them will shape the way development occurs in our region.
- **Entitlement.** We need to recognize that, in many ways, we are responsible for shaping the way our community will look (or not look) in the coming years, simply by the deals we facilitate.
- **Empowerment.** Despite having no apparent authority or power—we are, after all, simply intermediaries between buyers and sellers, landlords and tenants—we do have a significant amount of leverage in shaping where and in what ways development occurs in our own backyard.
- **The capacity to be change agents.** Our reach into development activity is far broader and deeper than that of our colleagues who do development work for Melaver, Inc. For every single green project they do, we literally touch one hundred

other deals that could potentially be green. The pace of diffusion of green practices, the speed with which green gains market acceptance, can to a large degree be affected by the work of brokers.

It has taken our brokerage division a long time to arrive at these fundamentals. In many ways, we are still not there, but we are taking steps to re-define brokerage. Sustainable brokerage plays a key role in the green movement that is rapidly becoming mainstream in this country. In the following sections we discuss how our business model for sustainable brokerage works, how it differs from traditional brokerage practices, how it has the capacity to be an agent for change toward green building practices, and, finally, how it actually can be profitable.

Defining Sustainable Brokerage

To talk about how sustainable brokerage differs from traditional or conventional brokerage, let's remind ourselves what traditional brokerage is about. First and foremost, traditional brokerage is about getting the deal done within the ethical and legal framework of the real estate profession. Often that is a tall order, in and of itself. Traditional brokerage is most often focused on a single bottom line—making the commission. It rewards the highest producers—the ones who sell the highest dollar volume and secure the largest commissions. Future success (high dollars, high volume) is predicated on similar past success. A quick survey of real estate business cards tells you this is so. Production is the top priority. Success is measured in dollars. Traditional real estate companies stress the importance of their agents and brokers belonging to professional organizations, a source of individual status and networking potential.

Community service, or whether the particular deal serves the interests of the community, is generally not a priority. Traditional brokerage serves the interests of brokers and brokerage clients, but doesn't typically consider the greater consequences a deal might have once it is accomplished. Traditional brokerage doesn't ask questions like: How might the development of this site impact the surrounding natural environment? How might the development of this site create good jobs or provide needed services for area residents? How much energy, water, and materials will the buildings that are built here consume? How much carbon dioxide will those buildings emit? What would it be like to work in this building? Could you ride public transportation or walk to this building? How will this transaction impact the community in the future?

Sustainable brokerage asks all of these questions and then some. Sustainable brokerage looks beyond the immediate impact of a deal (closing the sale, satisfying the client, making the commission) and considers the deal in a triple bottom line context. We work with our clients to create deals that are good for business, good for the communities where properties are located, and good for the natural environment.

Figure 8.1   A print ad for Melaver, Inc.'s brokerage division.

Sustainable brokerage also embraces a much wider concept of service, serving the aims and goals of the client in the wider context of the community.

Our communities are growing. How and where our clients decide to develop their land, construct their buildings, and build out the spaces they lease will have broad and lasting impacts. So we understand that the work we do as real estate professionals has consequences for the neighborhoods where we live and raise our families and for the communities where we—and our clients—do business. And as people involved in the communities where we live and play and work, and as professionals whose livelihood depends on community viability and repeat business, we're both entitled and obligated to do more—much more—than simply facilitate deals.

First and foremost, sustainable brokerage fosters a personal vision of community. It is about becoming educated and educating others as to the bits and bytes of knowledge we need to acquire to implement this vision. It is about facilitating connections among those players in the community—developers, owners, designers, government workers, academicians, folks from the non-profit world—who are most empathetic to furthering our vision of community. And it is about orchestrating the various parts so that vision becomes a reality.

The Process of Sustainable Brokerage

The business case for sustainable brokerage, like most anything related to sustainability generally, is a process. It is bottom-up driven, starting from within a brokerage organization and radiating outward into the larger real estate community and beyond. It is integrative and collaborative, involving brokers' close working relationships with developers, the financial community, and key end-users. It is educational and iterative, with each step in the process repeating and learning from and refining prior, more naïve earlier steps. And it is non-linear, characterized by two steps forward and one step back, since sustainable brokerage is still in its infancy and advancing a new paradigm while still, for the moment, engaging in the old paradigm. The business case for sustainable brokerage involves these steps:

1 Re-framing the notion of brokerage.
2 Taking ownership of and responsibility for the development deals brokered in one's community.
3 Becoming educated internally about sustainable practices.
4 Designating an internal champion to serve as the point person for galvanizing a sustainable ethos, disseminating information internally, and assisting with sustainable business development.
5 Developing a toolbox of information with materials, technologies, and case studies that will assist brokers with their day-to-day work.

6 Leveraging knowledge and social capital.

7 Creating a set of financial incentives that rewards brokers for brokering sustainable deals.

8 Marketing sustainable brokerage primarily through a category build, using outreach to share best practices with the brokerage community generally.

Let's consider each step in greater detail.

RE-FRAMING THE NOTION OF BROKERAGE

Climate scientist Joseph Romm, in his book *Cool Companies*, recounts a story about Pierre Wack, one of the executives in charge of Royal Dutch/Shell's strategic planning process in the early 1970s. Wack and his group started a process of visionary scenario planning. They not only anticipated the global energy crisis that subsequently ensued, but also planned the re-shaping of Royal Dutch/Shell's organization to make it more responsive to what they rightly forecast as a need to decentralize decision making to address different market conditions in different parts of the world. The work Wack put in motion enabled the company, which was the weakest of the seven largest oil companies in 1970, to become one of the two strongest within ten years. Impressive, certainly. But what does this have to do with sustainable brokerage?

Wack and his group faced a classic problem of change management within an organization. His group might have had a clear sense of the alternative market scenarios looming on the horizon, but how were they to disseminate what they knew was coming in a way that would stick? How could they keep their scenario planning from being viewed (and hence dismissed) through old paradigms or mental models of how the oil business then operated? Wack's reflections on this challenge are telling:

> I cannot overemphasize this point: unless the corporate microcosm changes, managerial behavior will not change; the internal compass must be recalibrated....
>
>   Our real target was the microcosms of our decision makers: unless we influenced the mental image, the picture of reality held by critical decision makers, our scenarios would be like water on stone....[2]

Wack's strategy was simple and effective. He and his planning group presented key decision makers with a business-as-usual scenario. They also pointed out the underlying assumptions of that business-as-usual scenario. It was clear to all: those underlying assumptions were simply untenable, depending on several "miracles" to occur simultaneously. As Romm notes, "once managers saw that their faith in the status quo was built on miracles, they were more receptive to new thinking."[3]

Pierre Wack was functioning as a broker in the most noble of ways—he was brokering knowledge, conveying his vision of the future. But to make that knowledge stick, he needed to first remove the old mental frame that would impede understanding of his group's scenario planning and then provide a new frame. Sustainable brokerage begins with just this type of re-framing. But re-framing brokerage, as we will see, is slightly different.

Some years back, Martin Melaver and other colleagues at Melaver, Inc. tried to make the case to our brokerage group that the conventional ways of doing real estate were simply indefensible: Land was (and continues to be) swallowed up for development at unprecedented rates, causing ever-greater sprawl and longer commutes (more time on the road and less time with family), and adding to ever-greater concentrations of carbon dioxide in the atmosphere. Water resources, even in the fortunate area of Savannah (which draws upon the large Upper Floridan aquifer), were being tapped beyond carrying capacity and were being compromised by saltwater intrusion. Buildings were obscenely inefficient in their use of energy. And so on. The argument, logically speaking, is compelling: doing real estate in conventional ways is simply untenable. But so, too, was this re-framing. To whom would brokers pitch these ideas? And why would anyone want to listen? What could we gain by pitching an idea that was a fundamental critique of what our potential clients practiced? Customers or clients may not always be right, but telling them point blank that everything they did was idiotic and made no sense was hardly going to help us drum up business. Ideology does not sell real estate. We were getting nowhere.

Re-framing would require, instead, contextualizing the issue in terms of market forces as well as personalizing it. For years, our colleagues doing development work at Melaver, Inc. were telling us waves of LEED product were coming. All we could see, at least initially, were the early pilot buildings our own company was constructing. And, as the biblical saying goes, you can't be a prophet in your own hometown. But then things began to change. Grubb & Ellis, one of the largest real estate firms in the world, approached Scott Doksansky, who heads up asset management for Melaver, Inc., to run a day-long seminar on sustainable management practices for brokers and property managers across the country. That got our attention. So too did an SIOR (Society of Industrial and Office Realtors) conference where the keynote speaker focused on green practices as the new standard by which all real estate would be conducted. Seeing other familiar and very middle-of-the-road trade organizations—NAIOP (National Association of Industrial and Office Properties) and ICSC (International Council of Shopping Centers) and BOMA (Building Owners and Managers Association) and ULI (Urban Land Institute)—devote conferences and articles to green building practices also made a profound impact. Attending USGBC Greenbuild conferences in 2006 (in Denver) and 2007 (in Chicago), with 20,000-plus attendees and an extensive exhibit hall, seemed worlds away from the almost quaint first Greenbuild conference in Austin (in 2002), which attracted only 3,000 conferees.

We were beginning to get the message. It really hit home when a large, publicly traded industrial real estate investment trust (REIT), AMB Properties, announced it was building 3.3 million square feet of warehousing to LEED standards in our home market. And our main brokerage competition in town had gotten the nod to do the leasing for this development. It was time to recognize not only that market conditions were changing fast, but also that our brokerage division had two choices: play a lead role in promoting LEED development in our area or lag behind as others took charge.

Why Broker Green Development Practices?

Market forces helped answer part of the question and provided the financial context for questioning the paradigm within which brokerage has worked for decades. Maybe we wouldn't lose business by promoting a new way of doing real estate. Maybe we would lose business by *not* promoting a new paradigm. Market forces, in many ways, served as the catalyst. Personalizing the issue helped us with the actual re-framing.

I can't help it. I'm the type of person who wears my heart on my sleeve. People say they can read me like a book, and they are probably right. But it's a curious thing about business—we feel that in order to be professional, we need to keep emotions and feelings at bay. Yet some of the most successful and innovative businesses and business models of our time have been fueled by passion and feeling.

Ray Anderson, as a result of reading Paul Hawken's book, *The Ecology of Commerce*, concluded that the petrochemical-intensive floor tile company he founded, Interface, made him a plunderer of the planet and a thief, stealing the futures of generations to come.[4] Hawken himself, who got his start running a natural foods store in Boston and then founded the high-end mail-order gardening business Smith & Hawken, notes that at the very moment he was being awarded the Environmental Stewardship Award by the Council on Economic Priorities, his company's practices were antithetical to sustainability.[5] Yvon Chouinard, founder of Patagonia, who likes to refer to himself as a reluctant businessman, describes his company's purpose as one that challenges conventional wisdom and "presents a new style of responsible business."[6] David Gottfried, recognized as the founder of the U.S. Green Building Council, argues that it should be the mission of environmental groups such as the USGBC "to change how we make money and define financial value."[7] Jeffrey Hollender, founder of Seventh Generation, believes that a business should be about "spiritual and moral and personal growth."[8] Gary Hirshberg, co-founder of Stonyfield Farms, frames his entire story of running his company in terms of one simple, profound mission: "The truth is that we've all got a limited time in which to do something useful on this Earth."[9] And so on.

Personalizing the issue—whether it's selling carpet tiles or outdoor sportswear or organic yogurt or real estate—is a fundamental aspect of a sustainable business. In terms of brokerage work, personalizing the issue means coming to terms with the fact that when we brokers are stuck in traffic in the morning coming to work, we have at least partial ownership of the problem by having brokered deals that foster unsustainable transportation issues. Or that when we see our well-planned, historic 18th-century town diminished by franchise blight, we are in part responsible for our unique town beginning to resemble a geography of nowhere.

Sustainable brokerage starts with caring about what happens in our own backyard: protecting our marshes and local fisheries, keeping nearby tracts of land from being clear-cut, preserving buildings (both remarkable and ordinary) that are a part of community lore and everyday use. Sustainable brokerage recognizes a sense of authentic community—not something that looks like a Hollywood backlot—and an appreciation for the blood, sweat, toil, and tears (not to mention time) it takes to make any-place our place.

Each of us in the brokerage division has a slightly different sense of the type of community that speaks to us, the kind of work we each would particularly like to foster. Some of us tend toward an agrarian setting, a cluster of village houses set along-side untouched open fields. Others prefer a more urban, in-fill environment, one that is intimate and seamlessly melds elements of work, live, and play. Some of us have a preference for a heterogeneous coastal village community, while others see them-selves as part of a mixed-use, mixed income, diverse community. Despite our indi-vidual preferences, however, we all gravitate toward phrases like "timeless design," "preservation," "authentic," and "community-oriented."

To make the professional work of brokerage both personal and close to home, we have begun to ask our brokers what type of community they would like to foster in their own back yard. That, in a nutshell, is the critical re-framing of brokerage work.

TAKING OWNERSHIP AND RESPONSIBILITY

Re-framing brokerage by personalizing its context goes hand-in-hand with taking direct ownership of the development deals one brokers. By definition, a broker is a person who arranges contracts for a fee or commission. A broker can make connec-tions, act as a go-between, negotiate, mediate, deliver messages, or represent. As real estate brokers, we are the people in the middle of deals, but it's hardly neutral terri-tory. We are far from unaffected by what transpires.

Real estate brokers have multiple points of entry into the building process, from the sale of undeveloped land to assembling parcels for development to facilitating the sale of buildings and leasing. We have the capacity, if we so choose, to seek out particular developers and types of development that fit with the sense of community we would like to foster. We also have the capacity—though we haven't exercised it all that often—to say "no" when it comes to representing certain developers and types of development that do not fit with our sustainable ethos.

At each point of entry into development deals, we also have the opportunity—if we choose to take it—to educate and advocate for green building and sustainability. (We'll discuss this in greater detail later in this chapter.) As sustainable brokers, we can also help our clients to both improve their financial bottom lines and have posi-tive social and environmental impacts. Sustainable brokers, because of this triple bot-tom line focus, are aware that the ways clients develop their land, construct their buildings, and build out the spaces they lease can have lasting consequences that affect, for better or for worse, the communities where we live and work. This aware-ness brings with it the recognition that what we do affects our own back yard—a place we might variously define as the block where we live, our neighborhood, our town, our county, our watershed, our state, our country, or our planet. It is for our own sakes and our families' and our clients' that we are entitled, even obligated, to do more than simply facilitate deals and collect commissions.

Most brokers probably don't view themselves as entitled (or empowered) to seek out particular clients and exclude others. But developers certainly feel free to select the locale and locations where they wish to build. And financial institutions decide whether to invest or not invest in various projects. Tenants make decisions every day about where (and where not) to locate. It seems odd even to pose the question, *why shouldn't brokers feel a similar sense of entitlement?* They don't have to be simple order-takers. In fact, they should not be.

BECOMING EDUCATED ABOUT SUSTAINABLE PRACTICES

Everyone in our brokerage division has taken a LEED preparatory workshop and has taken the LEED exam at least once. Half of our group is LEED accredited, which is sub-par for the Melaver organization as a whole, but still not bad when compared to other brokerage companies. But getting conversant with the LEED program is simply a baseline, a starting point. Education, at least for us, consists of three interrelated components: research, internal information sharing, and inter-departmental communication.

Quality information is as critical (perhaps more critical) in the commercial real estate industry as it is in any industry. Like traditional brokers, we are dedicated to staying informed of events and trends in our marketplace. We track demographic information, transactions, and economic growth rates; we locate comps and conduct market surveys just like everyone else. The sustainable brokerage difference is that we infuse all of this traditional information with the latest sustainable development news, trends, technologies, products, and processes. We conduct research on everything from financing tactics to who, what, where, and how different companies are developing high-performance green buildings. We've found cost-benefit analyses and type-specific studies (office, industrial, retail) to be particularly informative. Case studies, articles from print and online trade publications, and whatever sustainable development databases we can get our hands on are organized and compiled in a central, accessible location.

Sharing and discussing this information throughout the brokerage group is necessary for us to really learn from the flood of information, to let it sink in, to make sense. As a group, we share sustainable development news and information at our weekly staff meetings and monthly "Lunch & Learns" as well as through e-mails and informal discussions. The internal information sharing of sustainable brokerage is antithetical to the protective and competitive nature of traditional brokerage. Traditional brokers often treat information as proprietary, as a secret advantage they must guard closely and keep to themselves. As sustainable brokers, we figure that the more we talk to each other and share information, the more we all learn. And we believe that learning and a lack of internal competitiveness actually bolster our competitive advantage in the marketplace.

Our learning and our competitive advantage also are enhanced when we communicate and go on client calls with members of the other departments in the Melaver, Inc. organization. On any given call, one of our brokers might be providing office market expertise while simultaneously learning about a new product from the construction team, or a new source of green power from our in-house sustainability consultants, or a new process for stakeholder involvement from the development group. We try to find all possible opportunities for taking advantage of the depth and breadth of sustainable development knowledge held in our company.

Additionally, gaining on-the-ground exposure to LEED projects throughout the development process gives us a practical knowledge of the challenges, strategies, and innovations involved in green development and building. Inter-departmental communication rounds out our education and makes us better at what we do. As a sustainable brokerage group, we cannot only explain to a potential client what is entailed in developing a green project, we know the process firsthand and have our own performance metrics to provide as examples of what building strategies make sense (and which do not).

DESIGNATING AN INTERNAL CHAMPION

For a long time—way too long, as it turned out—we expected our brokers to stay informed about developments in green development (including building case studies as well as advances in materials and technologies), participate in our company's line-up of speakers, educate others in the community generally, and in the real estate profession specifically, about green practices, and advocate for sustainable legislation *on top of* their regular responsibilities representing clients. Forget it. Brokers—almost without exception—have an independent agency relationship with their parent company, which means that time devoted to deep, ongoing research into green practices, outreach, and advocacy is time taken away from earning a living. Maybe someday the market will change and, with it, the capacity of brokers to devote financially productive time to these efforts. But that's not the case at the present time.

Every brokerage division or company is organized differently, and so we hesitate to draw a line in the sand and say that every such business should have a full-time, salaried champion on staff to choreograph sustainable brokerage. There may indeed be companies out there where a head broker or agent is able to play this role, perhaps in addition to his or her ongoing daily brokerage responsibilities (which could easily be the case in a values-centric organization where values might trump financial productivity). Those situations notwithstanding, we would strongly recommend having someone within a brokerage division or company whose primary responsibility is to integrate the individual, professional skills of the brokerage team with the sustainable goals of the entire group.

The key responsibilities of a sustainability associate (or sustainability champion) are likely to include:

■ **Research.** Serve as the primary source for collecting information on green business case studies, materials, technologies, regulations, and policies;

- **Internal Communication and Education.** Serve as the primary source for disseminating information on green business case studies, materials, and technologies;
- **External Communication and Education.** Develop a set of toolboxes for distinct real estate product types (office, retail, industrial, etc.) that enable brokers to educate clients and/or potential clients on sustainable practices;
- **Education/Advocacy/Business Development.** Serve as the primary face of the company or division for pitching sustainable services to potential clients;
- **Business Integration/Business Development.** Serve as the primary link integrating brokerage services with other divisions of a real estate company (development, construction, property management, consulting, etc.) *or* as the primary link between a brokerage company and other companies that provide these services; and
- **Outreach and Advocacy.** Serve as one of numerous individuals within the organization who promotes sustainable practices through speaking engagements; involvement in civic task forces related to planning, promotion of sustainable policies and legislation, etc.

This list is not exhaustive, but it does provide a general flavor for the broad range of responsibilities that need attention. Someone needs to keep abreast of the almost daily changing landscape of green development. Someone needs to disseminate that information to others within the organization as well as provide a process for leveraging that information for business development. And so on. No wonder our original expectation of having the brokers themselves take responsibility for these tasks fell flat. It's simply a full-time job.

DEVELOPING A TOOLBOX OF INFORMATION

Think about it for a moment from a broker's perspective. You have a number of listings that you are diligent about servicing. Perhaps you're a tenant rep broker searching for locations in a broad geographical area. Meanwhile, you are working on adding some listings, partly through working a known database of potential clients, partly via cold calling. Deals in motion, potential deal listings you'd like to nail down, deal flow to feed the pipeline down the road. And now, in the midst of all of this, you are thinking: *How do I do all of this and pitch developing green to my clients and potential clients?* Clients will want to know: What does sustainability mean? What does it do for me? Can you prove these claims? Where else has this worked successfully? Do I need to change how I operate or how my brand looks and feels in order to be green? Will it cost more? A broker needs a toolbox of case studies and metrics that will ease a client's natural disinclination for doing business differently than he or she has done in the past.

Commercial real estate trade organizations such as Society of Industrial and Office Realtors (SIOR), Certified Commercial Investment Managers (CCIM), and Realtors Commercial Alliance (RCA) are realizing it is a matter of *when,* not *if,* sustainable development will become the industry standard. Education is critical. What we call our "sustainable brokerage toolbox" combines what we've learned from our three internal educational components: research, internal information sharing, and inter-departmental communication, and serves as the foundation for our outreach and client services. The toolbox resources allow us, as brokers, to be advocates for sustainable development in our professional associations as well as in the community at large. Our toolbox also enhances and broadens our already rich scope of client information and services.

Typically, when we broach the subject of sustainability with clients and potential clients, the first questions they ask are: *What is it? What will it do for me?* and *How much does it cost?* Our answers must be succinct, relevant, and helpful. With this in mind, the first item in our toolbox is a set of talking points that provide a brief yet convincing business case for high performance green building. Those talking points include:

1 An overview of the LEED rating system.
2 A list of the major REITs investing in and corporations building to LEED standards.
3 Risk and reward categories.
4 A sampling of cost and benefit data.

These talking points are intended to provide specific information that captures attention and gets a larger conversation started.

A number of our clients are skeptical of businesses that claim to be "green" or "sustainable." They feel like anyone could claim it and market a product as such without really having to back it up. We see this happening at both the local and national levels, and we think that as green becomes more mainstream, skepticism is warranted. Greenwashing and unsubstantiated claims of sustainable development hurt consumers as well as developers of truly sustainable projects and make it difficult to figure out what is what and who is really doing what they say they are doing. We view this confusion and skepticism as an excellent opportunity to dig into our toolbox and provide our clients and potential clients with information on LEED. LEED standardizes sustainable development process, practice, and performance criteria, and awards points for meaningful accomplishments over a broad set of criteria. The standardization aspect and broad scope, plus the fact that LEED is the most widely accepted and utilized high-performance green building rating system, appeal to skeptics because they see how LEED certification means something significantly more than claims of greenness, especially because certification is by an independent third party.

To be honest, our clients are primarily interested in the economic and social aspects of sustainable development. The environmental aspects are, for most, an added bonus rather than the main reason to change the way they do things. When we tell people BOMA International director Brenna S. Walraven predicted a competitive disadvantage in five years or less, maybe in as little as two years, for buildings that are not green and not operating efficiently,[10] people listen. When we tell them that everyone from Adobe Systems to Wal-Mart is building to LEED standards, they listen. They also listen when we talk about Genzyme's twelve-story, 350,000-square-foot LEED Platinum corporate headquarters in Cambridge, Massachusetts that uses 42 percent less energy and 34 percent less water than a comparable conventional building, or that after Toyota's customer services unit moved into a LEED Gold building, absenteeism fell by 14 percent. And when we cite the 2006 *McGraw-Hill Green Building SmartMarket Report*, which found that green buildings deliver 3.5 percent higher occupancy rates, 3 percent higher rental rates, a 7.5 percent average increase in building values, and a 6.6 percent higher ROI, they hear us loud and clear. These numbers from our toolbox often surprise the people we talk to. It is difficult for them to argue against the threat of becoming obsolete, or to argue against more efficient resource use (which means lower operating costs), healthier buildings (which mean higher productivity), and better investments.

Assumptions regarding the cost premium associated with building out spaces or developing land or buildings sustainably are typical. Our toolbox also includes information from the many studies that dispel these myths. Examples of basic talking points include:

■ In 2003, Gregory Kats, of Capital E energy consultants, released a study showing that the average construction premium for a sample of thirty-three LEED buildings across the country was 1.84 percent;[11]

■ In 2004, the U.S. General Services Administration (the agency that builds or leases millions of square feet for federal offices, courthouses, and special facilities) reported that the anticipated construction premium for new federal courthouses would range from a *negative* 0.4 percent for a "low-cost" LEED certified facility to a high of 8.1 percent for a "high cost" LEED Gold facility;[12]

■ In 2005, Turner Construction's Market Barometer study found that the average estimated cost premium for sustainable building is only 0.8 percent for a basic LEED certification;[13]

■ In 2006, real estate consultant Davis Langdon compared the cost of eighty-three buildings seeking LEED certification against one hundred and thirty-eight conventional buildings. The analysis concluded, "the cost per square foot for buildings seeking LEED certification falls into the existing range of costs for buildings of similar program type;"[14] and

■ In 2007, PNC Financial Group began a major green bank branch construction program. Their LEED certified branches cost PNC $100,000 less to build and take forty-five days less to construct than comparable conventional bank branches.[15]

Still, we do not convince everyone. But for those who want to learn more, we use the toolbox like a library, a library that acquires new materials on a daily basis. Information is sorted by product type: office, industrial, land, retail (with restaurant, hotel, and bank subsets), mixed-use, existing buildings, schools, healthcare, and residential. In addition to product types, we have created separate categories for LEED Information, General Business Case, Cost-Benefit Analysis, Financing, Investment, Policy, Green Development Specs and Guidelines, and Performance Evaluations, plus a case study database. Our library is currently expanding to include a comprehensive product/technology database. We are developing this database, comprised of field-tested products and technologies, with the help and expertise of our construction and consulting departments. Information directly from Melaver, Inc. projects about such aspects of green building as performance, operational costs, construction costs, design, or process innovations is particularly valued.

We use this library to compile information for clients and potential clients, depending on the scope or type of their project and the specific questions they need to have addressed. For example, a recent potential client who was interested in hiring us to list an office building that hadn't sold in the two years it had been on the market asked us about green renovations that might stir some interest. In addition to sending information on the costs and benefits of renovating office buildings this way, we also suggested (since there was going to be an investment in landscaping anyway) that the client invest in drought-tolerant and native plants and a more water-efficient irrigation system that incorporates rainwater capture. This information made us stand out from the other brokerage firms that were vying for the listing and gave the client a strategy to differentiate its building. We got the listing.

The final, and perhaps most important, component of our toolbox is the sustainable development services provided by the other departments of the Melaver, Inc. organization. We brought our consulting and development departments together with a bank client (for whom we are providing site selection services) to discuss the possibility of developing its flagship regional branch to LEED standards. In the site selection process, we are, of course, cognizant of meeting client needs. For example, this client wants to be where the most rooftops and growth, and highest incomes and traffic counts, are. But we are also aware that, in our region, some of those sites are exurban areas with no access to public transportation and high percentages of coastal wetlands. Serving clients is our topmost priority, so we aim to help them find sites that meet their business needs while mitigating these factors. And because we are interested in growing long-term relationships, we brought in the expertise from our other departments (our living, breathing library) to discuss the bank's flagship branch, which would be an urban infill project.

LEVERAGING KNOWLEDGE AND SOCIAL CAPITAL

Of all the elements comprising sustainable brokerage, this was the big *Ah-ha!* It should have been obvious, but it wasn't for a long time: For every sustainable development project Melaver, Inc. itself undertook, the brokerage division was touching a much greater number of development projects involving a much broader segment of the marketplace. The people in the development division are the innovators, but largely they preach to the already converted. We brokers, on the other hand, are in contact with the vast majority of people who haven't yet been exposed to sustainable ideas and practices. In slightly fancier language, for green building to hit a tipping point of broad market acceptance, the diffusion of information and innovation relies on brokers.

We have already seen this s-shaped diffusion curve in Chapter 1. First come the innovators, followed by the brokers who spread the word well beyond the small cadre of folks who have begun to innovate. Then come the early majority adopters, followed by the late majority adopters, and finally the laggards or resistors. The pace of acceptance is largely determined by how successful the brokers are at disseminating new ideas. A comparison of two s-shaped diffusion curves—one with a slow adoption rate, the other with a much faster adoption rate—can be seen in Figure 8.2.

Once the company as a whole got our heads around this basic concept, our priorities changed. Brokerage, earlier viewed as a small appendage of the company's overall operations, moved to the forefront as a critical focus for our overall sustainable agenda. We realized the touch points within the brokerage division, both quantitatively and in terms of diversity, were much richer than those provided by our development division. Our brokers met with a much greater number of potential developers and had access to a much deeper pool of financing sources. And there was a multiplier effect as well, since our brokers by and large had longstanding business relationships with other brokers well beyond our small community, and these brokers had a wealth of social contacts as well. It seems so obvious in retrospect—if an aspiring green real estate company is seeking to seed change, it needs to move beyond its own groupthink and bring into its scope of work those who are thinking and acting more conventionally.

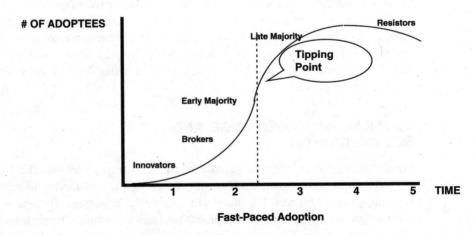

Figure 8.2   Two adoption curves for sustainability.

Martin Melaver is fond of quoting Donella Meadows about how to institute paradigm change:

> So how do you change paradigms? Thomas Kuhn, who wrote the seminal book about the great paradigm shifts of science, has a lot to say about that. In a nutshell, you keep pointing at the anomalies and failures in the old paradigm, you come yourself, loudly, with assurance, from the new one, you insert people with the new paradigm in places of public visibility and power. You don't waste time with reactionaries; rather you work with active change agents and with the vast middle ground of people who are open-minded.[16]

Our company as a whole may have been doing a reasonably good job of working with active change agents in the real estate profession, but it wasn't doing much with the "vast middle ground of people who are open-minded." That's where sustainable brokerage, we feel, has the capacity to shine.

We want our clients and potential clients to think about development differently. Sustainable brokerage integrates traditional brokerage services with sustainable consulting, development, construction, and property management services. It's a collaborative effort. We provide our clients with all of the traditional brokerage services like site selection, tenant and buyers' representation, landlord and sellers' representation, raw land assemblage, market and financial analysis, referral services, and retail, office, and industrial leasing and sales. But we also help our clients envision smarter, more sustainable developments and build-outs. When we help clients select a site, we direct them away from greenfield sites and toward sites with infrastructure, sites where development will create minimal environmental disturbance. Our retail leasing clients

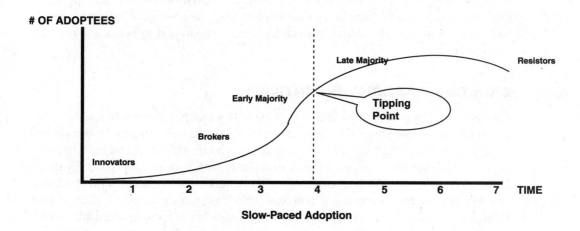

are exposed to information about LEED for Commercial Interiors as a way to build out their spaces to create healthier environments for their workers while saving money on operating costs.

The idea, still in its infancy, has been to include our brokerage sustainability associate on all business development calls. Her role, first, is to help educate our brokers so they can be more informed about sustainable practices with potential clients. Her role is also to develop the toolbox that facilitates how our brokers present sustainable ideas. And finally, her role is to be at the table on business development meetings, both as a resource in real time and to facilitate follow-up. If a potential developer client needs help guiding a project through the LEED process, our sustainability associate is able to connect this client with our consultancy group. If a potential developer client needs help building out a project to LEED specifications, our sustainability associate connects this client to our construction group. Or if a potential developer client needs help managing a project sustainably once it is complete, a hand-off is made to our property management division.

These practices are all standard for a vertically integrated real estate company, one that can develop, construct, lease, manage, and provide consultancy services for a project. What is different here is making the connection between our brokers' deep pool of social contacts and our similarly deep knowledge of and experience with sustainable practices. Sustainable brokerage is all about leveraging social and intellectual capital.

An interesting and not insignificant aspect of leveraging our intellectual and social capital is that it helps us compete favorably in the market. Often, our meetings with potential clients are vetting processes, as clients are evaluating our firm against competitive brokerage companies, some of which are regional or national in scope. The client is obviously looking for a brokerage group with the requisite knowledge and experience to successfully represent the project. Beyond that baseline, the client is looking for both tangible and intangible elements that serve to differentiate a particular brokerage firm (and by association, hopefully, differentiate the project). Our commitment to sustainability has amply served as a positive differentiator. It has not, as was initially feared, proven to be an impediment to being awarded an assignment.

CREATING FINANCIAL INCENTIVES

Let's be crystal clear about this: When working with a group of independent agents who are compensated solely by commission, a sustainable brokerage program absolutely has to be aligned with a well-designed compensation scheme. Otherwise, forget it. By hiring a full-time sustainability associate, a sustainable brokerage firm has basically made a statement about respecting a broker's precious time by not "wasting" it on enormous amounts of research. By creating a toolbox of information, a sustainable brokerage firm is further streamlining a broker's time, enabling him or her to focus on the deal. And by empowering the sustainability associate to make the pitch to potential clients, the broker is relieved of having to face an uphill battle alone. Having said all that, without a financial incentive the broker has little reason to push a sustainable agenda.

A caveat: We have a pilot financial incentive system in place for our brokers. But it is just a pilot program and has been in place for too short a time to evaluate its effectiveness. Our guess is that we will need to make considerable refinements to the program, or perhaps junk it altogether and replace it with an alternative system in the next few years.

The incentive system is simple. Melaver, Inc. pays a 10 percent referral fee for any third-party sustainable work that our brokers effect. If we serve as the developer or co-developer on a project, the broker who brought us the deal gets 10 percent of the fee. If we serve as the general contractor or provide consulting services, the same 10 percent referral fee applies. If a broker brings us a property to manage according to our Mark of a Difference sustainable property management guidelines, the company pays the broker 10 percent of all fees collected.

These fees, which are paid directly to the broker, are nothing to sneeze at. Here are some hypothetical examples:

- A broker brings in a development project—a 75,000-square-foot office building—that will cost in the neighborhood of $15 million to construct. At 4 percent, the development fee amounts to $600,000. The referral fee is $60,000.
- A broker brings in a developer who is looking for a local company to oversee the construction of that same 75,000-square-foot office building. At 10 percent, the contractor fee is $1.5 million. The referral fee for the broker is $150,000.
- A broker brings in our consultancy group to provide LEED consulting on a third-party development project (the same 75,000-square-foot office building). At one-half percent, the consultancy fee amounts to $75,000. The referral fee for bringing the deal is $7,500.
- A broker brings in our property management group to manage a development after it is completed. (Again, the same office building.) At $20 a square foot, the office building generates $1.2 million in gross revenues after a common area factor of 20 percent is factored in (75,000 × 80% × $20). Property management fees amount to 5 percent of rents collected, or $60,000. The referral fee to the broker is $6,000 annually. Given that property management contracts are often multi-year deals (three to five years, conservatively), this would entail fees to the company ranging from $180,000 to $300,000 and a broker referral fee ranging from $18,000 to $30,000.

At the time of this writing, our brokerage division has referred one small development deal (total fee of $25,000, broker referral of $2,500), one general contracting deal (total fee of $500,000, broker referral fee of $50,000), one consultancy deal (total fee of $60,000, broker referral fee of $6,000), and one property management deal (total fee of $30,000, broker referral fee of $3,000). Total potential revenue to the company amounts to $615,000 and broker referral fees amount to $61,500.

Again, we want to be very cautious in our discussion of financial incentives for sustainable brokerage. This is a program that we have not thoroughly vetted. It may not be rich enough to create the paradigm-changing traction we are seeking. It may, on the other hand, be too rich. It may create unintended consequences for some of our other divisions, where staff members are not compensated by such an incentive system. And this particular incentive program might well cause "gaming the system," a typical problem with most incentive-based compensation designs. It's too early to tell whether this particular program will be effective. What is clear to us, at least, is that some type of incentive system is called for.

SHARING BEST PRACTICES

The next step in becoming a sustainable brokerage involves passing along what we've learned to our clients and the community. Learning is great for us as individuals, but because we are interested in seeing real change in the way the real estate industry does things, outreach is critical—so critical that, although we don't really view it as marketing, it certainly functions that way.

It is not uncommon for agents and brokers of a traditional brokerage to participate in professional associations and industry conferences. Agents and brokers of a sustainable brokerage participate in professional associations and industry conferences too, but there is a difference. Instead of participating primarily to network, we also participate for the sake of outreach and discussion with our colleagues. We give presentations to groups such as appraisers about the value of sustainability; we chair sustainability task forces in city government and the local economic development authority. not because we need more lines to put on our résumés but because we want our colleagues and the community to think about development differently.

John Grant, in his groundbreaking work, *The Green Marketing Manifesto,* makes the point that as companies move from green to greener to greenest, they focus part of their marketing effort on the social level. (The two other tiers are the personal level and the corporate level.) The social component of moving up the green spectrum involves first fostering credible partners, then creating tribal loyalty around a brand, and then creating a paradigm-shifting product or service that gains credibility in the market by being normal.[17] (Grant's schema is discussed in greater detail in Chapter 10.)

Sustainable brokerage plays a particularly important role in the social level of marketing. Given the nature of the work, brokerage is a natural point from which to bring together credible partners. If nothing else, brokers foster connections between parties. Fostering partnering where there is strong philosophical alignment along sustainable lines is something at which a sustainable brokerage group should be adept. Since these partnerships have strong potential to extend over multiple deals over many years (that's why brokers always want to be "protected" by getting residual commissions in future deals, since they were responsible for cobbling together the initial partnership), a quasi-tribal bond between parties often evolves. As we have seen already in the context of s-shaped diffusion curves, brokers play a critical role—perhaps *the* critical role—in normalizing paradigm-shifting practices, enabling these practices to become immediately more mainstream and accepted by the general community.

Admittedly, our brokers are fairly new at this. It still feels a bit uncomfortable to be ahead of the curve, pitching sustainable practices to hardnosed, unsentimental, set-in-their-ways developers (as well as to the other brokers with whom we work). And we've been very careful in our public presentations to ensure that our talk does not exceed our walk. Our brokerage firm has been involved in brokering some of the first LEED leasing deals in the country, as a result of representing our company's various developments. But we are still in the process of becoming a sustainable brokerage firm—by learning more, talking to more and more people, refining our service offerings to best meet the needs of our clients, and growing our business along the way. But the very nature of our evolution is significant. Brokers, much more than cutting-edge developers, represent the mainstream. Our struggles are everyone's struggles. Over time, each person in the brokerage group has found an aspect of sustainable development that resonates with him or her. It could be a desire to preserve the quality and quantity of our coastal waterways or a concern for public health. Or it could be because we want our clients to remain competitive and think that high-performance green buildings are an important component of competition. It could be all of the above, or a million other things about sustainable development that make sense. We take what resonates with us and tell authentic stories to our clients and the community.

Competitive Advantage?

Michael Porter, considered an authority on competition and strategy, defines competitive advantage as the capacity of a firm to both create and sustain extraordinary profits (compared to rivals in a particular sector) over the long term. As Porter notes, there are two basic types of competitive advantage: the advantage that comes from being a leader in managing one's costs and the advantage that comes from differentiation.[18] Let's consider each briefly.

Pricing, in the traditional brokerage world, is not much of a differentiator. Most commercial brokerages charge similar commissions. Given the commodity-like nature of this business, brokerage firms are continually looking for value-added, fee-based services that can add to a company's bottom line while helping to differentiate one firm from the others. Many provide asset management services, appraisal work, and/or financing. None, to our knowledge, provides sustainable brokerage services. We are an early if not first mover in a burgeoning business.

Having said that, we don't feel this differentiation will last indefinitely—or even all that long. First, more and more green building product is coming on line, making it easier for even conventional brokers to market green product. Second, there is a widely held belief that, whereas today green buildings are a differentiator, tomorrow they will be the new standard. Everyone will be building to LEED standards. We hope so. If you take serious stock of the various reports emanating from the Intergovernmental Panel on Climate Change, you quickly begin to realize that if we all don't get on board quickly with a new way of doing business, there will be a lot less business for all of us to do.

A sustainable brokerage business such as ours is then faced with an interesting challenge: Do we try to be proprietary about what we know (whatever that is) in the hope of extending the time period of our competitive advantage and thus enhancing our revenue stream? Or do we share what we know with anyone who will listen, in the hope that this green movement will ratchet up all the more quickly but with the recognition that our own time frame of superior profitability will be cut short?

It's an interesting quandary, but at the end of the day it's a false choice. From a purely pragmatic point of view, the movement toward sustainability is happening so quickly that our own particular response to it can only be open sharing of what we know. If we were to be proprietary about what we know and do (we're not sure what that is, but no matter), we don't think it would slow down this green movement one iota. But being active, positive, and collaborative participants in this movement does help move the ball forward. We are doing what we can in our own backyard, which leads us to the more idealistic point of view.

Years		0	1	2	3
REVENUES/SAVINGS					
Net revenues from brokers			300,000	300,000	300,000
Total Revenues			300,000	300,000	300,000
EXPENSES					
Personnel			(60,000)	(60,000)	(60,000)
Rent			(50,000)	(50,000)	(50,000)
Administrative expenses			(20,000)	(20,000)	(20,000)
Total Expenses			(130,000)	(130,000)	(130,000)
Total Cashflow		0	170,000	170,000	170,000
Discount Factor		1.000	0.909	0.826	0.751
PV Cashflow		0	154,545	140,496	127,724
NPV	1,044,576				

Figure 8.3 Cash flow for a traditional brokerage company.

A sustainable brokerage business should spread the word as fast as it can—even to the extent of undermining its own very temporary competitive advantage. Peter Forbes, head of the Center for Whole Communities, notes: "I am concerned that we are 'saving' thousands of very beautiful and important miles of Maine coastline that will very likely end up under water because of global climate change."[19]

Forbes' point applies well beyond the Maine coast: Why have this specialized knowledge and experience of building green if there's literally no place to put that knowledge into action? As a seventy-year-old company, we like to think we have managed to stay in business by doing the right thing. By doing the right thing, we—all of us—will manage to stay in business. Survival itself is a sufficient differentiator.

The Business Case for Sustainable Brokerage

Let's return once again to our fictionalized company, Green, Inc., to which we now add a small brokerage division of five brokers and an executive office manager. The manager has a salaried position, while the brokers work solely on commission—a standard split whereby 60 percent of gross commissions goes to the broker and 40 per-

4	5	6	7	8	9	10 TOTALS	
300,000	300,000	300,000	300,000	300,000	300,000	300,000	3,000,000
300,000	300,000	300,000	300,000	300,000	300,000	300,000	3,000,000
(60,000)	(60,000)	(60,000)	(60,000)	(60,000)	(60,000)	(60,000)	(600,000)
(50,000)	(50,000)	(50,000)	(50,000)	(50,000)	(50,000)	(50,000)	(500,000)
(20,000)	(20,000)	(20,000)	(20,000)	(20,000)	(20,000)	(20,000)	(200,000)
(130,000)	(130,000)	(130,000)	(130,000)	(130,000)	(130,000)	(130,000)	(1,300,000)
170,000	170,000	170,000	170,000	170,000	170,000	170,000	
0.683	0.621	0.564	0.513	0.467	0.424	0.386	
116,112	105,557	95,961	87,237	79,306	72,097	65,542	

Years		0	1	2	3
REVENUES/SAVINGS					
Net revenues from brokers			300,000	300,000	300,000
Additional revenue from sustainable consulting					
Additional revenue from sustainable prop mgmt					
Total Revenues			300,000	300,000	300,000
EXPENSES					
Personnel			(60,000)	(60,000)	(60,000)
Rent			(50,000)	(50,000)	(50,000)
Administrative expenses			(20,000)	(20,000)	(20,000)
Sustainability associate					
Fee to broker for additional consulting revenues					
Fee to brokers for additional prop mgmt work					
Total Expenses			(130,000)	(130,000)	(130,000)
Total Cashflow		0	170,000	170,000	170,000
Discount Factor		1.000	0.909	0.826	0.751
PV Cashflow		0	154,545	140,496	127,724
NPV		1,364,906			

Figure 8.4   Cash flow for Green, Inc.'s sustainable brokerage division.

cent goes to Green, Inc. Out of that 40 percent revenue, Green, Inc. pays the salary of the executive office manager and the rent on its office space, and covers some basic elements of administration such as copying, stationery, and signage. We'll assume that each broker brings in gross commissions of $150,000 annually, of which they each retain $90,000, while Green, Inc. receives the balance of $60,000 per broker or $300,000 annually ($60,000 × 5 brokers). Gross revenues over a ten-year period amount to $3 million, while total costs over this same period equal $1.3 million, for net revenues over a ten-year period of $1.7 million. Discounted back to Year 0, the net present value (NPV) of this conventional set-up is just over $1 million. An overall view of this traditional brokerage set-up can be seen in Figure 8.3.

But Green, Inc. is not a conventional real estate company and so decides, five years into its overall commitment to sustainability, to bring in a sustainability associate to champion sustainable brokerage. There's a cost associated with that additional hire. But Green, Inc. discovers some benefits as well. The company also decides, as part of its overall incentive structure with its brokers, to provide a 10 percent referral fee for any additional work a broker brings to the company. As a result of this revised compensation structure, Green, Inc. gets one additional sustainable consultancy job annually, amounting to $75,000 in revenue. The company pays $7,500 as the referral fee on this business. Moreover, Green, Inc. gets some additional third-party property management assignments over this same time period, from companies that, because of introductions from the brokers, wish to have their assets managed sustainably.

4	5	6	7	8	9	10	TOTALS
300,000	300,000	300,000	300,000	300,000	300,000	300,000	3,000,000
		75,000	75,000	75,000	75,000	75,000	375,000
		75,000	112,500	150,000	187,500	225,000	750,000
300,000	300,000	450,000	487,500	525,000	562,500	600,000	4,125,000
(60,000)	(60,000)	(60,000)	(60,000)	(60,000)	(60,000)	(60,000)	(600,000)
(50,000)	(50,000)	(50,000)	(50,000)	(50,000)	(50,000)	(50,000)	(500,000)
(20,000)	(20,000)	(20,000)	(20,000)	(20,000)	(20,000)	(20,000)	(200,000)
		(60,000)	(60,000)	(60,000)	(60,000)	(60,000)	(300,000)
		(7,500)	(7,500)	(7,500)	(7,500)	(7,500)	(37,500)
		(7,500)	(11,250)	(15,000)	(18,750)	(22,500)	(75,000)
(130,000)	(130,000)	(205,000)	(208,750)	(212,500)	(216,250)	(220,000)	(1,637,500)
170,000	170,000	245,000	278,750	312,500	346,250	380,000	
0.683	0.621	0.564	0.513	0.467	0.424	0.386	
116,112	105,557	138,296	143,043	145,784	146,844	146,506	

Given the current status of referrals coming from Melaver | Mouchet to Melaver, Inc., we feel these revenue assumptions are reasonable to conservative.

The revised discounted cash flow for the brokerage division of Green, Inc. now shows gross revenues over a ten-year period of $4,125,000 and total costs of $1,637,500 for a net income over this decade of almost $2.5 million. Discounted back to Year 0 dollars, net income for the brokerage division of Green, Inc. is $1.36 million, a 36 percent improvement over a conventional brokerage operation. The cash flow for Green, Inc.'s sustainable brokerage division can be seen in Figure 8.4.

We have one more step in considering the business case for sustainable brokerage—integrating the performance of this division into the overall cash flow of the company as a whole. We last reviewed the overall performance of Green, Inc. in Chapter 4, which showed a net present value of cash flows over ten years of $2.3 million and a 28.38 percent internal rate of return (Figure 4.2). When we add the performance of sustainable brokerage into the mix, we find the numbers have improved, with a net present value of approximately $2.6 million and a return just above 30 percent (Figure 8.5).

There are still some overall costs we have yet to consider—primarily the legal costs associated with a green bottom line company and the costs of marketing. These topics will be taken up in Chapters 9 and 10, so we will be revising this spreadsheet a few more times. Nevertheless, what should become clear from the thumbnail financial analysis in Figure 8.5 is that sustainable brokerage has the capacity to ratchet up

Years	0	1	2	3
REVENUES				
Shaping values, chapter 1		21,300	21,300	21,300
Creating a culture of green glue, chapter 2		79,000	79,000	79,000
Green from the inside out, chapter 3		3,300	3,300	3,300
Developing expertise in LEED, chapter 4		0	0	0
Sustainable brokerage, chapter 8	0	0	0	0
Sale of Green, Inc.				
Total Revenues		103,600	103,600	103,600
EXPENSES				
Shaping values, chapter 1	(166,000)	(114,500)	(91,875)	(97,286)
Creating a culture of green glue, chapter 2	0	(57,600)	(64,631)	(72,717)
Green from the inside out, chapter 3	0	(78,490)	(22,381)	(22,998)
Developing expertise in LEED, chapter 4	(49,000)	(150,000)	(417,000)	(890,000)
Sustainable brokerage, chapter 8	0	0	0	0
Total Expenses	(215,000)	(400,590)	(595,888)	(1,083,002)
Total Cashflow	(215,000)	(296,990)	(492,288)	(979,402)
Discount Factor	1.000	0.909	0.826	0.751
PV Cashflow	(215,000)	(269,991)	(406,849)	(735,839)
NPV	2,621,324			
IRR	30.02%			

Figure 8.5   Revised discounted cash flow of Green, Inc.

general sustainable practices in the community and also prove profitable to the company as a whole.

Concluding Remarks

The reading I do in my spare time is usually not about business—I tend to read histories or non-fictional accounts of adventures such as Jon Krakauer's *Into Thin Air* or Amundsen and Scott's race to the South Pole. I do, however, find Yvon Chouinard's narrative of the founding of Patagonia compelling, perhaps in part because I like the gear the company produces and admire a businessman who devotes so much of his time to being outdoors. Chouinard makes a statement in his book that resonates with me: "I learned at an early age that it's better to invent your own game; then you can always be a winner.... I'd much rather design and sell products so good and unique that they have no competition."[20]

I don't think I'm all that different from most brokers. Or at least I didn't for most of my professional life. Brokers are not known for inventing our own games. In fact, we are about as conventional a lot as you will find.

4	5	6	7	8	9	10	TOTALS
21,300	21,300	471,300	538,800	548,925	550,444	550,672	2,766,640
79,000	79,000	79,000	79,000	79,000	79,000	79,000	790,000
3,300	3,300	3,300	3,300	3,300	3,300	3,300	33,000
0	0	2,116,500	1,176,475	1,222,971	1,261,446	1,303,442	7,080,834
0	0	150,000	187,500	225,000	262,500	300,000	1,125,000
						0	0
103,600	103,600	2,820,100	1,985,075	2,079,196	2,156,689	2,236,413	11,795,474
(103,490)	(110,605)	(118,765)	(128,125)	(138,866)	(151,190)	(165,335)	(1,386,037)
(82,016)	(92,710)	(105,007)	(119,150)	(135,413)	(154,117)	(175,625)	(1,058,987)
(24,858)	(26,997)	(29,456)	(32,285)	(35,538)	(39,278)	(43,580)	(355,862)
(57,000)	0	0	0	0	0	0	(1,563,000)
0	0	(75,000)	(78,750)	(82,500)	(86,250)	(90,000)	(412,500)
(267,365)	(230,311)	(328,228)	(358,310)	(392,317)	(430,835)	(474,541)	(4,776,386)
(163,765)	(126,711)	2,491,872	1,626,765	1,686,880	1,725,854	1,761,873	
0.683	0.621	0.564	0.513	0.467	0.424	0.386	0
(111,853)	(78,678)	1,406,597	834,788	786,942	731,931	679,278	0

But when the tsunami hit South Asia in 2004, we in this country were exposed to pictures of widespread natural devastation and untold numbers of people who suddenly found themselves without homes, without families. Those images spoke to me in a way that is hard to describe. I was struck then by how fortunate we are in this country, and how we need to make the most of what we have bountifully been provided. And so, in our mid-fifties, my wife and I decided to adopt a three-year-old girl from China, whom we named Caroline. Many of my friends and colleagues in the real estate world said we were crazy to start a family again, just when our two sons were headed off to college. My colleagues at work, however, seemed to understand and proved deeply supportive.

It's probably a cliché, but Caroline has been a gift to us, not the reverse. She is a constant, daily reminder that providing for others resonates with a sense of meaning and purpose that is beyond measure. It is probably no accident that our brokerage division began to embrace sustainability in earnest about the time my wife and I adopted Caroline. Here we were, inventing a new life at home, our own game as it were, despite the nay-sayers and skeptics around us.

So, too, our brokerage division has begun a journey that is largely of our own invention. It's not always easy. The other real estate professionals in our community were initially somewhat patronizing. No more. There's still some pushback from some of my broker colleagues. That's OK. We'll get there. We have to.

Sustainable brokerage combines the traditional real estate values of honesty, ethical behavior, and integrity with a strong stewardship ethic, a dedication to community service, and the realization that, both as individuals and as members of a company, we bear responsibility for what happens in our back yard. By our actions, we hope to leave a meaningful legacy for our children and grandchildren and generations to come. We want to set an example that differentiates us from a traditional brokerage company, both because we see it as a competitive advantage and because we believe it is the right thing to do for us all. We intend to broker deals that enhance the quality of life where we live.

The word "broker" comes from the old Spanish word *alborque*, meaning a gift given at the consummation of a business transaction. Our gift is to be the real deal to our clients: to facilitate deals of quality, deals that are meticulously researched and tirelessly nurtured and consummated with passion and integrity. Our gift is also to be the real deal to our community: to work in ways that consider the best interests of the place we call home. Our gift is to be the real deal to the next generation: to broker a paradigm change in the kind of legacy we leave.

Like most gifts that come from the heart, without strings attached and with no particular intent to gain by the giving, we're not sure how this gift of sustainable brokerage will be regarded. Time will tell.

NOTES

[1] E. O. Wilson, *Biophilia* (Cambridge, Mass.: Harvard University Press, 1986).

[2] Joseph J. Romm, *Cool Companies: How the Best Businesses Boost Profits and Productivity by Cutting Greenhouse Gas Emissions* (Washington, D.C.: Island Press, 1999), pp. 16–27.

[3] Ibid., p. 19.

[4] Ray Anderson, *Mid-Course Correction: Toward a Sustainable Enterprise: The Interface Model* (Atlanta: The Peregrinzilla Press, 1998), p. 7.

[5] Paul Hawken, *The Ecology of Commerce: A Declaration of Sustainability* (New York: HarperBusiness, 1993), p. xi.

[6] Yvon Chouinard, *Let My People Go Surfing: The Education of a Reluctant Businessman* (New York: Penguin Press, 2005), p. 4.

[7] David Gottfried, *Greed to Green: The Transformation of an Industry and a Life* (Berkeley, Calif.: WorldBuild Publishing, 2007), p. 105.

[8] Jeffrey Hollender and Stephen Fenishcell, *What Matters Most: How a Small Group of Pioneers Is Teaching Social Responsibility to Big Business, and Why Big Business Is Listening* (New York: Basic Books, 2004), p. x.

[9] Gary Hirshberg, *Stirring It Up: How to Make Money and Save the World* (New York: Hyperion, 2008), p. 195.

[10] "The Green Quotient: Q&A with Brenna S. Walraven," *Urban Land*, November/December 2007, p. 118.

[11] Gregory H. Kats, "Green Building Costs and Financial Benefits: A Report to California's Sustainable Building Task Force," 2003. Available from www.cap-e.com/ewebeditpro/items/O59F3259.pdf.

[12] Steven Winter Associates, Inc., GSA LEED Cost Study Final Report, 2004. Available from www.wbdg.org/ccb/GSAMAN/gsaleed.pdf.

[13] Turner Construction Company, Market Barometer: 2005 Survey of Green Building, 2005. Available from www.turnerconstruction.com/greensurvey05.pdf.

[14] Peter Morris and Lisa Fay Matthiessen, The Cost of Green Revisited: Reexamining the Feasibility and Cost Impact of Sustainable Design in the Light of Increased Market Adoption, 2007. Available from www.davislangdon.com/USA/Research/ResearchFinder/2007-The-Cost-of-Green-Revisited/.

[15] Harold Brubaker, "Green Are Its Branches," *Philadelphia Inquirer*, August 15, 2007. Available from www.usgbc.org/News/USGBCInTheNewsDetails.aspx?ID=3321.

[16] Donella Meadows, "Places to Intervene in a System," *Whole Earth Magazine*, Winter 1997.

[17] John Grant, *The Green Marketing Manifesto* (Chichester, U.K.: John Wiley & Sons, 2007), pp. 59–69.

[18] Michael E. Porter, *Competitive Advantage: Creating and Sustaining Superior Performance* (New York: The Free Press, 1985), p. 3.

[19] Peter Forbes, *What Is a Whole Community? A Letter to Those Who Care For and Restore the Land* (Fayston, Vt.: Center for Whole Communities, 2006), p. 13.

[20] Yvon Chouinard, *Let My People Go Surfing: The Education of a Reluctant Businessman* (New York: Penguin Press, 2005), pp. 10, 97.

THE FINE PRINT: LEGAL ISSUES IN GREEN BUILDING PROJECTS

ROBERT E. STANLEY, ESQ.
AND JUSTIN SHOEMAKE, ESQ.

SUMMARY

If the development and construction of green buildings may be said to have finally reached Main Street, the legal ramifications surrounding this rapidly burgeoning field are still in their infancy. In this chapter, Robb Stanley and Justin Shoemake, both LEED Accredited Professionals as well as legal counsel for Melaver, Inc., explore the myriad legal issues that, while standard real estate fare, take on unique twists in the context of going green. The first section looks at the initial development and construction phase of green buildings, focusing on the legal aspects of verification, risk management, and avenues for remedy should things go wrong. The second section considers the issue of permitting and incentives from a green perspective, highlighting how the fast-changing nature of local, state, and federal legislation directly impacts the legal exposure of green developers. The third section entails a brief look at some of the financial implications of a green development. In the fourth section, the authors consider the challenges and potential legal exposure posed by those who broker and market a green development. Sections five and six consider the standard and familiar elements of a real estate lease—tenant build-out, the economics, tenant operations, rules and regulations governing tenant behavior, and repairs and alterations—all from a radically new perspective. Section seven focuses briefly on the legal aspects of buying and selling a green property. In the final section, the authors consider the costs and benefits to a developer/owner of addressing these numerous green issues.

Suppose you own a piece of land you want to develop. You have someone perform a market analysis, and that person concludes that the best use of the property is as a multi-tenant office building. Now you have an idea, but how do you make it reality?

Developing real estate is no simple task, and you won't be able to do it alone. You will need an architect and an engineer to help you design the building and a contractor to build it. To attract tenants you will need a marketing strategy, and you must find a broker with knowledge of the local market. You will likely need someone's help to make sure that you are complying with local laws and that you get the approvals required to build the project. Perhaps most critically, you need money, either from a lender or an equity investor. Finally, you must have a property manager and service providers to help you operate the building once it is constructed.

In order for your project to be successful, you have to make sure that all of these parties understand what their responsibilities are. The overall responsibility, of course, is yours, and as you contemplate the literally thousands of individual actions required to build, lease, and operate your office building, you may begin to lose a little sleep or decide you don't want to be a real estate developer after all.

How can you manage all of this? You need to begin by putting everything in writing. Hire an attorney who can help you think through the issues and can also draft and negotiate the critical documents you will use to allocate obligations among yourself and the various parties helping you implement your contractual obligations. Through contracts you will assign tasks, provide for payments, allocate risk, and set deadlines. Once contracts are in place with every member of your team, every action that is required in order to complete your project should have been assigned to someone. You should begin to sleep a little easier.

Now suppose you are developing your building in 2008. You've seen *An Inconvenient Truth* and heard about peak oil. A water shortage might be looming in your region. You know that potential tenants of the building are concerned about traffic in the area where your project will be located. Your neighborhood just fought, unsuccessfully, against the installation of a landfill nearby. You start to ask questions like:

- How can I reduce energy use in the building? Should I incorporate daylighting? What types of energy-efficient equipment are available? Can I generate any energy on-site with wind turbines or solar panels?
- How can I reduce water use? Should I install a system that can reuse graywater for flushing toilets? Should I collect rainwater for irrigation? Can low-flow fixtures be installed?

- How will people get to the building? How might I encourage the use of public transportation? Could people travel to the building by bicycle?
- Should I install pervious paving? A green roof?
- How do I make sure that indoor air quality is healthy? How do I measure it? Where do I find less toxic, low volatile organic compound (VOC) products?
- How can I help tenants of the building recycle? Are there recycled materials I can use in the building's construction?
- If I make my building more "green," can I use that to my advantage when marketing the building to potential tenants or buyers?

As you seek answers to these questions, you begin to understand that building green concerns nearly every aspect of your project, from the design of the building to your construction budget, to how you market the building and what is expected of your tenants. This means, of course, that in every contract you have with a third party, you must address the concerns raised by the project's green aspects.

This chapter explores how developers can address the contractual and legal issues that arise in green building projects. Because many such issues must be dealt with in the context of the design and construction contracts and in leases to building occupants, our discussion will focus on those documents but will (at least briefly) address all the major legal and contractual aspects of a real estate development project.

Design and Construction

KICK-OFF

Once a developer has decided to construct a building, he or she must begin to think about what that building will look like. How much square footage should the building have? What will its intended use (or uses) be? How it will relate to the surrounding area?

Most important for our purposes, perhaps, is how the building will provide the services that tenants expect. We expect much more of buildings today than four walls and a roof. Among other things, buildings must provide easy access to electricity and water, maintain a comfortable temperature, provide for telephone and Internet communication, regulate humidity, and provide security and comfort for their occupants. To satisfy these demands, buildings incorporate hundreds of different materials and complex mechanical systems. Green building requires developers to satisfy these demands in ways that use fewer resources and produce less waste than conventional methods.

The best way to work through these problems is to hold, very early in the design process, a design charette—a meeting or meetings that include the architect, engineers, contractor, and other members of the project team. Design charettes accomplish a few things:

1 They allow the early communication of project goals to all parties. If a developer is planning to seek LEED certification, the charette is the time to discuss the LEED points it is feasible to seek. (If no one on the project team has any experience seeking LEED certification, the developer might consider hiring a LEED consultant to help with this exercise.) The project team should also discuss any pertinent legal requirements and, if the developer is seeking tax credits or other incentives, how those incentives will be obtained.

2 By bringing expertise from different disciplines into one room, charettes can be a good testing ground for design ideas. The recycled material an architect suggests as insulation may sound great, but the contractor may know from experience that it does not perform well. The composting toilet system an engineer proposes might save lots of water, but the property manager may be able to suggest alternatives that are easier to operate over time.

3 Integrated design—the recognition that building components are not independent of each other, and that efficiencies result when the components interact well—is a central principle of the green building movement. Design charettes bring this principle into practice. For example, how the building is oriented will affect how much sunlight it receives and how large the HVAC system should be, but you need both an architect and an engineer to advise in this area. Rainwater from the building might be used to irrigate landscaping; the charette allows your building engineer and landscaper to talk about how much water will be collected and needed.

Design charettes are advisable throughout the design and construction of a project, but a design team should emerge from the initial meeting or meetings with a clearer set of goals for your building, and you should gain a clearer sense of the issues you will need to address in the contracts with your architect, contractor, and others.

Possible Participants in a Design Charette

Developer

Architect

Mechanical or structural engineer

Landscape designer

Planner

Sustainability consultant

General contractor

Attorney

Broker

Marketing professional

Property manager

Any other parties contributing to, or whose work might be affected by, the design or construction of the project

Possible Participants in a Design Charette (*continued*)
Financing partners Key city officials Key members of a neighborhood association Permaculturist Tenants

FOUR DEVELOPMENT CONDITIONS

Developers may enter into formal contracts with architects and engineers prior to the design charette process and amend those contracts following the charette process or wait to enter into formal contracts until the charettes have clarified the project's requirements. In either case, developers must be certain their contracts with design professionals ensure that, at the end of construction, the following four conditions are satisfied:

1 The project's green features and improved performance can be verified;
2 The sustainability goals for the project's construction were actually met;
3 The risk that any green features of the building will cause problems with the building's performance has been minimized; and
4 The developer is adequately protected in the event that problems do arise.

We discuss each issue below.

VERIFICATION

For various reasons, including higher energy costs, higher prices for raw materials, and possible policy developments, demand for green buildings is likely to increase. More and more tenants will seek to lease space in buildings that are energy- and water-efficient, healthier, and more convenient to alternative transportation options. Some lenders have made commitments to invest in green projects and may offer incentives like reduced interest rates if a particular building meets their requirements. Buyers will pay more for buildings with lower operating costs that are more attractive to tenants.

In order to take advantage of any of these possibilities, a project owner must be able to validate the green features of its building. The most common way to do this is, and will likely remain, seeking certification from an outside party through programs like the U.S. Green Building Council's Leadership in Energy and Environmental Design (LEED), Green Globes, EarthCraft, and others. LEED is the best known of these programs.

Certification can require various steps, including registration of the project, testing of project systems, the submission of documents related to the different green aspects of the project, and responding to the comments of the reviewing authority. Someone must be responsible for these steps. If the developer lacks the in-house expertise to do so itself, then it must have a contract obligating another party to do so.

As mentioned above, one possibility is for the developer to hire a sustainable-building consultant specifically to navigate the certification process. A LEED consultant can offer advice on which points are achievable, complete the necessary documents, make sure testing is completed at the appropriate time, and alert the developer to points in the process where particular actions are required. (A project will earn a LEED point simply for having a LEED Accredited Professional on the project team.) In situations where the architect, engineer, or general contractor is qualified to provide services related to certification, it may not be necessary to retain an additional consultant. Whoever the party helping you with verification is, your contract with that party should be specific as to what its obligations are.

Sample Contract Provision for Certification Assistance

The following is a sample provision of a contract between a developer and architect where the architect has agreed to assist the developer seek LEED certification for a project:

"Architect shall be responsible for assisting Owner to seek LEED certification for the project. Without limiting the generality of the foregoing, Architect agrees to:

(i)   Attend design charettes and other meetings to specify goals for the project;

(ii)   Research the feasibility of the project's achieving LEED points;

(iii)   Consult with Owner regarding the design of the project and the pursuit of LEED certification;

(iv)   Provide such reports to Owner regarding the project's pursuit of LEED certification as Owner shall request;

(v)   Prepare a plan for LEED certification, including a list of LEED points to be sought, the strategy for achieving each such point, a certification schedule, and such other information as Owner shall request;

(vi)   Register the project as required for certification;

(vii)   Supervise the bidding of the project to prospective contractors, including providing such information as bidders may require in order to prepare a bid contemplating the requirements of the proposed LEED points being sought;

(viii) Supervise the construction of the project to ensure that the project is constructed in accordance with the requirements of the

---

Sample Contract Provision for Certification Assistance (*continued*)

LEED points being sought, including without limitation communicating with the general contractor and other members of the project team regarding such requirements, visiting the site as necessary in order to ensure or confirm compliance with such requirements, and responding to proposed changes in the project;

(ix) Submit an application for LEED certification for the project in accordance with the requirements of the USGBC, including without limitation all required supporting documentation;

(x) Submit such additional documentation as is required by the USGBC in connection with such application; and

(xi) Provide such other services as are reasonably required in order to assist Owner in achieving LEED certification for the project."

The above language is a sample only. Each building project is different, and the above language should be negotiated according to circumstances and the particular needs and concerns of the parties.

MEETING SUSTAINABILITY GOALS

A building cannot be certified, of course, if it does not actually meet the requirements of the certification program. Developers must ensure that their requirements are clearly outlined in the design and construction contracts.

Design Contracts

In design contracts, care must be taken with respect to how project goals are described. If a developer expects a building to use 25 percent less energy than a comparable building, for example, it may not be enough to simply include this requirement in the design contract and leave it up to the architect to come up with a solution. The architect might easily achieve a reduction in the building's energy usage, but the method for doing so might be unacceptable or might not include certain elements, such as increased daylighting, that the developer expected. If there are green technologies that the developer is not including initially but hopes to add later, then the design team should be made aware of those. Installing solar panels, for example, may be easier later on if their use is anticipated in the initial design. The developer's goals and expectations for the project should be communicated as fully as possible and, to the extent necessary, included in the design contract.

Many of the technologies likely to be included in a green building's design are relatively new, and it can be difficult to predict how these technologies will actually perform in a particular building. There are two methods that can increase the likelihood that a building design will meet the goals of the developer: building information modeling and peer review.

Building information modeling involves using software to predict a building's performance. Peer review allows a developer to take advantage of another design professional's experience with green building. Developers should consider requiring one or both in design contracts for green buildings.

One difficulty with peer review is that the reviewer will be earning a set fee and is unlikely to warrant a building's performance or assume liability if the reviewer's advice turns out to be wrong. A reviewer also is not hampered by budgetary constraints and might, therefore, tend to advise surer but more expensive alternatives than a developer's budget will tolerate. Despite these drawbacks, it can still be helpful, particularly if the architect on the project team has not previously designed a green building, to get the benefit of outside expertise.

Construction Contracts

Once a developer has a building design that is satisfactory, the construction phase of the project can begin. The most important goal when negotiating the construction contract, of course, is to ensure that the building is built consistently with the approved design. The contractor should be obligated to use only the materials and technologies specified in the design documents. Some aspects of the design, however, are likely to be changed during the construction process—materials may not be available when needed, problems with a technology included in the design might come to light, etc. In order for the developer to be sure that the design will remain green, a process for the careful review of any change orders should be put into place. The contractor should not be allowed to make any substitution or otherwise deviate from the plans for the project without the approval of the party responsible for ensuring that the project meets its sustainability goal, whether that is LEED certification or some other standard.

In addition to the finished product, the developer and contractor must pay attention to the construction process itself. The practice of green building relates not only to *what* is built but to *how* it is built. The construction phase of the project is the only opportunity to get some things right, such as the recycling of construction waste and the indoor air quality during construction, so items like these must be addressed in the construction contract. Failing to do so might cost a developer certification and make it impossible to meet promises made to tenants, investors, and others.

Once construction is complete, the building's systems should be tested to ensure that they perform as expected, and the building's property manager should be trained in the use of the systems. It is often advisable to produce a manual for use with a system. In many situations, LEED certification will require that the building's HVAC, lighting, and energy systems be commissioned in this way; in larger buildings this commissioning must be performed by a third-party commissioning authority. The construction contract should require that the contractor cooperate with the commissioning authority and, following the commissioning process, correct any work necessary to ensure that the building's systems function as desired.

Construction Contract Considerations

Listed below are LEED for New Construction prerequisites and points that relate to the construction process rather than a project's design. As it will be either impossible or, in some cases, highly impractical to achieve these points following the completion of construction, the construction contract and schedule must contemplate them.

SS Prerequisite 1: Construction Activity Pollution Prevention

EA Prerequisite 1: Fundamental Commissioning of the Building Energy Systems

EA Credit 3: Enhanced Commissioning

MR Credit 2.1: Construction Waste Management, Divert 50% from Disposal

MR Credit 2.2: Construction Waste Management, Divert 75% from Disposal

EQ Credit 3.1: Construction IAQ Management Plan: During Construction

EQ Credit 3.2: Construction IAQ Management Plan: Before Occupancy

Source: U.S. Green Building Council

PREVENTING PROBLEMS WITH BUILDING PERFORMANCE

Just because a building is green does not mean that it will otherwise perform well. LEED and other certification systems are green rating systems, but they do not measure building performance in every area. The fact that a building is LEED certified does not mean that it will be immune from problems.

In fact, incorporating green elements in a building's design, if not done carefully, may increase the risk of problems. Some have pointed out, for example, that LEED rating systems are not tailored to different regions of the country; their requirements are the same regardless of where a building is constructed, even though building elements that can achieve LEED points may work better in some places than others. In areas with higher average humidity, for instance, such as the southeastern United States, features like operable windows that increase ventilation may also increase the risk of mold and mildew. Building flush-outs, which are done to clean the building's air following construction, move large volumes of outside air through a building and may bring in more moisture than the building is designed to handle.

More broadly, green buildings are likely to incorporate new products and technologies, many of which have not been widely used. Most new products are experiments, and as Stewart Brand warns in *How Buildings Learn*, "[m]ost experiments fail."[1] New building products may not perform as well as advertised or may not interact well with other components of a building. A 1996 Florida Solar Energy Center study found that the risk of building problems increases as building complexity goes up, and green buildings are often more complicated than others (think of green roofs and on-site energy generation facilities).

To decrease the likelihood that problems will occur in a green building project, developers, among other things, should:

- Assemble a project team with green building experience;
- Hold design charettes that give the project team the chance to discuss potential issues;
- Give preference to products and technologies that have a satisfactory warranty and make sure that the warranty is in the correct party's name (unless the architect or another party is warranting the product's performance, warranties should normally be in the project owner's name);
- Do more than look at a product's promotional material—review technical data and research the manufacturer;
- Require the architect to only specify products for the use recommended by the manufacturer, ensure that the general contractor installs products in accordance with the manufacturer's recommendations, and prohibit any party from taking an action that could void a warranty;
- If an architect or engineer advocates the use of a particular product or technology in an unusual way, consider requiring them to warrant the performance of the product, at least for a certain period of time (this is particularly true if they claim that they have worked with the product successfully elsewhere);
- Have a building's design modeled prior to construction and if moisture control is a concern, consider having wall modeling performed;
- Have the building's design peer-reviewed;
- Put procedures in place that ensure thorough evaluation of change orders;
- Monitor the construction process carefully through on-site inspections and review of documentation from the general contractor;
- Perform commissioning of building systems following construction; and
- Consider using a single entity to perform design and construction functions. Design/build arrangements can make communication easier by reducing the number of members on the project team. It can also, by making one party responsible for a number of different functions, make it easier to assign responsibility in the event that problems do arise.

Even if all of these steps are taken, some buildings will inevitably have problems. In the next section, we discuss how the developer of a green building should protect itself in that event.

REMEDIES

Consider the following claims that have arisen out of green building projects:[2]

- An architect agreed that its design for an office building would achieve enough points for LEED Gold certification. The developer advertised that the building would reduce operating costs and provide a healthier indoor environment. Budget and scheduling concerns prevented the building from achieving Gold certification. The developer sued the architect, claiming that the architect had guaranteed the certification level.

■ After a green roof was installed on a building, water infiltration resulted in significant damage. Analysis concluded that a lack of structural stability caused the infiltration. The structural engineer claimed that no one had provided proper information related to the green roof. The engineer's liability had been limited contractually to $50,000, an amount insufficient to repair the damage.

■ A law firm hired an architect to design a sustainable office that would receive favorable press. The architect received promotional materials from manufacturers of green products and, based on these materials, incorporated the products into the design of the office. The local press investigated the products and claimed that they were not as sustainable as suggested by the promotional materials. The law firm claimed its reputation was damaged and demanded that the architect remedy the situation.

■ A county government required the recycling of construction materials. The architect on a project in the county agreed to provide construction observation services. The contractor decided during construction that the recycling obligation was causing delays and that prices for recycled material had declined. It began dumping construction waste in another state. When the county took action against the project's owner, the owner sued the architect for failure to properly observe construction.

Other claims have involved patent infringement, consumer protection laws, breaches of confidentiality, and mold issues. The range of possible claims from any building project is nearly unlimited, and developers always need to consider how to protect themselves. How are these considerations different for a green building project?

1. Nature of the Obligation

When a developer enters into a contract with a design or construction professional, the contract typically does not require the professional to perform its obligations perfectly. Rather, the professional usually agrees perform its obligations in a manner consistent with the degree care ordinarily exercised by other members of the profession in similar circumstances. This is called the "standard of care."

In a green building project, the standard of care may not be adequate to protect the developer's interests. Two of the above examples describe instances where an architect recommended green products that either caused construction delays or were not as green as advertised. If the architects in those cases had agreed to use only the degree of care ordinarily exercised by other architects in similar circumstances, they might not have been expected to thoroughly investigate the availability or environmental attributes of the products in question.

If additional risks arise out of the fact that a developer is constructing a green building, that developer may need to consider requiring guarantees or other assurances related to such risks from its design and construction professionals. If, for example, a developer knows that the tenant with whom it has signed a letter of intent will require that its premises be at least 30 percent more energy-efficient than premises in other buildings, then the developer may ask the architect to warrant that the building will meet that requirement. If 70 percent of construction waste must be recycled, then a developer may demand a specific guarantee from its contractor that this will be done and include measurement documentation and other reporting requirements in the construction contract.

Architects and contractors will, of course, resist provisions of this sort, and the form contracts published by the American Institute of Architects and commonly used in commercial transactions do not include many such provisions. Developers must be prepared to specifically negotiate warranties, guarantees, and the like where they feel those protections are necessary.

Architects may be particularly unwilling to warrant that specified products will not be defective (as opposed to whether the product was appropriate for use in the project). For product defects, developers will likely need to look to the product warranty (and should examine the warranty prior to agreeing to the product's use).

For a project where the developer is seeking LEED certification, many contracts will obligate either the architect or the contractor to complete documentation required for certification. These documents will require the party completing them to make certain statements regarding different aspects of the building's construction and performance, but developers should not rely on these statements as warranties that will run to the developer's benefit. (Architects or contractors concerned about this issue might consider including specific language in their contracts that their completion of LEED documentation is for certification purposes only and does not create any sort of guarantee in favor of the project owner.)

2. Damages

Ensuring that an architect or contractor has assumed the responsibility to do something, of course, does not automatically protect a developer. Developers must also pay attention to the damage provisions of contracts to ensure that, if a design professional defaults, the developer can recover all or some portion of this damage.

How should damages be calculated? In some cases the parties can predict the harm that would arise from a default with some specificity. A contractor might fail to do something and prevent a developer from obtaining a LEED credit, making it impossible for a project to obtain Silver certification and for the developer to get a discounted interest rate on its loan. An architect's design might fail to qualify the building for a tax credit. For situations like these or where damages could be predicted to fall within a certain range, parties to a contract might negotiate a liquidated damages provision whereby the design or construction professional would agree to pay a sum certain if it failed to achieve an obligation in its contract.

In most situations, however, it will not be possible to know in advance how much damage would result from a default. In this situation, the best a project owner may be able to do is ensure that the contract gives it the right to make a claim against the defaulting party and to beware of provisions that limit the damages that may be recovered.

3. Insurance and Bonds

Even if a contract says that an architect or contractor will be liable for damages, what if it cannot pay? Damages related to a large building project are potentially so large that many design or construction firms would not be able to pay them even if required to do so. Contracts typically address this difficulty in two ways—insurance and performance bonds.

There are many different types of insurance policies. Architects and engineers typically carry an errors and omissions policy that covers defects in design, while commercial general liability policies carried by contractors cover construction defects. Insurance policies come with many limitations, however. Of particular concern in the green building context, policies often insure only that the architect or contractor performs to the standard of care described above. If a design or construction contract includes obligations that exceed the standard of care, developers need to pay close attention to the extent of the insurance coverage carried by the professional. If the insurance policy will not cover the obligation and there is no way to amend or supplement the policy to include the desired coverage, then the developer needs to be aware of that risk before signing the contract with the architect or contractor.

Performance bonds are agreements by a third party to guarantee a contractor's performance on a project. Since some green building technologies are still relatively untested, the issuers of bonds, like insurance companies, may be reluctant to guarantee enhanced contractual obligations by the contractor on a green building project. This is likely to change as green building practices rapidly become mainstream.

Conclusion

Building construction is inherently risky, and developing a green building, because it can require a project owner to make additional promises to investors or occupants about a building's performance, can be even riskier. Since it can be difficult for a project owner to be made whole in the event that there are problems with the design or construction of a building, taking steps to prevent problems from occurring is critical.

Permitting and Incentives

Before constructing a building, of course, a developer·has to obtain the necessary permits. The permitting process can deeply affect a green building project in both positive and negative ways.

In many cases, existing building and development codes can stand in the way of making construction projects more sustainable. Local officials may be unwilling to approve technologies they view as risky or be bound by outdated laws. The re-use of graywater, for example, is often resisted by local building authorities. Development codes require a minimum number of parking spaces for any project, but the minimum might be more than the green building developer, anxious to develop in a dense area where building occupants could use public transit or walk to a building, would prefer to install. Building codes and homeowners' or community associations might prohibit wind turbines or other on-site generation facilities. Developers need to be prepared to persuade local governments and owners' associations that green building features are beneficial.

Problems like these should become less common over time; building codes and other laws are likely to be revised to incorporate energy efficiency, water efficiency, and other sustainability requirements. In fact, this trend is already underway. For example, many state and local governments now require that their own buildings be LEED certified. The District of Columbia and Pasadena, California, for example, go further, mandating that certain private development projects meet the requirements for LEED certification. California recently passed a law requiring electric and gas utilities to keep consumption records for all nonresidential buildings and, starting in 2010, for building owners to disclose that data and the building's EPA performance rating (a rating calculated by evaluating the building's attributes and comparing them to other similar buildings in a national survey) to prospective buyers, tenants, and lenders. Municipalities including DeKalb County, Georgia are exploring ways to ensure that when homes built prior to a certain date are sold, their plumbing fixtures are updated. In Chicago, homebuilders developing to the city's unique green program are fast-tracked through the permitting process.

Beyond simply dictating that privately owned buildings be more sustainable, more and more governments are offering incentives to encourage green development. At the federal level, tax deductions already exist for energy-efficient building equipment. Tax credits are available for certain renewable energy equipment. Various incentives to install energy-efficient and renewable energy facilities are also in place in California, Minnesota, Washington, New York, and other states. Seattle allows projects in certain areas to increase their size and height if those projects achieve a LEED Silver rating.

Since buildings account for such a large percentage of greenhouse gas emissions (approximately 40 percent of total emissions in the United States), green building would have to be a focus of any national climate change policy. A cap-and-trade program could give developers who construct very energy-efficient buildings the opportunity to sell energy savings on the open market. One expert has noted that carbon trading could account for from 3 to 8 percent of a building's cash flow.[3]

The legal landscape will change rapidly in the coming years. Developers planning to build green must be aware of that landscape's roadblocks and opportunities.

Financing

The emergence of green building does not appear, at this writing, to have significantly affected the legal issues related to financing commercial real estate, but we note here a couple of items to consider when negotiating a loan related to a green building project.

The first is appraisals. The amount an owner will be able to borrow against a green building depends on the appraisal. Green buildings, because of lower operating costs and expected market demand, will be more valuable than comparable buildings that are less efficient. To make sure this premium is considered, owners should try to have appraisals performed by appraisers with experience evaluating green buildings.

Other Examples of Laws and Incentives Relating to Green Building

Green Building Codes

Boston, Massachusetts—All new and rehabilitation construction projects over 50,000 square feet must earn at least 26 LEED points.

Montgomery County, Maryland—County-built non residential buildings must achieve a LEED Silver rating; private nonresidential or multi-family residential buildings must achieve LEED certification.

Santa Cruz, California—Residential and nonresidential projects must earn points from green building checklists. The checklist for nonresidential projects is based on LEED; the checklist for residential projects was developed independently.

Babylon, New York—New homes must be built to meet Energy Star standards.

Incentives

Corona, California—Green building projects are fast-tracked through the plan-review process, cutting permitting times by as much as half.

Arlington, Virginia—The county allows increased densities for certain projects meeting green building requirements.

State of Maryland—Awards grants for solar, geothermal, and wind installations and tax credits for green building projects.

Austin, Texas—The local utility company is one of many nationwide offering rebates for energy efficiency and renewable energy installations.

The second issue to consider is the role that financial incentives can play. In early 2007, New Resource Bank in San Francisco announced a program giving the developers of LEED buildings a reduction of one-eighth of a percentage point in interest rates and the opportunity to finance a higher percentage of a construction project's cost than traditional construction loans. Bank of America has committed to lend billions of dollars for green projects, a large portion of which can be expected to go to developers of green building projects. GE Real Estate is working to green its massive portfolio, and at least one real estate investment trust committed to investing in green buildings has been announced.[4] There are other examples. With so many lenders and investors seeking green building projects, lenders and investors may need to give developers incentives in order to meet their commitments.

All of this is good news, but from a developer's perspective it can also increase risk. Loan agreements that include incentives, for example, will also include requirements that the project meet certain standards. For each requirement, developers must be sure that they can either satisfy the requirement themselves or that another member of the project team will be obligated to do so. If requirements are not met, the penalty, if possible, should be the loss of the incentive rather than a loan default. Default under loan documents can spell the end of a project, and even the loss of an incentive may destroy a project's profitability, so the need to coordinate the requirements in a loan agreement with the requirements in other project contracts is critical.

Brokerage and Marketing

Developers building green projects will want to advertise that fact. This raises a couple of concerns. The first is that brokers and marketing professionals may not fully understand the product they are selling. This can make it difficult to communicate the project's benefits to a marketplace that increasingly expects projects to be more sustainable, and can also make it harder to sell to potential buyers and tenants that may be uncomfortable with new building technologies or unusual operating requirements.

The second concern is that the language used to market green projects should be chosen with care. Though the law with respect to green buildings is not well-developed, the federal government and some states have shown an interest in the marketing of products that are presented as beneficial for the environment. Marketing claims that do not meet certain requirements can be subject to challenge under state and federal consumer-protection laws.

The Federal Trade Commission (FTC), which regulates unfair or deceptive trade practices, has published guides related to the marketing of environmental products.[5] These guides do not have the force of law, but they do represent the FTC's position on certain issues and can be useful indicators of where legal boundaries at the state and national levels are likely to settle (California's law, for example, explicitly references the FTC guidelines).[6]

The guidelines should be reviewed in full prior to marketing a product, but very generally, marketing claims should:

1 **Be substantiated.** There must be a "reasonable basis" for making the claim. Research, tests, studies, or other reliable evidence must show that the claim is true.
2 **Be clear and properly qualified.** In the context of green buildings, it could be deceptive, for instance, to call a building "environmentally friendly" if the only difference between it and a typical building is the use of low-VOC paints. The claim should be limited to the particular component of the building that is in fact "environmentally friendly." Comparative claims, such as "save 50 percent on energy bills," should be clear as to the basis for the comparison.

3 **Not overstate environmental benefits.** It might be deceptive to advertise that a building component is recycled, for example, if it is also of questionable safety.

Objective, easily verifiable, specific claims are less likely to be challenged. Referring to a project's LEED or other certification can be a useful shorthand for communicating that a project is green. When negotiating contracts with brokerage and marketing firms, developers should address how those parties will market projects, both to ensure that the green attributes of a project are communicated to the public and to require that those attributes are presented in a manner that complies with the law.

The Green Lease

The operative contract between a landlord and a tenant is the lease. The lease agreement allocates risks and responsibilities, as well as costs, and controls the behavior of the parties. Without a well drafted and properly negotiated lease agreement, the developer of a green building may not be able to achieve and maintain the type of building desired.

To date, the desire to construct and lease space in green buildings (other than owner-occupied buildings) has been driven by the developer of the building, rather than the tenant. As such, the lease agreement must be primarily focused on regulating the tenant's behavior. As more tenants, particularly those tenants leasing significant amounts of space or who have particular leverage in a lease transaction, begin seeking out green buildings or demanding that buildings are constructed and operated more sustainably, lease agreements will begin to incorporate obligations and requirements to be met by the landlord with respect to the design, performance, and management of the building. For purposes of analyzing certain lease provisions in this chapter, however, we will assume that the landlord is the primary party promoting the green elements of the building and is seeking to regulate the tenant's behavior through the lease agreement.

Structuring a "green lease" must begin with an understanding of the design and performance objectives of the building. Specifically, the green features of a building must be identified to determine if any specific modifications to the lease agreement are required. Examples include the energy management system of the building; whether or not the building has a green roof, water management and reuse programs, mandatory or voluntary recycling programs, parking and integration with public transit; and the extent to which the developer desires to regulate the interior design and construction of the tenant's space in addition to the structure and common areas of the building. In some cases, the lease agreement will simply require the tenant to acknowledge the existence of these features or programs. In others, it will require the tenant to comply with certain landlord requirements, accept regulation of its operations to some extent, and pay the costs of certain of these features and programs as a component of its rent under the lease.

Once the landlord and tenant have come to agreement on basic terms such as the amount of space the tenant will lease, the rent the tenant will pay, and the length of time the tenant will lease the space, the parties can begin negotiating the specific provisions of the lease agreement, including 1) the tenant build-out, 2) the economic structure of the lease, 3) the costs associated with operating the building, 4) tenant operations, 5) the rules and regulations governing tenant behavior, and 6) repairs and alterations. Let's consider each of these provisions in greater detail.

TENANT BUILD-OUT

One of the first and most important elements of any type of lease is an agreement between the parties as to the nature of the design and construction of the improvements in the tenant's space. Depending upon the extent to which the landlord desires to incorporate the green elements of the building design into the individual tenant spaces, the lease agreement will need to be highly specific about the design of the space, as well as the manner of construction. Likewise, if the landlord is requiring the tenant to construct its interior improvements to a particular standard (e.g., LEED for Commercial Interiors), this requirement must be specifically set forth in the lease agreement, along with appropriate tools for the landlord to ensure compliance by the tenant with this requirement. Examples of specific provisions that may be set forth in a green lease include requirements for management of any waste resulting from the demolition and removal of existing improvements, as well as during the construction of new improvements in the space; the reuse of any existing improvements or fixtures; the nature and source (e.g., recycled content and/or local sourcing requirements) of the materials utilized in the construction of the improvements; requirements for the use of non-VOC emitting paints, stains, and sealants; standards for ensuring and measuring the indoor environmental quality of the tenant's space (as well as its impact on the remainder of the building); the type of furnishings and equipment that may be utilized in the space; and the incorporation of daylighting, individual climate control, etc., in the design of the tenant's space.

There are a number of tools that a landlord can use to regulate and enforce any particular requirements regarding the construction of the tenant's space. Most important is the right to review and approve the tenant's plans and specifications. Further, the landlord could require the use of a particular contractor acceptable to the landlord or reserve the right to approve any contractor selected by the tenant for the construction of the tenant's improvements, as well as the right to review and approve the construction contract with the approved contractor. The lease agreement may also contain a "work letter" or construction exhibit that prescribes the allocation of responsibilities between landlord and tenant, establishes timing and approval procedures, describes the specific materials and fixtures the tenant must use, etc., in connection with the construction of the improvements in the tenant's space.

Beyond regulating the construction of improvements to the tenant's space, the landlord may have an interest in regulating the type of furnishings and equipment brought into the space by the tenant. This may be as simple as requiring Energy Star® labeled

The Green Lease Work Letter

Among other things, a green lease work letter could include requirements such as:

- Tenant shall utilize only energy-efficient lighting fixtures, and all lighting (excluding safety lighting) shall be equipped with occupancy-based controls.
- Tenant shall certify the commissioning of all HVAC, lighting, and energy systems prior to the occupancy of the Premises for the conduct of Tenant's business therein and provide such commissioning data to Landlord.
- All wood used in connection with the construction of any improvements shall be FSC-certified.
- Tenant shall be required to utilize low-flow toilets, waterless urinals, and sensor-controlled faucets for all restrooms within the Premises.
- Tenant shall comply with Landlord's requirements for construction waste management and recycling, including any required sorting and storage of such materials.
- All paints, stains, sealants, and adhesives used by Tenant in the construction of improvements in the Premises shall be water-based, low- or no-VOC emitting products.
- Carpet systems shall meet the Carpet and Rug Institute's Green Label Plus certification.
- Fiberglass insulation used on site must be formaldehyde-free.

appliances and equipment, or as detailed as regulating the type of workstations and other office furniture the tenant is allowed to bring into the space, particularly with respect to the volatile organic compound (VOC) content of these items. Most lease agreements contain no provisions that regulate the tenant's furniture, fixtures, and equipment. As such, if a landlord desires to include these types of controls, the green lease will need to include specific language addressing the landlord's goals or requirements along these lines. Because this is not traditionally an area controlled by the landlord, these provisions may meet with some resistance.

Once the tenant's improvements have been constructed and tenant has taken occupancy of the space, the lease agreement dictates the rights and responsibilities of the parties over the term of the lease (and, in some cases, beyond the term—e.g., indemnity obligations, or covenants to pay any costs or damages that the party being indemnified may suffer that survive past the term of the lease). In the context of a green lease, there are a number of provisions that may need to be modified or enhanced. These include provisions addressing operating costs of the building, insurances, taxes, janitorial services, building operating hours, supply of utilities and the measurement of tenant's consumption of the same, recycling programs, smoking policies, and controls over subsequent repairs and alterations.

LEASE ECONOMICS

As an initial matter, the economic structure of the green lease must be examined, as this structure has a direct impact on the incentives created to build and operate high-performance buildings, as well as the return received by one party or the other from energy cost reductions. The typical economic structures for a commercial office building with multiple tenants fall into one of three categories: a so-called "gross lease," a "modified gross lease," or a "net lease."

Under the gross lease structure, the tenant pays a single rent to the landlord, which is inclusive of the tenant's share of the costs to operate and manage the building. This structure allocates the fiscal responsibility to the landlord to manage the costs of operating the building, as the amount of rent received by the landlord is fixed. In this case, the benefits of a green building, such as lower energy costs, would benefit the landlord. As a result, there is arguably little incentive for the tenant to participate in any program to reduce or control energy costs, as there is no change in the total costs incurred by the tenant to occupy the leased space if energy costs go up or down.

The modified gross lease structure represents the transference of some of the risk of controlling the operating costs of the building to the tenant. Here, the tenant's initial rent includes operating expenses for a specified year (typically the first year of the lease), called a "base year," or up to a particular dollar figure, called an "expense stop." Following the initial base year, in addition to the base rent, the tenant also pays the landlord for its share of operating costs in excess of the base year or expense stop amounts. In this case, the total amount of rent paid by the tenant increases as the cost of operating the building increases, including increases based on higher energy costs.

The so-called "net lease" is a different lease structure that leverages all of the costs of operating the building to the tenants. Under this structure, the tenants pay the landlord a base rent, plus an additional charge for costs of operating, maintaining, insuring, and managing the building. Here the tenants bear all of the risks that the building operating costs will increase; however, the entire benefit of any reduction in building operating costs accrues to the tenants, not the landlord. Historically, the net lease structure has been utilized predominantly in single-tenant and retail lease settings. While the net lease may provide some incentive for tenants to minimize operating costs, it does not directly reward a landlord for designing, constructing, and maintaining the most environmentally conscious and energy-efficient building.

Under either the gross lease or modified gross lease rent structures, the landlord has an incentive to control the costs of operating the building in order to maximize the rents received from the tenants, which in turn maximizes the value of the building. As such, this creates a financial incentive for landlords.[7] This incentive must be supported by the terms of the lease agreement.

Lease Provision for Sharing Energy Cost Savings

The following provision is designed to allow a landlord and tenant to share the cost savings of energy-efficient features installed by the landlord where the tenant either pays the landlord for utility costs or pays the utility provider directly. The landlord's portion of the savings would allow it to recoup the initial cost of such features, while the remaining savings make the economic terms of the lease more attractive to a potential tenant.

Energy-Efficient Features

(a) Landlord will install, at Landlord's sole cost and expense, energy-efficient lighting and HVAC systems, daylighting, and other measures designed to reduce the electricity usage of the building, all as more particularly described on Exhibit C attached hereto.

(b) Landlord will deliver to Tenant, on or before the first date Tenant occupies the Premises, a certification from a licensed professional engineer setting forth the projected monthly energy savings (the "Monthly Savings") as a result of Landlord's installation of the features described in subsection (a) above. Such savings shall be calculated based on the standard method of calculating energy savings in the metropolitan area in which the Premises is located.

(c) On the day that Tenant is first obligated to pay rent, and continuing throughout the remainder of the term of the Lease, Tenant shall pay to Landlord, as additional rent, _____ percent (___%) of the Monthly Savings in equal monthly installments.

With respect to calculating the savings from energy-efficient features, a tenant might insist that at least two engineers (one of which could be selected by the tenant) calculate such savings, with the calculations of the two engineers being averaged to compute the actual amount that the tenant will pay to the landlord.

Basing the amount the tenant must pay the landlord on actual energy savings may be problematic. Not all of a tenant's energy usage will be as a result of fixtures installed by a landlord. Much energy usage will relate to the tenant's own equipment and habits, and a tenant would have less incentive to use energy efficiently if energy savings would mean increased payments to the landlord under a cost-sharing arrangement.

Where the landlord of a multi-tenant building charges all tenants in the building a set amount for electricity usage, a tenant that installs energy-efficient equipment may have difficulty recovering its investment. Such a tenant may wish to negotiate a provision similar to the provision above but which, instead of requiring the tenant to pay an additional amount to the landlord, would give the tenant a monthly rental credit equal to all or a por-

Lease Provision for Sharing Energy Cost Savings *(continued)*
tion of the calculated savings resulting from the tenant's installation of such equipment.

COSTS OF OPERATING THE BUILDING

In leases using a modified gross or net rent structure, the definition of "operating costs" is typically fairly broad. Usually, it encompasses all costs incurred by the landlord to operate, maintain, repair, and manage the building. In some cases, there are negotiated limitations on the landlord's ability to pass through certain costs incurred by the landlord as operating costs. In the green lease, landlords must be very careful to both ensure that the definition of operating costs is sufficiently broad to encompass additional or unique costs associated with the green building and avoid limitations which may limit or prohibit the landlord's ability to pass these costs through to tenants. Some specific examples include the costs of maintaining a green roof, maintenance and repair of pervious paving systems, maintenance and repair of water management and collection systems, costs associated with on-site and off-site access to and use of public transportation (bus shelters, walkways, underground tunnels, lighting, discount programs for users of public transportation, etc.), specific products and procedures associated with green cleaning, operation of recycling programs together with the costs associated with storage of recyclable materials as a part of these programs, and the costs associated with on-site power generation systems. Further, the operating costs provisions of a green lease should be sufficiently broad to allow the landlord to amortize the costs of operational enhancements made to the building over the life of the lease that are designed to enhance the energy or water efficiency of the building, improve indoor environmental quality, etc., as a component of operating costs. Without the ability to pass through the cost of such enhancements to the tenants of the building (who will directly benefit from the same), there is little economic incentive for the landlord to make such investments.

TENANT OPERATIONS

The form of any lease agreement contains certain provisions regulating the operations of the tenant in the leased premises, as well as on the balance of the landlord's property. Along these lines, the cost of utilities such as electricity, water, sewer, and waste removal are usually included in the definition of operating costs or dealt with through specific provisions in the lease. In the green lease these provisions must specifically address issues like the energy management system for the building, energy conservation measures such as occupancy-controlled lighting, building operating hours, limitations on the tenants' ability to control the temperature in their spaces, as well as

A Definition of Operating Costs

A typical broad definition of Operating Costs might read as follows:

For the purposes of this Lease, "Operating Costs" shall mean and include all expenses, costs and disbursements of every kind and nature relating to or incurred or paid in connection with the ownership, management, operation, repair, landscaping, and maintenance of the Building, including but not limited to the following: (1) wages, salaries and other costs of all on-site and off-site employees engaged either full or part time in the operation, management, or maintenance of the Building; (2) the cost of all supplies, tools, equipment and materials used in the operation, management, and maintenance of the Building; (3) the cost of all utilities for the Building, including but not limited to the cost of electricity, gas, water, sewer services, communication services, and power for heating, lighting, air conditioning, and ventilating; (4) the cost of all maintenance and service agreements for the Building and the equipment therein, including but not limited to security service, garage operators, window cleaning, elevator maintenance, HVAC maintenance, janitorial service, waste disposal and recycling service, telecommunications services, interior and/or exterior landscaping maintenance and customary interior and/or exterior landscaping replacement; (5) the cost of repairs and general maintenance of the Building; (6) amortization of the reasonable cost of acquisition and/or installation of capital investment items and/or capital improvements made by Landlord (including security and energy management equipment), amortized over their respective useful lives, which are installed for the purpose of reducing Operating Costs, promoting safety, or complying with governmental requirements; (7) the cost of casualty, rental loss, liability, and other insurance applicable to the Building; (8) the cost of trash and garbage removal, air quality audits, vermin extermination, and snow, ice, and debris removal; (9) the cost of legal and accounting services incurred in connection with the management, maintenance, operation, and repair of the Building; (10) all taxes, assessments, and governmental charges attributable to the Building or its operation; and (11) the cost of operating the management office for the Building.

temperature controls for the common areas of the building, and programs to individually measure a tenant's energy or water consumption. The landlord may also reserve the right to designate the provider of certain utilities, including electricity (green power) and water, as well as for removal of waste and recyclable materials. Finally, the lease may also include a specific declaration of the landlord's intentions with respect to seeking LEED certification for the building and a requirement for the tenant's cooperation.

If some or all of the building's energy or water needs are met through on-site energy generation programs, the lease should specifically address the landlord's right to provide these services to the tenants and to pass through the actual costs of producing the same to the tenants. This may also give rise to concerns by the tenant regarding the interruption or reduction of these services by the landlord, and the lease may need

<hr>

Declaration of Landlord's Intent

What follows is an example declaration of Landlord's intent with respect to LEED Certification.

LEED Certification. Landlord may design, construct, and/or maintain the Building in accordance with the principles of LEED (Leadership in Energy & Environmental Design), a voluntary rating system utilized to mitigate the impact of development on the environment, and may obtain certification from the U.S. Green Building Council that the Building complies with standards set forth in LEED. Tenant shall comply with requirements specified by Landlord for the construction, alteration, operation, maintenance, repair, and replacement of the Premises, which requirements may include, without limitation, (i) the use of recycled, rapidly renewable, or other materials, or the non-use of other materials, in Tenant's improvements, fixtures, equipment, and personal property; (ii) the installation of utility facilities or lighting fixtures specified by Landlord in the Premises; (iii) the compliance of the Premises with energy or water conservation; lighting; ventilation; waste disposal or reduction; and other design or operation standards for the Building; and (iv) the maintenance of a minimum level of indoor air quality in the Premises. Tenant shall take no action which shall cause Landlord to lose the certification described above. Tenant shall reasonably cooperate with Landlord in connection with Landlord's efforts to obtain or maintain such certification, including without limitation the provision to Landlord of utility bills and other documents relating to the functioning or maintenance of the mechanical systems installed within the Premises. Landlord shall have the right to construct, alter, maintain, repair, replace, and operate the remainder of the Building in such a manner as Landlord deems necessary in order to obtain or maintain the Building's LEED certification.

<hr>

to address issues of landlord liability in this regard. The landlord will obviously desire to incorporate provisions that relieve the landlord from any liability for the interruption or reduction of these services. The tenant will desire to incorporate language that creates an incentive for the landlord to ensure consistent levels of service and provide a remedy to the tenant in the event the landlord fails to do so.

If the landlord has a recycling program for the building, the green lease should contain provisions permitting the costs of such a program (including the cost of reserving areas of the building for the interim storage of recyclable materials) to be included in operating costs. Beyond this, the landlord may desire to require tenants to participate in such programs as an obligation under the lease. In such an event, the lease should specifically describe the tenant's obligations and provide remedies for the landlord in the event the tenant fails to perform such obligations. These might include monetary fines, loss of services, an obligation to reimburse the landlord for performing the tenant's obligations, and ultimately, the right of the landlord to terminate the lease.

RULES, REGULATIONS, AND OTHER CONTROLS ON TENANT BEHAVIOR

In the context of a green building, the landlord may have an increased need to regulate the tenant's behavior, depending upon on the nature of the overall design of the building and the landlord's goals with respect to energy usage, indoor air quality, waste reduction, etc., and the lease agreement should reflect this. Examples include regulation of the types of furniture and equipment a tenant is allowed to bring into and utilize in the premises, prohibition of smoking in or around the building, and monitoring requirements related to indoor air quality issues such as carbon dioxide and ventilation efficiency. These provisions range from rules and regulations prohibiting certain noxious uses and activities (including smoking in or around the building), to limitations on the tenant's ability to perform any repairs or alterations to the leased premises without the landlord's prior consent. A green lease may also include specific janitorial specifications that define the materials, procedures, and protocols for cleaning the building in a sustainable manner.

A green lease may have increased reporting requirements for a tenant. A typical lease might require a tenant to report to the landlord its gross sales from the premises, or perhaps provide periodic financial statements on the tenant's business. A green lease could also require the tenant to report to the landlord its energy consumption data, monitor readings related to indoor air quality issues (e.g., carbon monoxide levels and ventilation efficiency), and equipment efficiency. Such requirements in a lease must be drafted to first require the tenant to collect and maintain such data, as well as to provide it to the landlord on a periodic basis or upon the landlord's request. Failure to do so may result in a tenant default. Obtaining this information could be very helpful to the landlord in terms of ensuring that desired indoor air quality standards are being met, confirming that tenants are complying with energy management programs, and collecting data about the overall performance and efficiency of the building that could be used for certification or marketing purposes.

Examples of Green Cleaning Specifications

The following are examples of green cleaning specifications, some or all which may be included in a green lease:

- Use only products and equipment considered environmentally safe and that are phosphate free, non-corrosive, non-flammable, low-VOC emitting, and fully biodegradable.
- Use only "green" products as defined by Green Seal's GS-37 standard.
- Minimize the use of products that release irritating fumes.
- Use products that are packaged ecologically.
- Utilize paper products that are 100% recycled content.

Examples of Green Cleaning Specifications (*continued*)

- Utilize biodegradable or compostable garbage bags.
- All vacuum cleaners will be of a type that can remove 99.97% of harmful particles, including dust, mold spores, and most microscopic respiratory irritants and allergens. If disposable vacuum bags are used, they must be replaced in accordance with the manufacturer's directions once filled and cannot be re-used.
- Tenant will provide Material Safety Data Sheets (MSDS) to the Landlord for all janitorial supplies provided by the Lessee.
- Tenant will follow good housekeeping practices, including proper disposal of open food and drink, toner cartridges, and any other items that may attract pests, damage Landlord's property, or threaten health.
- Where recycling programs are in effect, Tenant will properly separate recycling material from trash and use the appropriate receptacles for disposal.
- When the Tenant provides light bulbs/tubes under the terms of the lease, the Tenant will establish a lighting recycling program for spent lighting.

REPAIRS AND ALTERATIONS

One of the most important areas to be addressed in the green lease is the ability of the tenant to make repairs, alterations, or improvements to its leased premises. Like the initial construction, the landlord has a vested interest in the materials and procedures utilized by the tenant to make any changes in its space over the term of the lease. Most leases make the tenant's ability to perform any repairs, alterations, or improvements to the leased premises subject to the landlord's prior approval; however, leases often contain provisions that allow tenants to make certain alterations or improvements to the premises without the landlord's consent if they are of a specific nature or are estimated to cost less than a certain amount to perform.

The landlord of a green building must be particularly wary of such exceptions. For example, the cost to repaint, replace carpet, or install wall covering in a tenant's premises may well cost less than the threshold established in the lease to trigger the landlord's consent rights, but the landlord may have significant concerns about the *type* of paints, carpets, sealants, or adhesives used, as they may have a significant impact on the indoor air quality of the building or may not meet the landlord's standards for recycled content (in the case of products like carpet and wall covering). To address these concerns, the green lease may simply prohibit any repairs or alterations by a tenant without the landlord's prior consent. Alternatively, the green lease could contain specific requirements, whether in the body of the lease or as separate exhibit or handbook, for such alterations and improvements, including a list of permitted and/or prohibited products.

The green lease could also contain specific requirements for the contractor performing any such work on behalf of a tenant and rules and regulations for how and when such improvements may be performed to minimize any adverse impact on the occupants of the building. As well, the landlord may prefer to retain responsibility for maintenance of the mechanical systems serving the tenant's premises (particularly the HVAC system), rather than attempt to regulate the tenant's performance in this regard. The landlord may even elect to require that all improvements by the tenant to the premises meet a particular standard, such as LEED for Commercial Interiors.

The so-called casualty provisions of a lease addressing events like fire, wind, and water damage may also have different implications in the context of a green lease. Specifically, the landlord may want to retain control over the reconstruction of the tenant's space following a casualty so that the landlord can ensure that the space is rebuilt in the most sustainable manner. Further, the landlord may reserve the right to reconstruct the tenant's space following a casualty to a higher standard, from an environmental standpoint, than the original premises.

The Double-Edged Sword of the Green Lease

The use of the green lease by a landlord may have some unintended consequences. For example, if the lease declares the landlord's intention to seek LEED certification for a building and the landlord is not successful in obtaining it, a tenant may want to have a remedy against the landlord for such failure, including the right to pay reduced rent or possibly to terminate the lease. Similarly, a tenant may want recourse against a landlord if the landlord desires to impose certain standards on the tenant with respect to indoor air quality, yet fails to impose or require compliance by other tenants in the building with the same standards.

As well, other seemingly innocuous provisions in the lease may take on new meaning in the context of the green lease. For example, most leases contain a "quiet enjoyment" paragraph. This paragraph typically provides that so long as the tenant pays the required rent and abides by all of the other terms and conditions of the lease, the landlord will not disturb or allow others to disturb the tenant's use and enjoyment of the premises. It is possible to imagine the tenant using such a clause as a sword against the landlord if, for example, the indoor air quality of the building declines or is not as represented by the landlord.

Finally, when the tenant, rather than the landlord, is driving the green leasing process, a landlord can expect to be held to a higher standard for the features and performance of the building than is typical, and the lease may contain certain tools the tenant can use to ensure that the landlord meets these standards, including the right to perform for the landlord at the landlord's expense, the right to reduce or withhold rent, and the right to terminate the lease. This could pose new challenges in the financing arena, as lenders and capital partners will also be required to understand the dynamics of operating a green building and the obligations being placed on a landlord/borrower in this context. As well, the landlord may have certain obligations to provide green janitorial services to the building, to offer a recycling program, to upgrade the building's electrical, plumbing, or HVAC systems for higher efficiency, or to report data to the tenant regarding the building's operations and energy efficiency. When a new building is to be constructed or renovated, the lease may grant the tenant significant rights in determining the overall design of the building and the engineering of the various systems within the building. As more tenants begin to specifically seek out green buildings, the green lease will become a requirement and the provisions of the lease may work to impose requirements or limitations on both the landlord and the tenant.

Building Operations and Maintenance

Once a developer has completed a green building, received LEED certification, and leased the building to tenants, it will still need to ensure that the building actually operates as intended (and as the developer advertised). In some situations a developer must ensure that the building continues to operate satisfactorily in order to maintain LEED certification. The LEED for Existing Buildings program (formerly known as LEED-EB), for example, requires that a project be re-certified at least every five years. In order to achieve re-certification, a building owner must provide performance data for every year following the original certification.

The need for operational issues to be addressed in leases has been discussed above. Property managers and service providers must also be made aware of the building owner's goals for the ongoing operation and maintenance of the project. Contracts with those parties must ensure that their obligations are performed in a manner consistent with the owner's goals and that, where applicable, the requirements for maintaining the project's certification are fulfilled.

Buying and Selling

Developers (particularly green developers) may have many goals for a project, but one goal nearly everyone has is to make a profit. Evidence continues to accumulate both that green buildings are more valuable and that the market has recognized this value in terms of higher purchase prices. The price spread between green buildings and conventional buildings will only increase as energy prices continue to climb.

Buyers, of course, need to be sure that the value is really there. LEED or some other certification helps confirm that a building is green, but it will not give a complete picture. If certification was obtained when the building was constructed, for example, it will not help a buyer confirm that the building has actually used less energy than other buildings over time. Certification also would not preclude the possibility that alterations have been made to the building that are less sustainable or green than the initial improvements.

Energy and sustainability audits will likely become a typical part of a potential buyer's investigation of a property. Buyers may request information regarding how energy and water are managed, what materials have been used to make alterations, how air quality is measured, whether programs are in place to encourage the use of public transportation or bicycles, and what percentage of the building's waste is recycled. Sellers will need to be able to provide all this information.

A large part of being green is simply being aware of how much energy a building uses, how many resources it requires, how healthy it is, and where there is room for improvement. To truly protect the value of its asset, however, the owner of a green building must not only know these things but must document them as well.

The Costs and Benefits of Green Counsel

Developers should strongly consider hiring an attorney to help prepare the various contracts discussed in this chapter. As we have discussed throughout, because developing a green building raises issues that do not exist in a conventional building project, each contract is likely to be more complex in the green-building context, some, of course, more than others. Legal costs are therefore likely to increase somewhat.

The variables that can influence the amount of legal costs are nearly unlimited and include the complexity of the project, the level of experience that participants in the development process have with green building, and whether aspects of the project change during the development process. In an effort to provide some idea of how legal fees might increase in a green building project, however, we offer the following example:

Suppose a developer is building a 70,000-square-foot office building on previously vacant land. Total budget for the overall development of the project is estimated to be approximately $15 million, with a developer fee of around 4 percent, or $600,000. The developer hires an attorney to assist with the acquisition of the land, financing of the construction, permitting of the project, negotiation of the design and construction contracts, and leasing of the building. For a conventional project, the budget for legal fees will amount to slightly less than 1 percent, or $135,0000, broken down as shown in Table 9.1.

Legal feels associated with acquisition and financing costs should not increase by more than 5 percent or so if the project is a green building. Legal fees to negotiate the design and construction contracts could be expected to increase by something closer to 10 percent, however, given the issues related to certification, construction practices and the like that we have outlined above. Legal costs related to leasing can also easily increase by 10 percent or more, since a new lease form may need to be developed,

TABLE 9.1 LEGAL FEES FOR A CONVENTIONAL PROJECT	
ASPECT OF WORK	**PROJECTED LEGAL FEES**
Acquisition and Financing	$0.43 per square foot ($30,000)
Design, Construction, and Entitlement	$0.30 per square foot ($21,000)
Leasing	$1.20 per square foot ($84,000)
Total	**$1.93 per square foot ($135,000)**

and tenants can be expected to resist or negotiate provisions related to the green features and operations of the building.

Applying these percentages to the numbers in the previous example, the budget for legal fees changes as shown in Table 9.2.

The total increase in legal fees in this example is $12,000, or 8.88 percent greater than fees anticipated as part of a conventional non-green development project. In the overall scheme of our fictionalized green development project of $15 million, the added cost amounts to eight basis points, or less than one-tenth of one percent.

The benefits of these increased legal costs, though difficult to quantify, can be significant, and include the successful construction, certification, leasing, and sale of a project; the avoidance of violations of law; the achievement of tax credits and other incentives; and the prevention of lawsuits.

Concluding Remarks

The development, financing, operation, management, and leasing of a commercial building is a complex process; this is especially true in the case of a green building project. There are a large number of stakeholders involved in and affected by a developer's decision to develop a green building. Underpinning the relationships between

TABLE 9.2 PROJECTED LEGAL FEES FOR A GREEN PROJECT			
ASPECT OF WORK	**PROJECTED LEGAL FEES**	**GREEN BUILDING PREMIUM**	**NEW PROJECTED LEGAL FEES**
Acquisition and Financing	$30,000	5% ($1,500)	$ 31,500
Design, Construction, and Entitlement	$21,000	10% ($2,100)	$ 23,100
Leasing	$84,000	10% ($8,400)	$ 92,400
Total			**$147,000**

the developer and these various stakeholders (lenders, capital partners, architects, contractors, and tenants, to name a few) are a number of legal agreements, each of which must be modified in certain material respects to communicate, advance, and protect the developer's goal of developing a green building. These agreements must be drafted in a manner that clearly communicates the developer's goals, properly allocates the rights and obligations of the parties, and provides the necessary flexibility for the developer to regulate the conduct or performance of the various stakeholders through the evolving process of developing and certifying a green building. As such, a critical element of the developer's strategy will be engaging an attorney who is knowledgeable not only of the typical legal issues involved in this complex process, but who also has a thorough understanding of green building technology and the requirements for certification of a green building, and who can utilize this knowledge to effectively guide the creation and negotiation of agreements with all the necessary parties.

As the demand for green buildings in the marketplace continues to grow, the need for knowledgeable attorneys will rapidly increase. Developing professional expertise in this area takes a significant investment of time and study. To date, we have gained our expertise through a number of different channels that include attending courses on the LEED certification process and green building practices, studying for and taking the exam to achieve the designation of LEED Accredited Professional, participating in design charettes, getting involved in industry organizations and trade shows, developing lease and contract language, and in-the-trenches negotiation of various types of agreements. As the number of green building projects grows, experience will reveal new issues and new areas for study, possibly resulting in an entirely new sub-specialty of real estate development law. Developers hiring an attorney should thoroughly investigate that attorney's qualifications to guide them through the green building process.

NOTES

[1] Stewart Brand, *How Buildings Learn* (New York: Penguin Books, 1994), p. 54.

[2] Claim examples courtesy of Victor O. Schinnerer & Company, Inc. The examples have been simplified to isolate the aspects of the claims related to green building. As of this writing, none of the examples had actually resulted in litigation.

[3] Tyler J. Krutzfeldt, managing director of Mont Vista Capital, quoted in "Carbon Trading and Portfolio Approaches Seen as Opportunities for Owners and Investors," Green Real Estate News, December 2007.

[4] "In Green Realty He Trusts," CoStar Group Realty Information, Inc. 2008.

[5] 16 C.F.R. Part 260.

[6] California Business and Professions Code Article 7, Section 17580(a).

[7] Jerry Whitson, "Green Lease," *Environmental Design + Construction*, July 17, 2006.

MARKETING SUSTAINABLE DEVELOPMENT: A MILLION SHADES OF GREEN

DAN MONROE AND LISA LILIENTHAL

SUMMARY

Taking a long view of real estate development is rare in a world where short-term gains typically rule. Couple that with a lack of data about long-term returns, and the misperception that green requires a premium and the business case for sustainable development might not sound particularly persuasive. Pioneering, as Melaver, Inc. has done, has its advantages. But being the first-mover can also be challenging if your ideas are too far ahead of the curve, however grounded they might be in the right thing to do. Balancing that message—that green is both right and smart—was a challenge for Melaver, Inc., its outside marketing firm, and its outside public relations people. But the payoff is clear: today the company is widely regarded as an expert with deep roots in sustainable development.

In this chapter, Dan Monroe, a principal in the marketing firm Cayenne Creative, which is responsible for Melaver, Inc.'s marketing work, and Lisa Lilienthal, an independent contractor who oversees the company's public relations work, focus on the evolving challenges of green marketing. Section one sets the scene of a burgeoning green movement occurring against the backdrop of weakening brands and a growing need for values, in many ways an ideal setting for a green bottom line company. Section two looks at the early work of Melaver, Inc. and its need for a communications strategy for internal branding purposes. Sections three and four consider the broader marketing and public relations needs of a green development company as it begins to construct (and create expertise in) green projects. Section five looks at the actual costs of such a multi-year marketing strategy. The chapter concludes with projections regarding the future challenges of green marketing.

Setting the Green Stage

The market for green building and smart development is growing—some would say it is even reaching a tipping point. And it doesn't start or end with green building. You can't pick up a newspaper or buy a cup of coffee or a gallon of gas without being bombarded with the evidence of climate change. While skepticism and misinformation typified the early days of the green movement, today there is much more consensus around the idea that green is good.

Trend information confirms that consumers are looking for companies to take the lead. Consider, for example, a June 2007 survey by The Climate Group. In this study, consumers reported that they wanted companies to address climate change, but few were able to identify a single brand taking the lead on the issue. The top five U.S. brands perceived to be "green" included General Electric, Toyota, BP, Ford, and Honda. (These were, admittedly, companies touting their green-ness in national television advertising.)

About two-thirds of respondents thought companies should play a bigger role in tackling climate change and said they make purchasing decisions based on that sentiment. Researchers divided this consumer group into three "tribes": the concerned but pessimistic "campaigners" (27 percent), the feel-good "optimists" (17 percent), and the "seen-to-be-green" followers (9 percent). Interestingly, The Climate Group's research also shows a gap between what these consumers *expected* companies to be doing about climate change and what they *perceived* companies to actually be doing. In the U.K., for example, 69 percent of respondents drew a blank when asked to name climate-friendly companies; the percentage was slightly higher in the U.S., at 74 percent.

And yes, the skeptics still have a say. Of respondents who do not consider climate to be a factor in making purchasing decisions, the report drew a distinction between the "unwilling" (12 percent), who are accepting of the issue but not prepared to act, and the "rejecters" (16 percent), who confidently reject the issue and feel informed enough to do so.

The Natural Marketing Institute (NMI) also sheds light on current consumer attitudes with its annual "Top Ten Trends" research into the LOHAS (lifestyles of health and sustainability) marketplace. According to NMI research in late 2007, consumers are ready for what was termed a "deeper values experience." In the past decade, researchers tell us, we've seen an explosion of consumerism some call the "new luxury." Brand experiences that had before been the province of the elite have become available to the masses. For example, the luxury brand Prada and its iconic handbags were once associated only with the very wealthy. Now Prada is within the grasp of the middle and upper middle class consumer. But with the raising of such a bar comes the raising of our expectations. Consumers who have come to expect a high-level brand quality experience at every touch point are now looking beyond the physical and emotional brand dimensions in search of an experience of *fundamental core values*. Now, it's not enough that a luxury car features leather seats. Today we seek luxury *hybrid* cars with leather seats.

What this research suggests is that consumers are equating a company's values with the products it produces. From luxury hybrid cars to couture dresses made from organic and sustainable fabrics, it is not enough simply to have it all—consumers also want to *feel better* about having it all. Consider that luxury ecotourism is the fastest growing market in the tourism industry, or that cause marketing programs are exploding among consumer companies as sourcing, materials, trade practices, and social causes become a part of the brand experience—we want fair trade coffee, Rainforest Alliance bananas, and clothing and shoes made by people who have been paid a fair wage. Locally grown and sustainable foods are sought both for their regional appeal and premier quality. The influence of the USDA Certified Organic seal on foods has increased significantly since 2004. Because of the Internet, consumers are able to research the origins of products and understand, on a level unimaginable a mere decade ago, the relative social and ecological impact of any given product.

So what does this latest chapter say about the brand experience in the United States? In the earliest days of commerce, what people bought was, by and large, locally sourced (and probably pretty sustainable and organic, too). Local farmers served the communities in which they lived, and craftsmen built from materials that were available locally. Exotic materials like silk and tea were imported and available to wealthy consumers. When you traveled across the country, you had new and different experiences in dining and shopping.

Enter the industrial age and the ensuing automobile age. Railroads and trucks made transport over long distances feasible. No longer were you limited to local products and the seasonally available food grown in your community. The early days of national retail were still fairly simple in terms of brand experience. It was a time when brand attributes were more directly related to the quality of materials and workmanship. You could, for example, order from the Sears, Roebuck catalog and understand exactly what you would get. The wealthy and well-traveled could buy a Chanel suit as an investment in design, workmanship, and timeless style, and it would come from Coco Chanel's atelier in Paris. As we entered the electronics age, gadgets from overseas were the new exotics, pricey and full of sex appeal.

You know the rest of the story. As the world has grown smaller, the brand experience has become both more homogenous and more complex. The "blanding of brand" means that regional and even international boundaries have been erased; you can eat at McDonald's no matter where you are in the world. The fancy electronic gadget you once coveted is now ubiquitous, cheap, and probably not that well made. A Chanel suit is still an investment in design, workmanship, and timeless style, but now it's a mass-produced status symbol that can be bought at the mall in just about any major city.

That's where the complexity enters in—the marketplace has become such an overwhelming cacophony of hype that consumers are having a hard time matching their values with a brand. If it's not local, if they can't count on quality of materials and workmanship, and it's not special (but instead something they can buy anywhere), what should they look for? Enter the era of labels—where brand is connected to self, and you use your consumer purchasing power to confirm your social status.

So where are we today? The conclusion by the Natural Marketing Institute that consumers are now looking for a *deeper values experience* suggests that maybe the brand experience has come full circle, with a twist. Antibiotic- and hormone-free grassfed beef seems like a good way to protect yourself against E. coli and mad cow disease. Tomatoes and strawberries taste so much better when they're in season and grown nearby. Price sensitivity and labels are still important, but we want to *know* what we are buying. We're still looking for a toy manufacturer we can trust, and foreign car manufacturers are still making the best high-performing hybrids, but we are increasingly coming around to the idea that local, sustainable, and organic equal safe, nutritious, and dependable.

What does this mean for marketers? Brand is more complex than ever, but a few things are clear, and they translate to the real estate marketplace. The green building phenomenon means that consumers are looking to pursue new high standards in the places where we live, work, and play. Demand for sustainable development is growing. We've spawned something very new. And it's very much alive.

But, again, it's a noisy world out there. Consumers are bombarded with messages and left to decipher their relative truth or relevance with little help. That's where marketing and public relations expertise is critical, in helping a company interpret its brand for the marketplace and—importantly—filtering consumer feedback for the benefit of brand development.

In the 1990s when Melaver, Inc., first started developing green projects, *sustainability* wasn't the buzzword it is today, and the communications challenge was 1) to understand internally the clear direction of the company, 2) to develop a market for green by wooing tenants, persuading lenders and partners, and educating municipal officials, and 3) to set in motion a paradigm change in the community and beyond for a more sustainable mindset. We learned that it required a consistently persuasive message, and that often the sustainability story was not persuasive enough on its own. Fair enough. The real estate development market rightly deserves good aesthetics, sound financials and, of course, location, location, location. These lessons serve us well today, as we work to cut through the clutter of an increasingly crowded green marketplace. We also learned that the communications strategy for a company aspiring to be truly sustainable evolves and deepens as the company's own commitment and capacity to be green evolve and deepen. Early work tends to focus on internal communication—"internal branding" is what we sometimes call it—that is meant for staff members and not the general public. As the company's work evolves, communication shifts to specific targets having to do with direct users of the buildings a real estate company develops. As that expertise matures, communication shifts to broadening the message to the wider audiences the company wants to influence.

John Grant, in his compelling book *The Green Marketing Manifesto*, develops a three-by-three matrix to provide a comprehensive study of green marketing techniques. Along the top are three broad types of green marketing objectives, ranging from narrowly defined green criteria to much deeper and far-reaching ones. Along the side are three levels at which marketing occurs: the personal, the social, and the company level. Each box in the grid calls for a different marketing strategy having to do with the level of a company's commitment to green (setting new standards, sharing responsibility, pushing innovation) as well as the way in which marketing that commitment gets expressed (communicate the message, collaborate with others, work to reshape the culture). Grant's matrix can be seen in Figure 10.1.

Although this matrix was not available when we first began working with Melaver, Inc., it is particularly useful in allowing us to analyze our work with the company as it has evolved over time. More specifically, we can identify three key phases occurring over a seven- to ten-year period, each involving different communications challenges, opportunities, and strategies. Phase 1, probably best located in box A3 of Grant's matrix, involved brainstorming sessions within the company and focused on internal alignment around a set of core values as well as green practices. The early chapters in this book (Chapters 1 through 4) speak well to this early phase in the company's green evolution. Phase 2, best located in boxes A2 and B3 in Grant's matrix, involved communications strategies focused on specific green projects, which in turn involved drawing into the fold credible partners as well as attracting key users of those projects. The middle sec-

	A. Green	B. Greener	C. Greenest
1. Public Company & Markets	Set an Example	Develop the Market	New Business Concepts
2. Social Brands & Belonging	Credible Partners	Tribal Brands	Trojan Horse Ideas
3. Personal Products & Habits	Market a Benefit	Change Usage	Challenge Consuming
	Set new Standards Communicate	Share responsibility Collaborate	Support Innovation Culture Reshaped

Figure 10.1 Grant's green marketing grid[1].

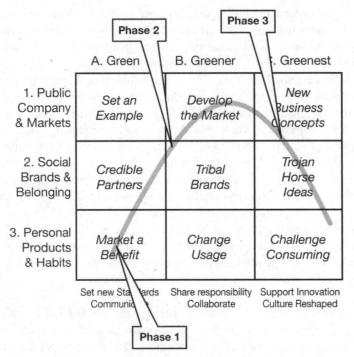

Figure 10.2   Phased evolution of Melaver, Inc.'s approach to green.

tion of this book (Chapters 5 through 7) addresses this phase in detail. Finally, Phase 3, captured in column C, represents current work for the company, as it tries to reinvent how a real estate company can remain viable by not only promoting less consumption but by producing less product. The evolution of Melaver, Inc. over time can be seen in Figure 10.2, which shows a trend line overlaid on Grant's matrix.

Each phase, as we will see throughout the remainder of this chapter, carries with it distinct communications strategies. In Phase 1, the focus was largely on working within the organization, but at an individual level, to reframe the core purposes of its business and to enable staff members to become more comfortable with a "less-is-more" value proposition. In Phase 2, the focus has been one of sharing the company's enthusiasm for green with others, entailing educating the marketplace, sharing information in a non-proprietary way with any and all, and being as broad and inclusive as possible in bringing other stakeholders into the fold. In Phase 3, in motion at the present time, the focus seems to be on linking forces with a network of change agents throughout the field to emphasize the strong linkage between green buildings and practices and notions of sustainable community. Not every company involved in green will experience this same trajectory. There are not only a million shades of green out there, but just as many routes to a green marketing strategy. Our long-term work with

Melaver, Inc., however, enables us to take an in-depth view of how one green marketing strategy evolves over time. Let's consider this evolution in greater detail.

Phase 1: Getting Organized

Melaver, Inc.'s road to green began well before we arrived on the scene, starting back probably in the late 1980s as family shareholders wrestled with how to practice real estate differently. This slow evolution carried into the 1990s, as management and staff zigged and zagged in various directions, taking steps toward becoming greener without really managing to integrate the various pieces into an overarching sense of purpose or executional focus.

Cayenne Creative was brought into the picture to help Melaver, Inc. with two inter-related issues: clarifying the value proposition of the company for all staff members and creating coalescence around the notion of sustainability.

The first of the two issues may best be understood in the context of a seminal article by Michael Treacy and Fred Wiersema, entitled "Customer Intimacy and Other Value Disciplines."[2] Treacy and Wiersema argue that successful businesses by and large have one value discipline at which they excel. They provide superb, customer-intimate service (for example, Ritz-Carlton Hotels), or they are known as product innovators (Apple), or they are highly profitable as a result of being operationally lean and efficient (Dell, Wal-Mart). On occasion, albeit rarely, successful companies are able to focus on two of these three value propositions. No company, the authors contend, performs well if it spreads itself thinly across all three disciplines. The staff at Melaver, Inc. felt precisely this sense of being spread thinly across value disciplines, executing none well. The company was ostensibly in the business of providing service to its clients (mostly tenants). But it also seemed interested in being a product innovator (through its green orientation) and interested in creating operational efficiencies (through developing high-performance buildings). At which value discipline was it supposed to excel?

Compounding the confusion over value disciplines was the specific focus on being green. The shareholders of the company had become convinced this was the direction to go, as had senior management. But the staff was still puzzled about, and not a little fearful of, the changes this orientation seemed to imply. Cayenne was thus brought in to help in a process we call brand engagement. In the words of Martin Melaver, senior management wanted to introduce this "new" Melaver to the employees in a way that would help integrate brand sensibilities into the behaviors of the company. And they wanted us to do it at their annual corporate retreat.

We found ourselves asking what was then and still is a challenging question: "What the heck is a sustainable real estate company?" Fully knowing we were entering uncharted waters, we began where brand should always begin—with an understanding of corporate values. We conducted a series of workshops with various employee groups within the company. We learned that the four Melaver corporate values were Ethical Behavior, Learning, Service to the Community, and Profitability. We also learned that, although the employees had a general understanding of these values, recall was not immediate or second nature. So one of the first things we set out to do was to rephrase the values, to simplify them into something mnemonic. The result was: "Do right. Learn. Serve. Grow." The use of verbs gave the values immediacy and the sense of a call to action.

With those values in mind, our design team explored a visual identity system, creating a graphic toolbox that assigned one of four signature colors to each value. We further created a brand "manifesto" to inspire Melaver employees and encourage reflection. Next, we put our minds to work on how best to introduce an overall brand to the entire Melaver corporate family. With the retreat in mind, we sought to create a series of elements, both passive and interactive, to help engage the entire organization in what was to become the Melaver brand.

Element #1: Brand video. The point of corporate retreats is to gather the team, inspire them, and send them off ready to accomplish great deeds. In creating the video, we sought to kick off the part of the retreat with which we had been entrusted with a piece that would inspire, excite, or—at the very least—cause reflection and—at its most powerful—raise goose bumps.

Element #2: Brand engagement exercise. We landed upon the idea of using the cognitive device of a terrarium to help bring the Melaver values to life. It made sense. It created a personal exercise in stewardship—micro-environmentalism, if you will. In the course of our design exploration, we had associated each of the values with a color. In packaging and presenting the terraria, we further related each value to a natural material to be used in the exercise. For example, we associated "Do right" with a taupe color and, subsequently, with the idea of foundation, with stone. We packaged gravel for the base of the terrarium in one of four boxes that neatly fit inside the terrarium, emblazoned with the words, "Do right." Doing right is, after all, truly the foundation of Melaver, and that upon which all else is built. We associated the value "Learn" with soil, and with a deep, rich brown. Soil in this exercise reflected the feeding of the company, just as soil feeds plants. Again, we packaged the soil for the terrarium in a nicely designed box emblazoned with "Learn." We associated "Serve" with watering, with service to the terrarium, with a pale blue, and enclosed a bottle of water in a blue-colored "Serve" box. And lastly, we associated profitability, or "Grow," with the plant itself and the color green. Serendipitously, the plant we selected for the terrarium and which we wrapped carefully in our green "Profit" box—the "prayer plant" (*Maranta leuconeura erythroneura*)—has, as part of its markings, what might be seen as a loose interpretation of the Melaver logo. The effect of the exercise was galvanizing. To this day, every Melaver employee can recite the corporate values with ease. And the terraria can still be found throughout the corporate offices (see Figure 10.3). Remember earlier, when we were talking about *alignment* and *authenticity?* These are the sorts of lessons that cement those corporate values and help Melaver maintain that alignment and authenticity.

Element #3: A succinct statement of purpose. After much jawing and circular arguments about the focus of Melaver, Inc., the entire staff finally recognized that

Figure 10.3   Core values "packaged" in a terrarium.

it was and would continue to be a service-focused company, serving the land and community of which it is a part. The central mechanism for this service was a sustainable orientation that called for providing cutting-edge, innovative products in its developments—innovations that would assist the company's clientele consume less and be more organizationally efficient. The company indeed touched upon three disciplines— service, innovation, and efficiency. But its primary focus was service to land and community. The light bulb went off in staff members' heads as we shaped a statement of purpose for the company. One sentence, but it took a full day of discussion at the company's annual retreat to shape it fully: *To envelop our community in a fabric of innovative, sustainable, inspiring practices.*

Element #4: Internal collateral materials. As a follow-up to the company's retreat, Cayenne created a series of posters built around the company's four core values. In addition, a series of four photographs—stand-alone shots that when combined provided a story of the company—was created, each one built around a core value and containing visual clues or "inside jokes" that only a staff member would know and understand. The point here was clear: the branding was internal, not meant for an outside audience. The company had to get the walk right before it would do any talking to the outside world.

Element #5: CommPost materials. We recognized in the course of learning about sustainable development (as part of the process one Cayenne employee became LEED accredited) that sharing of information was vital to this very new sort of endeavor. No single person or company knew all there was to know about sustainable development. Exchange of information with groups outside the company was important, exchange of information within the company even more so. CommPost is the idea of an open network of information exchange. As part of the retreat, we created journals for each employee. The journals featured perforated pages where the removed portion could be used to exchange ideas with others, and each employee was provided with a personalized stamp and inkpad. (See Figure 10.4.) The journals may never have been used in quite that way, but the point of exchanging information was not lost on the Melaver corporate family. The name CommPost, however, has stuck as the name given to Melaver's communications team, a team charged with, among other things, coordination and dissemination of information through the company's intranet site.

There are a few important things to note as a result of this initial phase of work with Melaver, Inc. For one, while the company had done considerable work in prior years shaping a sense of shared values as well as a powerful and empowering bottom-up culture, its value proposition still needed to be clarified and better communicated. Put another way, staff members needed to better understand how their shared sense of values connected to the business case for creating value. In the early years of the LEED program (2000–2005), there was not only a strong perception but also a great deal of truth to the idea that green cost more. Staff members, early on, needed to be convinced that the company, with its sustainable orientation, was not heading down a path that was "flavor-of-the-month" trendy or financially dubious.

Another critical piece in this initial phase concerned the degree to which it was strictly internal. Marketing played a critical role in framing the notion of sustainability for staff members, enabling them to share and exchange knowledge and information with one another that would later shape the company's delivery of products and services. But at this point public relations was not at the table. The company knew it was not even close to being ready for prime time. Lots of internal work—training,

Figure 10.4   CommPost stamp and journal.

measuring its own environmental footprint, learning the ins and outs of the LEED program, etc.—was called for first.

THREE FUNDAMENTALS

Three fundamental "musts" helped us lay the groundwork for messaging that makes it through the clutter. As with most successful marketing and public relations campaigns, you'll notice that it's not just about catchy slogans and slick ad copy. We owe every bit of our success to the fact that Melaver, Inc. is a mission-driven organization with a clear commitment to sustainable development. Critical to our marketing success are top down support for the project, alignment with the organization's mission, and authenticity and transparency. Let's consider each factor separately.

Top-down support for the project. The management team at Melaver, Inc. has a clearly defined mission that informs how every deal is brokered, developed, leased, managed, and built. Support for green building initiatives starts in the boardroom at Melaver and it's apparent in every management and participatory decision the company makes. There is no confusion in messaging and execution—what the company says is consistent with what it does. They're walking their talk.

Consumers can see the difference between companies that simply talk the talk and those that walk it. Sometimes it's subtle—how the company talks about itself on its website, how others talk about the company, how the receptionist answers the telephone, how its offices are different. Those implicit cues cannot be faked—they are either part of a company's fundamentals and genetic makeup, or they're not.

You'll know you've achieved a level of consistency when you can think of your company, your people, and your projects as one piece cast from the same material. If the market were able to slice through the entire aggregation, the cross-section would all be of a single consistency—a single color. And it darn well better be green.

Alignment with the organization's mission. The shaping of deep alignment was probably one of the single most critical outcomes of our Phase 1 work with Melaver, Inc. That early work brought home to all of us how senior management's perceptions of alignment and actuality can indeed be rather skewed, even (perhaps especially) in the context of a values-centric organization where there is a strong supposition of value congruence. Our early Phase 1 work made it clear that you can never take that alignment for granted, even in later iterations of a green company's evolution.

Alignment is critical to effectively communicating the benefits of green building. It is critical to have a communications team on board that is genuinely behind the effort and not just mouthing lip service in the interest of a consultancy contract. It is critical to have other partners—legal, HR, IT, general contractors, engineers, landscape and building architects—aligned with the green effort as well. Perhaps the most critical piece of early aligned partnering for Melaver was the U.S. Green Building Council (USGBC) and its LEED program. The commitment to the LEED program as the gold standard for development sets a clear mandate. This well-defined, well-executed, mission-driven approach distinguishes companies like Melaver, Inc. in a market where competitors may not be developing to the same standard or have the same mission-driven approach. When the USGBC launches a set of standards for a new type of development it is not surprising to see Melaver involved in a pilot project. For example, when USGBC launched LEED, Melaver's Whitaker Building project was one of the first to be certified. Abercorn Common was the first LEED certified all-retail center in the country (and a pilot project in the LEED Core and Shell program). The company has just completed its renovation of a 19th-century historic home that is in the LEED for Homes pilot project, earning a LEED Platinum rating in the process. At this writing, Melaver is engaged in the development of Sustainable Fellwood, a low-income housing development in Savannah, Georgia that's an early entrant into the LEED Neighborhood Development pilot program. The bottom line is that, rather than a formulaic, project-of-the-month approach, Melaver exhibits a long-term commitment to the sometimes steep learning curve that comes with pioneering. The genuineness of initiatives such as these cannot be faked in the marketplace.

Authenticity and transparency. There was a time when tenancy and purchase decisions followed a rather formulaic approach. Market trends, cost per square foot, location, size and accessibility, and, to a lesser degree, operating costs were all spreadsheet entries in what was, at the time, a tried-and-true approach to making a buying or leasing decision. But the marketplace for green building is an increasingly sophisticated one. In addition to those tried-and-true traditional metrics, potential tenants and buyers interested in sustainable developments consider different project attributes when making their decisions. Buildings are perceived in terms of their relative health. In addition to the obvious savings accrued through energy efficiencies, employers are looking at harder-to-measure benefits such as decreased churn, lower absenteeism, and greater productivity. In a crowded field where insincere greenwashing abounds, authenticity and transparency make communication simpler and more effective. You don't have to fabricate benefits—the potential market already understands those. You simply have to show how you're delivering them.

Phase 2A: Green Marketing 101 for a Different Type of Product

You can build it, but they won't necessarily come. If you are a sustainable developer, you're not only competing against other sustainable facilities—you are also competing against the existing inventory of conventionally developed real estate. While marketing green products and services does have unique challenges, it is still, in the end, marketing. And it still boils down to solving three basic problems: what, who, and how, or, in marketing parlance—message (what you want to say), target (whom you want to reach), and means (how you will deliver your message to your target). Knowing what your project really is and creating an identity for it comprise challenge number one. Deciding whom you want to tell about it is the second hurdle. Finally, you want to know how to go about telling them. Message. Target. Means.

FIRST THINGS FIRST: WHO ARE YOU AND WHAT IS THIS PROJECT?

Again, in marketing parlance, what's your brand? Some of the hardest work you'll do is in identifying the inherent nature and personality of your company or your project—in other words, your brand. But this work is vital. Knowing what your company represents and creating a compelling brand around it are keys to maintaining a longstanding position in any market. The same is true of each project you undertake And knowing brand at the beginning saves all sorts of identity crises on the back end. Every message you take to the market is informed by what you know about the brand of your company or project.

WHY BRAND?

Remember that your project is truly an expression of the company or companies behind it. The public begins to associate projects with developers. Having a strong corporate brand will help ensure success over the long run. Development of a corporate brand is important business and is done in much the same way as we've been discussing for projects. In the case of Melaver, we were asked initially to define a visual identity for Melaver, Inc., and then to do the same thing for the company's various projects.

CASE STUDY: BRANDING ABERCORN COMMON—A TALE OF TWO TARGETS

Target #1—Tenants

Branding Abercorn Common Shopping Center to potential tenants was one of Cayenne Creative's first project-related challenges for Melaver. Here we had the chance to extend the Melaver, Inc. developer brand into a key project that was decidedly green. The primary target for our message didn't necessarily get sustainability yet. This was, after all, before large retailers like Wal-Mart were greening their businesses, and there were still a lot of misperceptions in the market about green. Would it cost more to rent there? Would the quality suffer? Would a potential tenant's hair salon be located next door to a bunch of hemp-wearing, patchouli-scented, hippy-dippy candle makers, or to a nationally advertised deli? Secondly, because Melaver was truly in new territory with the LEED certified development of a 100 percent retail project, the budget for marketing was necessarily tight and focused.

In creating the initial marketing materials, our primary inspiration—but also our challenge—was setting. The development was something of a conundrum. On the positive side of the ledger, it was located at the prime main-on-main retail location in Savannah, Georgia, with a significant traffic count. Formerly the site of a shopping complex developed by the Melaver family when it was in the supermarket business, its *bona fides* as a successful and desirable retail location were unquestioned. When the original shopping center was developed in the late 1960s, it was located along the corridor of southern-moving sprawl beyond the city's urban core. But over the subsequent forty years, the location became more a part of the city's urban built-up area, despite a general ambience that suggested franchise-row suburban blight. The Melaver team had conceived Abercorn Common to evoke design elements of Savannah's historic downtown, while remaining true to its shopping center roots. From a design standpoint, the site sat at the crossroads of an older historical city center and a newer suburban one.

Even more challenging was the fact that while Melaver, Inc. knew this project would likely be the first LEED retail shopping center in the country, the company decided that the design should probably not scream "green" but should have the look and feel of what tenants and shoppers were more accustomed to and yet, at the same time, serve as a demonstration project everyone could learn from. In fact, an early concept of incorporating a green discovery center into the shopping center was jettisoned for a plan that called for the entire Abercorn Common to serve as a walking, living educational center with informational signage and visual clues as to what was different about it. The project was intended to be what John Grant refers to as a "Trojan Horse": "making new green stuff...seem normal."[3] How do you create a marketing strategy for a product that is radically different but looks the same? How do you pitch a uniquely differentiated product to a universe of retail tenants that is likely uninterested in and largely uninformed about the differentiation you are providing?

The visual language we created for Abercorn Common Shopping Center incorporated imagery and language associated with the history of Savannah, Georgia. The name, Abercorn Common, lived deeply in the history of Savannah and incorporated a sense of "common space." We created a look that lived in earth colors and a feeling of warm, historic value. Very little true advertising was done for Abercorn Common, but the little we did featured natural imagery and spoke to tangible tenant benefits such as the creation of a natural working relationship between big box retailers and small boutiques, or the value of the traffic count. We talked very little about the sustainable aspects of the development because, though it was a part of Melaver's brand, the target wasn't ready for it. Our target, the retail community, is a notoriously "hard," bottom-line-focused group. Again, the process is basic—message, target, means:

- **Message:** The most important promises we could make to the retail community were related to location and co-tenancy. So our marketing talked to traffic counts and anchor tenants.
- **Target:** Our target was potential retail tenants—bottom-line-focused and needing to be reassured about the value of the location to them. They were not, for the most part, interested in green.
- **Means:** We advertised in trade publications that talked directly to our target such as *Shopping Centers Today*. We also created branded materials to enhance the presence of the leasing team at conferences where they would encounter a large concentration of potential tenants. We created a trade show booth and handout materials using the graphic toolbox we used to express the Abercorn Common brand. Those same graphics were used in a website that was focused primarily on

In Savannah, Georgia, shoppers from a
41-mile radius are drawn to a single road.

{ Abercorn Common – blooming this coming spring
at the corner of Abercorn Street and White Bluff Road. }

Figure 10.5   A typical ad for Abercorn Common.

informing potential tenants. A typical ad we used to convey the Abercorn Common brand can be found in Figure 10.5.

Target #2: The Public

Once Abercorn Common was completed and leased, we were tasked with helping to inform the shopping public of what made Abercorn Common different. We created signage and a self-guided tour that helped shoppers understand why choosing to shop at Abercorn Common was actually benefiting the environment. The information signs were provocatively worded. Here are some examples:

> *Rainwater Capture.* Why use drinking water to water all these trees? Rainwater is perfectly fine. Every year, about 5 million gallons of rain falls on the roofs and is collected and used to irrigate our landscaping.
>
> *Not Your Ordinary Parking Lot.* Aren't all paved surfaces the same? Nope! Over an acre of this parking lot is "pervious" (the white sections). What does that mean? Just pour a glass of water on a pervious section of the parking lot and see what happens. Regular asphalt carries pollutants like car oil into local watersheds. Pervious paving absorbs water and filters it through the ground. Bottom line: 25 percent less storm runoff. Cheaper and cleaner.
>
> *Not Just Any Old McDonald's.* Are you sure that's a McDonald's? Yep! This McDonald's looks different because it is the first LEED certified McDonald's in the world. More daylight, less energy and water usage, and better indoor air quality. To learn more about LEED (Leadership in Energy and Environmental Design), visit www.usgbc.org/LEED.

We created keyed hardscape maps, and walking tour maps, so curious shoppers could tour the property and see the efforts that were made in creating a sustainable property. (You can see the Walking Tour Map in Chapter 5.) It's important once again to note that the marketing process was key. Note how very different the marketing proposition for the public is from the marketing proposition for the tenants:

- **Message:** Shopping at Abercorn Common is anything but common.
- **Target:** The shopping public, curious about this "LEED" business they keep hearing about.
- **Means:** On-site signage and maps.

HOW TO BRAND

Coming to understand a brand is a process. Even after years of working with Melaver, Inc., we still work intently at understanding this company, what it feels comfortable messaging out there in the public eye, what it prefers to keep quiet about, how it prefers to work. For instance, as a bottom-up, shared-leadership company, Melaver, Inc. views all of its staff members as the face of the company. This probably dates back to its time in the supermarket business, when cashiers and baggers were trained to understand that as the last points of contact with the customer, their roles in many ways were the most critical for the business. This notion of everyone-is-the-face-of-the-company means, among other things, that Melaver staff members want to personally own the marketing materials that we create with them (not *for* them). That's a challenging—and in many ways inefficient—charge for an outside marketing company to deliver. And yet, that is a critical part of the company's persona, its brand. As in so many things with this company, the "how" or process is as critical as the "what" or final result.

Probably the hardest, most important branding work we've done for Melaver, Inc. was done in the first six months. During that time, we conducted numerous workshops and "brandstorming" sessions to define the Melaver brand. The company was not interested in the slightest about our creating an image for them. On the contrary, what was desired was the authentic company, warts and all. All parts of the company participated in these early brainstorming sessions on branding, and some of them were no-holds-barred scary, in the sense of staff members being brutally honest with one another. That, too, fits the company's brand, which is largely about the shaping of lasting community and the disparate jab-and-counter-punch give-and-take of conflicting perspectives trying to find good enough resolutions to complex problems. Consider, for a moment, the various challenges an ethos of sustainability poses: In the interest of reducing global emissions of carbon, we need immediately to move away from the use of coal, which leaves nuclear power as the current widely available alternative. Not a particularly palatable choice. Does a company such as Melaver, Inc. focus its efforts on creating denser, vertical urban developments in and around its home market of Savannah, even though rising sea levels over the next half century are likely to tax the resources of the 75 percent of Americans living within one hundred miles of a coast? Or does this company instead look to create community in more rural environments inland, even though the footprint of such non-urban environments is a multiple of that found in a major city? Working in such a complex arena as sustainability calls for a penchant for dialectical discourse and debate, where very few solutions are obvious or easily implemented. That dialectical sensibility is also part of the Melaver brand. Trying to capture this and other elements of the company has informed everything we've undertaken with this company, from developing the website, to creating pieces handed out at conferences, to designing the posters that greet visitors to the Melaver offices.

BRANDING YOUR PROJECT

You may or may not decide to work with an advertising or marketing agency to help create your brand. Perhaps it is my own bias here, but I do feel that it is often easier for an outsider to see the persona of your project or company more clearly than you can. That notwithstanding, whether you work with professionals or not, asking these questions provides a useful foundation for exploring the brand of your project or the brand of your company:

- What is the inherent nature of this project?
- Is it a housing development? Retail? Office?
- What makes it different from other projects out there?
- What do people need that this project provides?

Naming Your Project

What is the name of your project? Remember—this will become associated with the project for its life. You'll want to do your homework here. There are lots of ways to research possible names. Here are some things to consider:

- Make sure it's memorable.
- Steer clear of parts of names that are already in use, if possible.
- Check to see if the Web address for the name you select is available.
- Consider the descriptive words you use (Village, Heights, Center), and attempt to use words that are unique and not often used.
- Involve lots of minds in the process.

Strategies for name selection abound. It's best to involve lots of minds and generate as many candidates as possible. At Cayenne Creative, we involve the entire staff in naming projects. We literally cover the walls with sheets of paper, and everyone adds to the ever-growing list. In the end, when we've created as many candidates as possible, we arm each member of the firm with a colored marker and everyone votes for their favorites by placing a small dot beside the names they like. In this way, we are able to winnow the list. Here are some strategies you can use for finding potential names:

- Study the natural history of the area.
- Research the names (both Latin and common) for plants and animals native to the area.

- Explore the settlement history of the area. Were there Native American tribes? Are any names associated with the tribes appropriate? Are the names of any of the first settlers in the area appropriate? Explore the legacy names of the area and determine if any of the old names are ready to reappear.
- Look into the history of your own company and your founders. Are any of these names appropriate?
- Observe the terrain of the area where your project is sited. Are any of the names of terrain features (such as rivers, lakes, mountains) appropriate to your project?
- Try a free association exercise. Select a random starting point unrelated to your project and free associate. Write down the words you come up with. Use these to free associate further.
- Consider the end use of the project and the feeling you want people to have when they first hear of the property. Explore words that have that tone and feeling.

Brandstorming Exercise #1

Often, it takes seemingly offbeat questions to see the familiar through a new and eye-opening lens. If this project were a human being, how would you describe it? This is an exercise we often use with products, services, companies, or projects. Start with the sex of the project and go from there. Is this project male? Female? Old? Young? What would this project drive? What does it wear? This may seem like a little bit of craziness, but what you end up with is solid gold. What you are trying to isolate is the inherent nature of this project—how it "feels." Knowing this gives you a sense of the tone you will use to market the project.

For example, a project featuring high-end condos may "feel" female, mid-fifties, upscale—the sort of project that would drive a Cadillac, look best in a formal gown, wear pearls, prefer an upscale seafood restaurant to a sushi bar, drink Manhattans, and whose favorite movie was *Out of Africa*. Or this same high-end condo project may "feel" male, mid-forties, upscale—the sort of project that would drive a 1960 "scoop-side" Corvette, feel most comfortable in worn khakis, Italian loafers with no socks, and a silk crew neck, drink nothing but sparkling water, and whose favorite movie was *Citizen Kane*.

We're talking two *very* different projects.

Brandstorming Exercise #2

A variant on brandstorming exercise #1, this approach similarly seeks to get at the essence of a brand by reframing the question in ways that you can touch and feel. The process of putting the spirit or sensibility of a business into tangible words can be a frustrating one, though stick-to-it-tive-ness often yields good results. Gary Hirshberg, CE-Yo of Stonyfield Farms, describes his own team struggling through frustrating meetings before landing on its key slogan: "You just can't fake this stuff."[4] So what are other things that feel like this project? Here's a list you can use:

- Beverage (could be alcoholic, could be Kool-Aid)
- Vehicle (car, boat, bicycle, Segway)

- Personality (celebrity, living, dead, doesn't matter—could be George Washington, could be Isaiah Washington)
- Architecture (could be the Pentagon, could be the Sydney Opera House)
- Chair or seat (could be a park bench, could be an Eames chair)
- Social event (could be a wine tasting, could be a family reunion)
- Desk object (something you might find on somebody's desk—a picture, a paperweight—be specific)
- Music (band, album, type—be specific)

Again, the point of the exercise is to help identify the feeling of your project's brand. A project that "feels" like an 18-year-old single malt Scotch is very different from one that "feels" like hand-squeezed lemonade.

Blending Authenticity and Inspiration

What is the promise this project makes to the community in which it finds itself? Answering this question helps identify an overarching message you can take to market. If you partner with advertising agencies, design groups, or public relations firms, having done this homework, you can expect better outcomes. The trick, particularly when it comes to green marketing, is to balance the presence of outside professionals with people within the company who feel a passion for what they do. Most (if not all) of the time, sustainably oriented companies have deep ties to place and people, and they see their specific products and services in the context of a *zeitgeist* having to do with how their particular business is integrated into the overall fabric of the community: creating fair, stable, local, and sustainable working conditions, helping reduce waste, promoting education and the exchange of information. By the same token, sharing excitement for one's business or project and the promise it holds out for the community has to be balanced by the capacity to deliver on those ideals. A well-managed brandstorming process blends integrity with aspiration.

Getting Graphic

Once you've done the work to understand the attributes of your project's brand and what it represents to the community in which it finds itself, you will want to develop a strong visual identity for the brand, one that frames all of your communication regarding the project. Visual identity typically consists of a logo or logotype, a primary and secondary color palette, a corporate typeface to be used in correspondence, and other complementary typefaces as needed.

Your branding toolbox should include a strong graphic identity that tells your story and provides a consistent link between brand and project. The first things you will need are: a strong logo design and a professional corporate suite consisting of stationery, business cards, note cards and envelopes, shipping labels, and perhaps a press kit. Implicit in the design of these things is a color palette appropriate to the brand you've identified, a usable family of fonts, and a set of graphic devices and rules that will ultimately inform all outward expression of the brand. Many an endeavor began with a logo designed by "a friend of a friend who's pretty good with a Mac." Avoid the temptation.

While they seem to be adequate at the time, those logos and designs look more and more amateurish as time goes by and can come back to haunt you. Also, improperly designed logos can end up costing you money. You'll want a logo that can be reversed out (white on a color or black), that can be printed in a single color, and that looks good even when it's pretty small. Amateurish logos frequently look lousy printed small or require too many colors to print. Remember this is foundational work for the brand of a place you're pouring your heart and dollars into. Do not be tempted to cut this corner. (We will discuss non-conventional compensation schemes later in the chapter.)

SHAPING YOUR MESSAGE

An important question too few companies ask internally is, "Why do we need to deliver this message?" Ask yourself this: "What is the outcome I hope for from this communication?" For example, are you seeking to lease space to tenants? Are you seeking simply to inform? Here are a number of outcomes you may want to consider:

- **Awareness** – You want the target to be aware of your project/company.
- **Action** – You want the target to visit/lease/purchase/try your project.
- **Relationship** – You seek to establish a relationship with the target by having them call you or visit your website.
- **Education** – You seek to educate your target regarding sustainable development.

It's important to know up front what you want your outcome to be, and to establish a measurement tool to gauge the success of your campaign. For example, deciding at the onset that you want to generate one hundred inquiries a month gives you a target. If the target seems easy to hit, raise the bar. If you seem way short of your mark, try to determine if your initial goal was reasonable.

How you establish your goals depends on what you want your outcomes to be. For example, if you goal is to have people regularly visit your website, you will want to use engagement tools on your website that allow visitors to opt in. An example of such an engagement tool is offering the visitor a monthly update on all LEED changes. If that's something of interest, the visitor must offer up an e-mail address. Equipping your site with relatively simple analytics allows you to verify how often that same person returns to your site. Conversely, knowing the outcomes you seek can help you define your goals. Goals force us to measure results, and measuring results over time lets us know how we're doing. Setting specific measurable goals and determining how you can gather performance metrics are vital to effective marketing, whether you're doing it in-house or using an outside partner.

DEFINING YOUR TARGET

Before you spend a dime marketing your development project or creating communications around your sustainable effort, identify your target market. To whom are you talking? What are their interests? Are they early adopters of sustainability and, therefore, sensitive to communication that over-promises with a hyped-up green message? Or are they more likely to skew the other way—distrustful of environmental concerns and anxious about the costs associated with sustainable development?

Different targets require different tactics. When we sought to create communications for Abercorn Common, you'll recall we used a very different strategy to talk to potential tenants than we did to speak to the general public. We knew, for example, that potential tenants had pro formas within which they needed to operate. We initially projected that $20 a square foot would either make or break the deal. Their primary concern was the bottom line. (We were right about the focus on the bottom line. Our projections about the rental rate were understated by almost 100 percent.) Prospective tenants' first questions centered on how much the space would cost per square foot and whether the property justified paying that amount. The public, on the other hand, was open to a message about sustainability. In each case, we chose messaging strategies relevant to their respective targets. In *Shopping Centers Today*, we talked to the potential for success in locating in the center. Meanwhile, the public relations efforts of Lisa Lilienthal spoke to the "green" message and garnered a great deal of positive public attention from the local press and government, very likely boosting our ability to charge higher rents. Both efforts worked toward the common goal of ensuring success.

Once you've identified specific targets, knowing everything you can about them is key to a viable communication strategy. Create a profile of your target by gathering every piece of information you can. Create a focus group of people representative of the target. Take them to lunch. Sit them around a table. And pick their brains. If you think you can, capture the session on video or audio so you can come back to it. If that's not possible, have a really good note-taker on hand. You want to know who these people are, what they think, what they are predisposed to believe about you and your project, what they read, what shows they watch on TV. Your goal will be to discover what they are like, how open they will be to your message, and some really powerful ways to deliver it. Look for common behavioral traits, common likes and dislikes, demographic commonalities. For example, does your target attend any conferences where you might meet a large contingency of them? Do they subscribe to specific trade publications? Are they typically male? What do they like to do in their spare time? Do they tend to be hockey fans? The more you know about them, the more likely it is you will create messaging that reaches them. Relevance is key to your messaging strategy. Ask yourself, "What is the single most important message I can take to this target?" Another way to think about this is, "What is the single most important promise I can make?"

As we went through this process with Abercorn Common, we even crafted different messages for different potential retail tenants. For example, the message to the smaller retail tenant was, "We will enhance your business through a unique, strategically located property." The message to the larger, big-box type retailers, however, was, "We can fit your formulae, save you some on operational expenses, and enhance your perceived goodwill to the community." We developed a graphic language that incorporated a background texture that felt like aging paper, a classic font called Mrs.

Eaves, and some graphic flourishes that had a traditional feeling. We used these same elements in advertising pieces, in collateral (handouts to prospects, for example), and for trade show booth components that Melaver took to the International Council of Shopping Centers conventions. We used headlines like "New Savannah shopping that feels like shopping Old Savannah," and "We'd like to invite you to a nice heaping helping of a 90,000 traffic count." Clearly, there was no mention of sustainability, at least in the early phase of lease-up. Why? Sustainability was not initially a relevant message to our target. Daniel Esty and Andrew Winston, in *Green to Gold*, remind us that we shouldn't get so caught up in the technology that we forget to make the business case. Prospective clients may be interested in the green attributes of a product or service, but the green button should not be the first (or even second) button to push.[5]

The message is only a blueprint for a piece of communication. In other words, you wouldn't craft a print advertisement with a headline that reads, "We can fit your formulae, save you some on operational expense, and enhance your perceived goodwill to the community." In fact, you might divide that message into three different advertisements and craft a headline around each one. If you're working with an outside partner, narrowing your message will increase the efficacy of the piece of communication. Always, always, always craft your message so that it is relevant to the people you are targeting. With Abercorn Common, we found that much of our messaging took place face to face between the Melaver brokers and the potential tenants. We did very little true advertising but invested a good deal of effort brainstorming how the brokers needed to talk about Abercorn Common. We recognized early on that too much talk of sustainability turned them off. We chose to keep the sustainable sell soft, and focus our discussions on the financial benefits of the property as they related to the tenant.

MARKETING COMMUNICATION STRATEGY SIMPLIFIED

Key questions that need to be addressed in shaping a message include:

1 What do you hope to accomplish from this communication?
2 To whom are you talking, and what are their motivations and concerns?
3 What is the single most important message you want them to take away from the communication?
4 How can you support this message?
5 What is the tone of your message?
6 How should you deliver this message?
7 When do you need to be out with the message?

Beyond the specifics of targeted messaging, the shaping of an overall marketing communication needs to be clear. Know *what* you are communicating and *why*. Clearly identify your target(s). Know as much as you can about these targets. Develop messages relevant to each target.

GREENING YOUR MARKETING COLLATERALS

Put some time and energy into thinking about what you will produce, in terms of advertising collaterals. How will a piece be used, and what will happen to it at the end of its useful life? Does it need to be produced in the first place? Sustainably oriented companies are often begun on a shoestring, with little or no money for marketing materials, and by necessity have to be clever about how they get the word out. Can it be cobbled together from materials that were previously used elsewhere, as TerraCycle does by packaging its liquid plant fertilizer in discarded soda bottles?[6] Does it have a use beyond the momentary marketing need? Can it be recycled? These sorts of questions may not be in the best interests of marketing and ad agencies looking for high-dollar spends from their clients. But these are the sorts of questions any green bottom line company is asking these days. And an agency partner that truly gets the issue of sustainability and the goal of reducing both waste and consumption understands that its value add comes in the ways a green message is communicated effectively and with integrity.

For example, when we worked on a communication piece for the USGBC's annual Greenbuild conference—a piece designed to make the sustainable development community more aware of Melaver, Inc.—we screen printed our messaging on 12″ × 12″ pieces of corrugated cardboard scavenged from around Birmingham, Alabama. We took an example of the piece to Martin Melaver. He read it, loved the writing, loved that it was printed on found materials. But then he asked: "What do I do with it after I've read it?" After much gnashing of teeth and wringing of hands, we came up with an innovative solution. We used large alligator clips and pads of 100 percent recycled paper to turn the cardboard piece into a clipboard, arguably a very useful piece for attendees of a conference intent on taking notes at meetings. Moreover, it was easily disassembled into its component parts, and all of the parts could be recycled or re-used. The piece was a success, and it cost next to nothing to produce.

Greening What You Print

Greening what you print can be a challenge, but it is becoming easier. There are a number of environmental certifications related to the printing industry. Paper companies and printers are making an effort to produce products that are, for example, carbon neutral or made with green energy. Green Seal certification of paper stocks means that they contain a minimum of 30 percent post-consumer fiber. The Forest Stewardship Council (FSC) is a highly respected international organization that encourages the responsible management of forests. FSC-certified printers use mostly recycled papers or papers that were manufactured from wood harvested from well-managed forests. Other certifications to consider are chlorine-free processing and alternative fiber (e.g., cotton rag, hemp, banana) content. High post-consumer recycled-content papers have also come a long way, in terms of the quality of the sheet and the corresponding quality of the printing. You may also want to look at printing on a 100 percent recyclable plastic polymer "paper." Biodegradable inks are fairly common in the printing community, but not ubiquitous, and not every printer is able to use them and use them well. Ask your printer to use soy-based inks instead of petroleum-based inks.

Here some things you may want to ask your printer:

- Have you replaced harsh varnishes with aqueous coatings?
- Do you have experience printing with soy-based inks?
- Have you eliminated using alcohol as a solvent?
- Do you recycle?
- Can you FSC certify my project?
- Can you certify that the paper you're using was manufactured carbon neutral or with green energy?
- Is the paper you are using Green Seal certified?
- What is the percentage of post-consumer fiber?
- Does the paper contain sustainably produced alternative fibers?

Take care to evaluate green claims from paper manufacturers, who are fighting an age-old reputation as egregious polluters. Some have come further than others in greening their practices. Lastly, be aware of which certification logos you are allowed to print on your collateral. Which of the certification entities will allow you to use their logos changes from time to time, so ask your printer to find out which ones apply and might be used on your final piece.

WORKING WITH AN AGENCY

Developing a Marketing Budget

In general, marketing firms work on a retainer that is charged against an hourly rate, based on a project scope that you both have agreed upon ahead of time. Additional costs for out-of-pocket expenses such as production, printing, shipping, travel, and other activities that contribute to the work getting done are usually billed as incurred.

An alternative is for agencies to bill on a per-project basis. This may be appealing if you don't have a large budget, and it works especially well if you have someone on your team with some marketing know-how who can manage the relationship with the agency and serve as a go-between to get the project done. Public relations agencies (discussed later in this chapter) also work on retainer or on a per-project basis.

Here are some budgeting guidelines:

- Advertising production (what you pay the agency to produce the advertising) should be no more than 30 percent of your overall advertising budget. NOTE: As we will see later on in this chapter, Melaver, Inc., being heavy on word-of-mouth and leery of advertising, tends to be an exception to this rule.
- Advertising media should be at least 70 percent of your budget. Media here is broadly defined as the means by which you take your message to market. For example, if you were to use a billboard, the cost of printing and placing the board would be considered production. The fee for using the board would be considered media. For direct mail, the agency's fee for design and getting the piece printed would be

considered production. The fee charged by a "mail house" would be considered media. For print advertising, everything the agency charges to produce the ad, from photography and design to any color proofs they have made to verify the look of the ad, would be considered production. The cost of placing the ad in the magazine would be considered media.

■ Your marketing and communication budget for a development project should be in the neighborhood of 1 percent of your overall construction budget. Melaver, Inc.'s budget, as we will see later, tends to run a little higher, at about 1.4 percent, but most of that spend is devoted to outreach, advocacy, and education.

Cost Guidelines
Agency fees can be expensive. Here are some rough cost guidelines in 2008 dollars:

■ **Television:** No less than $15,000 per spot.
■ **Radio:** No less than $1,000 per spot.
■ **Print advertising:** Expect to pay a minimum of about $6,000 for a campaign of three ads; that will probably not buy you original photography or artwork. With regard to print work, photography and illustrated art are owned by the individuals who produce them. Consequently, fees for using those materials are often high. The quality of those materials is almost always directly related to what you pay for it. Great photography is expensive.
■ **Collateral:** Prices for printed pieces generally start around $10,000 on the low end and can run into hundreds of thousands of dollars, depending on complexity and quantity.
■ **Logos:** Logos range from a few thousand to tens of thousands of dollars.
■ **Hourly fees:** Creative time charged by the agency typically starts at $100 an hour and goes up from there.

Don't be afraid to ask what things will cost, and be willing to say how much money you have budgeted for a project. All of that being said, many agencies want to work with clients who do good work and who allow them to do good work. In short, there is no "blue book" for agencies like there is for cars. There is no manufacturer's suggested retail price for a printed collateral piece. Your best bet is to negotiate from your budget. Tell them what you have and what you want to accomplish. Let them be creative. Let them create work they are proud to show others. You'll get the best work for the best price that way.

Alternative Compensation Strategies: Don't hesitate to offer creative compensation solutions. Barter is still alive and well. If you're constructing office space, there's nothing an advertising or design firm likes better than cool digs. Offer rent concessions on future tenancy. Offer co-branding or sponsorship opportunities. Your goal should be to establish a relationship. If the agency wants to work with you, they will collaborate to work out a creative compensation strategy.

Try Not to Micro-manage the Agency

Lastly, and most important—allow your agency to do what it does best. Try not to micro-manage the process. Having said that, if our experience with Melaver, Inc. serves as a basis for making generalizations about marketing for green entrepreneurial companies, there is likely to be a huge degree of collaborative work between an agency and its green client. Admittedly, while our work with Melaver, Inc. has been gratifyingly ground-breaking in many ways, they do tend to drive us nuts from time to time. Such clients tend to be deeply concerned that their messaging will somehow be inauthentic or grandiose, and so tend to spend a significant amount of time—as they should—making sure that the messaging is dead-on right. Finding a balance that is collaborative without being micro-managerial is simply one in a long line of balances that a sustainable company tries to manage well. An agency taking on a green client needs to be fully prepared for and accepting of such a close, often intense relationship.

Finding an Agency in Eight Steps

If you choose to work with an agency, here's a simplified method for locating one:

Step 1—Consult your local chamber of commerce and various business guides to identify advertising and design firms in your area.

Step 2—Consult advertising/design trade publications to see which agencies are doing breakthrough work, and which ones seem to have an interest in sustainability. Here's the short list:

- *Communications Arts* (Order the advertising annual, at least.) https://www.commerce.commarts.com/secure/subscrib.asp
- *PRINT* https://secure.palmcoastd.com/pcd/document?ikey=0768WID01
- *HOW* http://www.howdesign.com/magazine/

Step 3—Find the closest ADDY competition and peruse the work of the winners.

Step 4—Make a list of agencies whose work you'd like to see.

Step 5—Invite each agency on your list to share its work with you. Evaluate the work based on what you've already learned in this chapter:

- Is the message clear?
- Does it talk to the target for whom it was intended?
- Is the means appropriate to the target?

Step 6—Ask your trusted corporate partners. They may know something about the firms on your list. They may also know of firms outside your area that are doing good work.

Step 7—Have a short list of agencies provide proposals. The sort of things they're going to want to know are your budget and what, at a minimum, you'd like to accomplish with the budget. Try to give them tangible goals, such as "I'd like to generate fifty tenant prospects in two months." Ask them in the proposal to answer the following:

1 How do they operate? What is their process? Do they utilize creative briefs? Signed estimates? How will they keep you informed?
2 How will they bill you? By project? Progressively? Do they expect a retainer?
3 What are their rates? They should be able to give you a rate sheet.
4 Who will be your point of contact? You will want a single point of contact. They will expect the same from you.
5 How much experience do they have with development? You will probably want to work with a team that understands real estate development.
6 What is their position regarding sustainability? Have they worked with FSC-certified projects? Soy inks?
7 Are they aligned with your sustainable orientation or are they simply mouthing platitudes in order to garner business? Have they turned away potential business because a possible client did not square with the agency's values? Would they serve as a valuable resource in researching alternative and innovative ways of doing things?
8 How well does the agency communicate? Judge by how they listen as well as by how they talk and present.

Step 8—Based on their proposals, narrow down your short list to one or two. Meet with them offline. Go to lunch. Have drinks. Get to know them. The partnership between you and your agency should be tight. Used properly, it will be a critical asset.

You get the idea. There is no set formula for an agency request for proposal (RFP). You can do as much due diligence as you deem necessary. In the end, however, what really matters is the quality of their thinking and of their work and, perhaps most importantly, their capacity to be a collaborative partner in your mission to be a green company.

Phase 2B: Public Relations—Gaining Credibility and Support

The best and most robust communications plan includes advertising (and components like direct marketing) and public relations. Public relations includes the activities you engage in to create goodwill for your organization in the communities where you are doing business. Success can come in the form of an article in the local newspaper or industry magazine, an invitation to speak at a trade group, or a chance to work on a project with a well-regarded business partner. This "third party endorsement" reassures your target that you are credible. There is often the temptation to call public relations "free advertising," but that's really a misnomer. Public relations expertise doesn't come free, and an experienced and well-connected professional is essential to the effort.

Ideally, your public relations team and your marketing/advertising team will work together so that you have a consistent and well-executed message and strategy. The research on your message and target, described earlier in this chapter, is also a great foundation for developing a successful public relations campaign. The work you will do to create and manage your green reputation is important, and you should take care to guard those credentials once earned.

Typically, public relations programs include media relations (developing relationships with journalists who cover your organization and its projects), as well as community relations (developing relationships with organizations that are a part of your business community or your industry). Your public relations plan should also include a crisis management plan so that you are prepared to deliver bad news if and when it happens.

Green public relations, like green marketing, is unique in that it deals with a highly sensitized target. And, like green marketing, organizational issues such as top-down support, alignment with mission, and authenticity and transparency are critical. There is a key difference between advertising and public relations that makes this commitment to authenticity critical. When you place an ad, you pay for the opportunity to secure a space for content that you will develop and deliver. You have complete control over your message. When you conduct public relations activities, you or your representative pitches your story to a potential target via a filter—a newspaper editor, a magazine writer, a television announcer—and that person takes your news and edits it to fit within the story he or she is telling and in the context of the particular news outlet. Therefore, the only chance for control is on the front end, and is dependent on the story idea you take to the media person and the working relationship you establish together. Understanding this key difference between advertising and public relations will help you develop the best strategy for each.

When you think about public relations for a sustainable project, you need to keep in mind a few rules of thumb.

RULE 1: GREENWASH DOESN'T WASH

The target for a green product or service is sensitive to over-the-top claims and hype about what a product or service can deliver relative to its competitors. The term "greenwash" is commonly used to describe the practice of making green claims without the ability to deliver on them, and while that is always a bad idea, this target is particularly sensitive to these claims. Avoid exaggeration, take extra care to be authentic, and your target will reward you.

In December 2007, the environmental marketing company TerraChoice[7] gained national press coverage for releasing a study called "The Six Sins of Greenwashing," which found that 99 percent of 1,018 common consumer products randomly surveyed for the study were guilty of greenwashing. Here's a list you don't want to find your name on:

- Sin of the Hidden Trade-Off: "Energy-efficient" electronics that contain hazardous materials. 998 products and 57 percent of all environmental claims committed this sin.
- Sin of No Proof: Shampoos claiming to be "certified organic," but with no verifiable certification. 454 products and 26 percent of environmental claims committed this sin.
- Sin of Vagueness: Products claiming to be 100 percent natural when many naturally occurring substances are hazardous, like arsenic and formaldehyde. Seen in 196 products or 11 percent of environmental claims.
- Sin of Irrelevance: Products claiming to be CFC-free, even though CFCs were banned 20 years ago. This sin was seen in 78 products and 4 percent of environmental claims.
- Sin of Fibbing: Products falsely claiming to be certified by an internationally recognized environmental standard like EcoLogo, Energy Star, or Green Seal. Found in 10 products or less than 1 percent of environmental claims.
- Sin of Lesser of Two Evils: Products in a category with questionable environmental benefit, e.g., organic cigarettes or "environmentally friendly" pesticides. This occurred in 17 products or 1 percent of environmental claims.

RULE 2: DON'T LET YOUR TALK EXCEED YOUR WALK

Similar to greenwash is the practice of overstating the green attributes your product or service delivers, or using language to imply that you have achieved more than you actually have. It is important that you make realistic claims. Keeping your talk in line with your walk is an important factor in winning over your target. This issue is particularly germane in the context of the green building trade, where there is often an extended time between the planning of a project on the front end and its eventual execution, sometimes years later. Developers (and Melaver, Inc. is no exception in this regard) tend to have this sense that the building they have conceived and planned already exists well before it is actually constructed. They want to talk about the various green features they have specified, the level of LEED certification it is likely to garner, the potential social programs they have planned to tie into the overall project. The short and skinny on this is simple: Don't talk about it until it's a done deal.

RULE 3: AVOID ME, TOO! SYNDROME

Marketing, in general, is about distinguishing your brand from others. Green marketing, in particular, is about articulating not only *how* your product or service is different, but *why*. Green claims are popping up everywhere, and it's tempting to raise your hand and say, "Me, too!" Instead, take the time and care to conduct the branding exercises described earlier in this chapter, align your message with your mission, and you'll avoid coming off as a "Me, too!" when it comes to being green. The unique attributes you assign to your brand will help you distinguish your message in your public relations program as well as in your advertising. A complex twist on this caveat, however, is the fact that mission-driven companies such as Melaver, Inc. are actively encouraging others to get on board with the sustainability movement. Real estate is, after all, local, and green companies in one locale find strong synergies with one another in helping with the general category build of green development. The key here, I think, is twofold: First, celebrate the work of erstwhile competitors—since the attitude is that you are in different markets by and large, and even if you are not, you are pursuing the same overall goals of reduced consumption and waste. Second, accentuate the particular nature of your own sustainable orientation and stay true to that vision.

How and When to Use Public Relations

A public relations strategy should be a part of your organization's business plan, not just a tool to employ every time you sign a new tenant or have a grand opening (although those are great times to use PR). When developed proactively and in sync with your marketing and advertising plan, public relations can create goodwill that will help you build incremental awareness for your company and your projects, enhance your reputation in the business community, promote a cause or issue, and attract good employees or potential partners. You might consider letting your targets know when you:

- Hire a new employee.
- Sign a new deal, lease, or tenant.
- Acquire new property, particularly if you have green aspirations for a conventionally developed property.
- Achieve LEED certification for a project.

- Receive recognition from your community or your industry for your commitment to green building.
- Post quarterly or annual earnings.
- Book one of your experts to speak at a meeting or be interviewed in the press.

MEDIA RELATIONS

A well-designed public relations plan includes media relations, or strategic outreach to print, Web, radio, and TV media. The toolkit includes the basics: press releases, fact sheets, and photography that illustrate your product or service. In keeping with your green message, consider creating an electronic press kit and use the Internet for distribution whenever possible.

In designing Melaver, Inc.'s press materials, we have taken care to reflect the company's green mission in how we deliver our news. News releases are distributed online, and high-resolution photography is maintained on a website where media can download the images they need (an FTP site—file transfer protocol—is the technology we currently use). When we want to hand deliver materials, such as for media attending a trade show, we create electronic press kits and store documents and images on a "jump drive," or USB-compatible memory stick. Not only do we conserve resources, we also provide the media with a reusable tool.

In addition to traditional media relations (you write a press release, send it to an editor, and hope that it results in a story), you might have the credentials you need to reach out to media in a collaborative fashion and establish someone on your team as a *subject matter expert*—someone the media knows they can call upon to provide insights or a quote on news or trends. Because green building is a new and developing field, media in your community or in your industry may be looking for people who have experience and a story to tell.

Start by developing a bank of story ideas that can be marketed to media as bylined articles (articles published under the byline of your subject matter expert). To do this, gather editorial calendars for the publications that cover your organization. You can typically find them online or get them from an advertising representative for the publication. Next, research trend information available from trade organizations or from market research firms. If you are promoting a green building project, the U.S. Green Building Council is a great resource. The organization tracks national trends, and your local chapter may also have information about trends in your area.

Once you have an idea of what is being written about and the trends that will contribute to future news, combine that information with what you know to be your expert's core competency. Ideally, you will emerge with something to say that should be compelling enough to get an editor's attention. Put it in the form of a letter to the editor who handles the publication or the section of the publication you're targeting, and write persuasively about your subject matter expert's unique perspective. A well-

Figure 10.6   Media materials.

Notice the flash drive in the lower right imprinted with the Melaver, Inc. logo. Electronic press kits delivered via a flash drive are convenient and sustainable. You won't print press materials or need CDs or DVDs to share photographs, and the drive is reusable time and time again.

crafted pitch will start you on the way to building a good relationship with editors, and, over time, you may be called upon to comment on other trends and events in your industry or community. .

Once you have secured an opportunity, your subject matter expert and your public relations consultant can collaborate on the article, taking care to keep in mind what most appealed to the editor who responded to your pitch.

SPEAKING ENGAGEMENTS

Another tactic for generating positive press is to position your spokesperson as a subject matter expert for speaking opportunities. In every industry and in every community, there are a number of opportunities for passionate, knowledgeable, and articulate speakers. Trade shows, industry conferences, and service organization meetings (e.g., Kiwanis and Rotary Club) provide an opportunity and a target. By studying the conference or meeting themes and determining how you can add value, you can develop a calendar of speaking opportunities that increases your presence in the marketplace.

Similar to developing your pitch to an editor, write a letter to conference organizers or respond to their requests for proposals, identifying how your expert(s) can make a unique contribution to an event's program.

This has been a particularly successful tactic for Melaver, Inc., as the company has taken the somewhat atypical approach, particularly for a company of its size, of developing a whole in-house speakers bureau of various staff members able to speak on different core competencies. While there was some initial concern that such a wide coterie of public speakers might confuse the public as to who was the face of the company, this approach has flipped that concern on its head by making the statement that the entire staff is the face of the company. Messages do differ from speaker to speaker, as there are speakers from different ideological as well as professional perspectives. Nevertheless, the overall message various staff members deliver is remarkably consistent from speaker to speaker—and is perhaps made more potent by the fact that the drive and passion within the company come from all points on the compass and not just the CEO. All of the contributors to this book have, at one time or another, spoken at local, industry, or trade organizations about their work with the company. They are effective, in part, because they believe personally in the mission of the company, and that commitment is apparent in their delivery.

Events like Greenbuild, the U.S. Green Building Council's annual conference and trade show, organizations like the Urban Land Institute and CoreNet, and local chapters of the U.S. Green Building Council all provide great venues for stories about how green building is affecting our communities.

SPECIAL EVENTS

Creating relationships in the communities where you do business is enhanced when people have a personal experience from which to draw their opinions and attitudes. People are particularly curious about green building and, in general, want to know more about how your project will contribute to the health of the community. Developers who engage with a community open themselves up to the opportunity to establish positive relationship that will serve them well when questions or problems arise. Special events, whether they be town hall meetings, project overviews held at a local community center, or even hard-hat tours of your work site can bring you face to face with the people who can influence your ability to successfully develop in the community.

This strategy was particularly successful for Melaver, Inc. in leasing Abercorn Common and introducing the center to the Savannah community. An on-site leasing center also serves as a green building education experience and the point of contact for the myriad tours given to school groups and professionals eager to see some of the development's key green features. The center has also served variously as a venue for the district's congressional representative to launch a new energy policy initiative, the site of the community-wide showing of the documentary *Kilowatt Ours*, and a showcase to convince a local school to build its new middle school to LEED standards.

A CRISIS COMMUNICATIONS PLAN

There's every chance that you won't need it, but if your business does experience a crisis that has implications for any of your key targets, having a crisis communications plan in place will ensure that you do the best possible job communicating the facts of the situation. The golden rule of crisis communications is that it is crucial in a crisis to tell it all, tell it fast, and tell the truth. If you do this you will have done all you can to minimize the negative effects of the situation.

By definition, a crisis would be something unanticipated that threatens the integrity or reputation of your company, usually because it would bring on adverse or negative media attention. The crisis could be anything from a legal dispute to an accident, a fire, or a manmade disaster that could be attributed to your company. Fortunately, we have seen only one instance of this with Melaver, Inc., when a small flurry of letters to the editor of the *Savannah Morning News* complained about the parking spaces set aside for hybrid cars at Abercorn Common. The SUV crowd wrote in to say that they found these privileged parking spaces to be insulting, while several members with handicapped license plates wrote in to say that they felt their own interests were being ignored. A team of us from Melaver met quickly with the latter group (the former were not interested in meeting with us) to hear their concerns and make the necessary adjustments.

A crisis communications plan is a document that details the chain of events, the key spokespeople, and how they can be reached in an emergency (home and mobile phone numbers). Your plan should address:

1 Who will assess the situation and brief the management team.
2 Who will develop and deliver a public statement. (Typically this is your CEO or the highest ranking member of your leadership team who is available, with the help of your PR team.)
3 What the public statement will say. Even if you do not yet have complete information, it is often important to issue a statement to that effect: "We do not yet have the information we need to answer your questions but are working on it and will have a conference call/press conference/issue another statement by XX time." Make sure everyone who answers the phone has a copy of the statement.

4 Follow through. Even difficult news, including accepting fault for an accident or admitting to a lawsuit that has been filed, is better managed when you use the same authentic, transparent voice you use for your other communications.

Making Your Shade of Green Work for You

Marketing may seem like the most intimidating part of what you do. Don't be intimidated. More importantly, don't ignore it. Pursue your marketing with the same deliberation you use in planning your projects. Focus your thinking. It always boils down to the same set of criteria—target, message, means. So here's the breakdown:

1 Know who you are and what your project is really about.
2 Know to whom you are talking and what they are like.
3 Use public relations to gain community support.
4 Do your homework.
5 Believe in what you're doing. Deeply. Don't try to be something you're not.
6 Keep doing what you're doing. The world needs more of you.

Putting Some Dollar Figures on All This Marketing Sense

So how much does all of this cost? As a rule of thumb, a company should allocate roughly 1 percent of its overall project budget for marketing expenditures. This is not too far afield from Melaver, Inc.'s actual marketing expenses over an extended period of time (five years), although accounting for this is a bit unwieldy since in the past the company has lumped project-specific marketing expenditures in with general marketing expenses.

Let's consider once again the fictionalized company we have been drawing upon throughout this book, Green, Inc., and take a look at the breakdown of its marketing expenses (which are based upon those of Melaver, Inc.). As you may recall, Green, Inc. has 20 staff members, an asset base of $100 million, and gross revenues of $12 million. Out of that $12 million in gross revenues, the company is targeting an expenditure of around 1 percent in standing annual marketing expenses, amounting to $120,000, plus a certain amount earmarked for marketing specific development projects each year. Since those project-specific marketing costs are accounted for in the pro formas of each development, we will leave those out of our analysis and simply focus on ongoing marketing costs for the company as a whole.

In the early years of Green, Inc., the company focuses its marketing efforts internally: holding a retreat built around the communication of core values, developing collateral materials that support work done at the retreat, creating a website that expresses the values of the company. Also early on, the company makes a multi-year commitment to support the capital fund drive of an important local environmental group. Senior management chalks this up to youthful exuberance on its part, identifies the financial commitment as cause-related marketing, and makes the decision that in future years, the company should shy away from cause-related marketing in favor of direct philanthropic giving. Another early move was a multi-year commitment of advertising signage at the local airport, something Green, Inc. is uncertain it will continue once the contract expires. The original intent of this ad was to make a statement linking Green, Inc. closely with its hometown. It's virtually the only advertising Green, Inc. engages in, and senior management wonders if it should renew the contract.

After the first few years of devoting its overall communication budget to marketing, Green, Inc. begins to focus more and more of its efforts on PR, primarily on outreach, advocacy, and education as it relates to green building practices. Several staff members get invited to speak on panels or to make keynote addresses at conferences, first locally and in the region, then nationally. From doing about two dozen talks annually after its first two green projects are completed, staff members at Green, Inc. evolve into doing about fifty per year, which entails a significant amount of travel time and expense. The company also begins to get involved in local, regional, and state political issues, things like curbside recycling, municipal commitment to ICLEI (International Council for Local Environmental Initiatives) standards, a statewide comprehensive water plan, and state subsidies for use of alternative energy. The overall breakdown of communication expenses for Green, Inc. over a ten-year period can be found in Table10.1.

There are a few things worth looking at in this ten-year spend on communications. First, the initial spend in Year 1 is devoted largely to internal communication, as well as the company's main vehicle for expressing itself to the outside world (its website). Advertising amounts to about one-sixth of the total spent. Early commitment to cause-related marketing tapers off once a large multi-year commitment is paid off. And while the majority of early year dollars is devoted to marketing line items, as the company matures, more and more resources are devoted to communicating what the company knows and cares about to the outside world. Average annual spend on communication is above industry standard, at 1.3 percent of gross revenues (instead of 1 percent), amounting to approximately $36,000 a year more ($156,000 versus $120,000) than is typical.

Every company will, of course, judge for itself not only the amount it will budget for overall communication needs, but also how it will allocate this overall amount to specific categories (e.g., advertising, cause-related marketing). It's difficult even for us to analyze critically the expenditures provided by Green, Inc., since it is rare for outside agencies to be included in a client's budgeting process. We do feel that this type of budget, weighted toward sharing information with the outside community, fits the mission-driven nature of this particular organization.

TABLE 10.1 GREEN, INC.'S COMMUNICATIONS SPEND OVER TEN YEARS

YEARS	1	2	3	4	5	6	7	8	9	10
Annual marketing spend										
Internal branding (retreat)	10,000									
Collateral (cards, photographs, brochures)	20,000	2,000	2,000	2,000	10,000	2,000	2,000	2,000	25,000	2,000
Web development and maintenance	50,000	12,000	12,000	12,000	30,000	12,000	12,000	12,000	50,000	12,000
Advertising, general		25,000	25,000	25,000	25,000	25,000	25,000	25,000	25,000	25,000
Cause-related marketing (including pro bono consulting)	30,000	30,000	30,000	30,000	30,000	10,000	10,000	10,000	10,000	10,000
Project-specific marketing (budgeted as % of development budget)										
Annual public relations spend										
Speakers bureau and article placements		18,000	24,000	36,000	48,000	48,000	48,000	48,000	48,000	48,000
Internet headlining			18,000	18,000	18,000	18,000	18,000	18,000	18,000	18,000
Speakers bureau travel and expenses		20,000	30,000	40,000	40,000	40,000	40,000	40,000	40,000	40,000
Educational tours of projects										
Total communications spend annually	110,000	107,000	141,000	163,000	201,000	155,000	155,000	155,000	216,000	155,000
Communications spend beyond market	10,000	13,000	(21,000)	(43,000)	(81,000)	(35,000)	(35,000)	(35,000)	(96,000)	(35,000)
Total marketing spend annually	110,000	69,000	69,000	69,000	95,000	49,000	49,000	49,000	110,000	49,000
Total PR spend annually	0	38,000	72,000	94,000	106,000	106,000	106,000	106,000	106,000	106,000
% spend on marketing	100.00%	64.49%	48.94%	42.33%	47.26%	31.61%	31.61%	31.61%	50.93%	31.61%

Years		0	1	2	3
REVENUES					
	Shaping values, chapter 1		21,300	21,300	21,300
	Creating a culture of green glue, chapter 2		79,000	79,000	79,000
	Green from the inside out, chapter 3		3,300	3,300	3,300
	Developing expertise in LEED, chapter 4		0	0	0
	Sustainable brokerage, chapter 8		0	0	0
	Green marketing, chapter 10				
	Sale of Green, Inc.				
Total Revenues			103,600	103,600	103,600
EXPENSES					
	Shaping values, chapter 1	(166,000)	(114,500)	(91,875)	(97,286)
	Creating a culture of green glue, chapter 2	0	(57,600)	(64,631)	(72,717)
	Green from the inside out, chapter 3	0	(78,490)	(22,381)	(22,998)
	Developing expertise in LEED, chapter 4	(49,000)	(150,000)	(417,000)	(890,000)
	Sustainable brokerage, chapter 8	0	0	0	0
	Overmarket costs for marketing, chapter 10		(35,800)	(35,800)	(35,800)
Total Expenses		(215,000)	(436,390)	(631,688)	(1,118,802)
Total Cashflow		(215,000)	(332,790)	(528,088)	(1,015,202)
Discount Factor		1.00	0.91	0.83	0.75
PV Cashflow		(215,000)	(302,536)	(436,436)	(762,736)
NPV		2,401,349			
IRR		27.94%			

Figure 10.7   Updated cash flow analysis of Green, Inc.

Years		0	1	2	3
REVENUES					
	Shaping values, chapter 1	0	21,300	21,300	21,300
	Creating a culture of green glue, chapter 2	0	79,000	79,000	79,000
	Green from the inside out, chapter 3	0	3,300	3,300	3,300
	Developing expertise in LEED, chapter 4	0	0	0	0
	Sustainable brokerage, chapter 8	0	0	0	0
	Green Marketing, chapter 10	0	0	0	0
Sale of Green, Inc.					
Total Revenues			103,600	103,600	103,600
EXPENSES					
	Shaping values, chapter 1	(166,000)	(114,500)	(91,875)	(97,286)
	Creating a culture of green glue, chapter 2	0	(57,600)	(64,631)	(72,717)
	Green from the inside out, chapter 3	0	(78,490)	(22,381)	(22,998)
	Developing expertise in LEED, chapter 4	(49,000)	(150,000)	(417,000)	(890,000)
	Sustainable brokerage, chapter 8	0	0	0	0
	Overmarket costs for marketing, chapter 10	0	(35,800)	(35,800)	(35,800)
Total Expenses		(215,000)	(436,390)	(631,688)	(1,118,802)
Total Cashflow		(215,000)	(332,790)	(528,088)	(1,015,202)
Discount Factor		1.000	0.909	0.826	0.751
PV Cashflow		(215,000)	(302,536)	(436,436)	(762,736)
NPV		10,112,215			
IRR		44.11%			

Figure 10.8   Final discounted cash flow analysis of Green, Inc. with sale.

4	5	6	7	8	9	10	TOTALS
21,300	21,300	471,300	538,800	548,925	550,444	550,672	2,766,640
79,000	79,000	79,000	79,000	79,000	79,000	79,000	790,000
3,300	3,300	3,300	3,300	3,300	3,300	3,300	33,000
0	0	2,116,500	1,176,475	1,222,971	1,261,446	1,303,442	7,080,834
0	0	150,000	187,500	225,000	262,500	300,000	1,125,000
103,600	103,600	2,820,100	1,985,075	2,079,196	2,156,689	2,236,413	11,795,474
(103,490)	(110,605)	(118,765)	(128,125)	(138,866)	(151,190)	(165,335)	(1,386,037)
(82,016)	(92,710)	(105,007)	(119,150)	(135,413)	(154,117)	(175,625)	(1,058,987)
(24,858)	(26,997)	(29,456)	(32,285)	(35,538)	(39,278)	(43,580)	(355,862)
(57,000)	0	0	0	0	0	0	(1,563,000)
0	0	(75,000)	(78,750)	(82,500)	(86,250)	(90,000)	(412,500)
(35,800)	(35,800)	(35,800)	(35,800)	(35,800)	(35,800)	(35,800)	(358,000)
(303,165)	(266,111)	(364,028)	(394,110)	(428,117)	(466,635)	(510,341)	(5,134,386)
(199,565)	(162,511)	2,456,072	1,590,965	1,651,080	1,690,054	1,726,073	
0.68	0.62	0.564	0.513	0.467	0.424	0.386	0
(136,305)	(100,907)	1,386,388	816,417	770,241	716,748	665,476	0

4	5	6	7	8	9	10	TOTALS
21,300	21,300	471,300	538,800	548,925	550,444	550,672	2,766,640
79,000	79,000	79,000	79,000	79,000	79,000	79,000	790,000
3,300	3,300	3,300	3,300	3,300	3,300	3,300	33,000
0	0	2,116,500	1,176,475	1,222,971	1,261,446	1,303,442	7,080,834
0	0	150,000	187,500	225,000	262,500	300,000	1,125,000
0	0	0	0	0	0	0	
						20,000,000	20,000,000
103,600	103,600	2,820,100	1,985,075	2,079,196	2,156,689	22,236,413	31,795,474
(103,490)	(110,605)	(118,765)	(128,125)	(138,866)	(151,190)	(165,335)	(1,386,037)
(82,016)	(92,710)	(105,007)	(119,150)	(135,413)	(154,117)	(175,625)	(1,058,987)
(24,858)	(26,997)	(29,456)	(32,285)	(35,538)	(39,278)	(43,580)	(355,862)
(57,000)	0	0	0	0	0	0	(1,563,000)
0	0	(75,000)	(78,750)	(82,500)	(86,250)	(90,000)	(412,500)
(35,800)	(35,800)	(35,800)	(35,800)	(35,800)	(35,800)	(35,800)	(358,000)
(303,165)	(266,111)	(364,028)	(394,110)	(428,117)	(466,635)	(510,341)	(5,134,386)
(199,565)	(162,511)	2,456,072	1,590,965	1,651,080	1,690,054	21,726,073	
0.683	0.621	0.564	0.513	0.467	0.424	0.386	0
(136,305)	(100,907)	1,386,388	816,417	770,241	716,748	8,376,342	0

We do need to update the discounted cash flow for Green, Inc. that we have been tracking throughout this book, since the company (based on Melaver, Inc.'s own actual expenditures for green communications) is spending a bit more than conventional companies to spread the word about sustainable practices. As can be seen in Figure 10.7, even factoring in this "premium" for green marketing, the overall performance of Green, Inc. is still strong, with a $2.4 million positive net present value (NPV) and an internal rate of return (IRR) of nearly 28 percent.

It's worth noting that when a sale of Green, Inc. is factored into the analysis in year 10, the net present value of all green investments exceeds $10 million and the internal rate of return increases to 44.1 percent. See Figure 10.8.

Phase 3: The Green Marketing Strategy for Paradigm Change

Jacquelyn Ottman, in her groundbreaking 1993 study *Green Marketing*, identifies seven winning strategies that are still cogent and relevant today:

1 Do your homework. Understand the full range of environmental, economic, political, and social issues that affect your consumer and your products and services now and over the long term.
2 Create new products and services that balance consumers' desires for high quality, convenience, and affordable pricing with minimal environmental impact over the entire life of your products.
3 Empower consumers with solutions. Help them understand the issues that affect your business as well as the benefits of your environmentally preferable technology, materials, and designs.
4 Establish credibility for your marketing efforts.
5 Build coalitions with corporate environmental stakeholders.
6 Communicate your corporate commitment and project your values.
7 Don't quit. Continuously strive for "zero" environmental impact of your products and processes; learn from your mistakes.[8]

It is this last point that we wish to focus on, since it suggests a constant movement toward greater standards of sustainability. We have seen that progression in our work with Melaver, Inc. The company has moved from Phase 1 work—a focus on doing its own homework, shaping its business culture around a set of core values and practices, and developing a communications strategy that is internally focused—to Phase 2 work—the development of numerous cutting-edge LEED projects, bringing along a set of broad stakeholders, and shaping a communications strategy around expertise, integrity, and passion. And the company is still on the move, continuously striving to minimize its own (and others') impact on the environment.

What does this next phase in the evolution of a green development company look like, as it moves, according to John Grant's schema, from column A ("green") to column B ("greener") to column C ("greenest")? The broad outlines seem to include:

- Raising the bar on the high-performance nature of its developments, with particular attention to strategies for dramatically reducing carbon emissions;
- Broadening engagement with a wide group of stakeholders, to facilitate greater traction in the market for green development; and
- Linking the social with the environmental by integrating social justice programmatic elements into the fabric of green structures.

John Grant argues, and we would agree with him, that this last turn of the screw into Phase 3 (column C) work involves a shift from a social order that is destructive to one that is revitalized. Marketing, in Grant's view, can play an important role in synthesizing conflicting ideas to form a greater whole, by 1) developing communication strategies that facilitate collaborative networking; 2) developing communication strategies that make radically green business models seem hip or normal; 3) developing communication strategies that wean us from consumptive practices. It will be an interesting journey, to say the least, as we move into this next phase of green development work.

NOTES

[1] John Grant, *The Green Marketing Manifesto* (Chichester, U.K.: John Wiley & Sons Ltd., 2007), p. 69. Reproduced with permission.

[2] Michael Treacy and Fred Wiersema, "Customer Intimacy and Other Value Disciplines," *Harvard Business Review* (Cambridge: Harvard Business Press, Jan.–Feb. 1993).

[3] John Grant, *The Green Marketing Manifesto* (Chichester, U.K.: John Wiley & Sons Ltd., 2007), p. 233.

[4] Gary Hirshberg, *Stirring It Up: How to Make Money and Save the World* (New York: Hyperion, 2008), pp. 94–95.

[5] Daniel C. Esty and Andrew S. Winston, *Green to Gold: How Smart Companies Use Environmental Strategy to Innovate, Create Value, and Build Competitive Advantage* (New Haven: Yale University Press, 2006), pp. 125–131.

[6] Gary Hirshberg, *Stirring It Up: How to Make Money and Save the World* (New York: Hyperion, 2008), pp. 154–159.

[7] htttp://www.terrachoice.com

[8] Jacquelyn A. Ottman, *Green Marketing: Opportunity for Innovation* (New York: J. Ottman Consulting, Inc. 1993, 1998), p. 49.

CONCLUSION: INSCRIBING KEY LESSONS LEARNED INTO THE FABRIC OF A GREEN BUSINESS

MARTIN MELAVER

Everything I know about sustainability can be summed up in this story about my Uncle Scribe.

Uncle Scribe (he's really my great uncle, the brother of my paternal grandmother, Annie Melaver) was an acerbic man, extremely well-read, sharp-witted, difficult, contrarian, someone who would take any side of a debate and beat you down with arguments as to why you were dead wrong, someone who would take the cutlery and water glasses at Thanksgiving and demonstrate—to my grandmother's horror—the proper way to carve the turkey. He was the occasional lawyer for our family grocery business in Savannah, Georgia, and one of my favorite relatives. Years ago, as I was preparing to go off to college to pursue a general liberal arts education, I dropped by Uncle Scribe's house to give him my new mailing address and to promise I would write when I could. I think he was glad to see me. In any case, he was ready with advice.

"Martin," he said, "it's great that you are pursuing a general humanities education. But there's one thing you must never forget."

"What's that, Uncle Scribe?"

"Martin, pursuing a general liberal arts degree is a good thing. But you must never forget to specialize in something. You need to have some type of focus and expertise."

"Yes, Uncle Scribe." That was him all right: contrarian to the max, hitting them where they ain't.

Years later, as I was heading to graduate school, I paid a similar visit. Same set of circumstances. Same Uncle Scribe, ready with advice.

"Martin," he said, "it's great that you have decided to pursue a graduate degree and specialize in one narrow field of endeavor. I think that's wonderful. But there's one thing you must never forget, something more important than your narrow field of study."

"What's that, Uncle Scribe?"

"You must always remember that despite your expertise in a particular field, you need to have a broad knowledge of disciplines and fields of inquiry. Always keep in mind the big picture."

"Yes, Uncle Scribe."

Time passed. I finished my graduate studies, married, had kids, and had begun to shape our family business into a sustainable real estate company. Managing that business, Melaver, Inc., was time consuming and I didn't make time to see Uncle Scribe much. He was aging quickly, and I knew I needed to visit with him, even as I didn't want to see him in declining health. I saw him shortly before he died.

"Martin," he said, "it's great that you have gotten a general humanities education and I think it's also wonderful that have also studied a specific field, and I'm fascinated by this sustainable business you are running, but you need to remember what's really important."

"What's that, Uncle Scribe?"

"You must never let yourself weigh over 150 pounds. Once you do, it's simply impossible to return to where you once were."

I see so much of what we do as an aspiring sustainable business through the lens of this story. There is the need to focus on the particular and narrow, the practices that result in the construction of green buildings, one at a time. The central chapters of this book provide what I hope is a wealth of detail into those practices, details we might think of as comprising a set of instructions on how to build a watch. What are the key take-aways from the details of building a green product? While each of my colleagues might compile slightly different lists, I think we would all agree on the importance of these points:

- Be clear about "why."
- Be clear about "what."
- Be clear about "who."
- Be prepared for resistance, often in the guise of the false belief that green costs more.
- Be focused on a deeply collaborative, integrative process.
- Be prepared for fast-changing circumstances in which today's cutting-edge technologies, materials, and processes are replaced by more efficient and effective strategies.
- Document and measure everything you do.
- Always let your walk precede your talk.
- Be as open and transparent as humanly possible.
- Be open to lifelong learning.

A few comments about each point: It's important from the very inception of a green building project to recognize why you are developing this way. Is it because senior management is mandating such an approach? Perhaps the root cause of the undertaking is because your company has a values-centric orientation (such as ours). Perhaps the root cause is based on a belief that superior economic returns over the long term can be garnered by building green (reduced costs and increased revenues in the short term; reduced liability and exposure, and enhanced reputation and prestige over the long term). Perhaps your company's approach is based on the anticipation that all buildings will soon be built to green standards. Whatever the case, understanding why your company is deciding to build green is critical to the undertaking, since it informs so many other downstream decisions.

One of those critical downstream decisions is determining what type of green building should be built. Is a company looking for outside, third-party verification such as is provided by the LEED program, or is it sufficient to develop a green building without the verification? (We strongly feel independent accountability is important.) If a certain standard of LEED certification is sought, which level and why? Our company, for instance, is particularly focused on the carbon emissions associated with our activities, and so makes decisions based largely on energy efficiency and consumption. Other companies, however, may believe it sufficient, at least for the moment, just to deliver a LEED certified building. Having clarity about the goal makes the actual tasks of development and construction flow much more smoothly.

That clarity around goals leads to a critical focus on the various stakeholders who are brought into the development process. Alignment of partners is critical. Having a stakeholder on board who is likely to be a stumbling block in the process of building green can be extremely costly. Recognizing those potential stumbling blocks in advance is important. Use of force-field analysis, which lists the various stakeholders in a project and evaluates each as either a driver or resistor of company goals, is an effective business tool that helps clarify the "who."

Despite best efforts, there will always be people involved in the project who resist the green approach. For the most part, that resistance is expressed in financial terms, with skeptics claiming that it costs more to do a LEED project. As we hope we have demonstrated in this book, it does not. On the contrary, it has been our experience that our sustainable orientation has facilitated returns above market benchmarks. Of course, beneath the resistance that is expressed in financial terms is a more fundamental resistance having to do with change. Fortunately, given the speed of market acceptance of green building practices, it is best to go where this journey is taking you, and leave the naysayers behind.

Those who do come along on the journey are in for a deeply collaborative process, as this book has tried to demonstrate. The authors of the various chapters range across numerous disciplines—human resources, environmental studies, finance, real estate development and construction, brokerage, law, marketing, and public relations—all of which are critical to the overall process. But just having multiple fields at the table from the inception of a green project is not sufficient. Unless the individuals have the capacity to let go of some of their expertise to enable others to provide insights, true leveraging of talents is limited. Some of our best insights on engineering challenges have come from our design people. Some of our most significant financial savings have come from our engineers. Some of our most significant marketing insights have come from our finance team. It pays to leave professional hats and egos at the door before entering a collaborative environment.

The strategies a collaborative team will come up with to develop a green, high-performance building are constantly in motion, ratcheting up in efficiency while tapering downward in price. Some of the cutting-edge approaches in our early green projects such as Abercorn Common, built for roughly $175 per square foot, have found their way into more recent green projects such as Sustainable Fellwood, a mixed-use, mixed-income neighborhood development that, at $85 per square foot, is bringing green construction techniques to the neediest people in our community. It's a win-win phenomenon, enabling green practices to spread rapidly across our culture, as well as deeply into those segments of our society most profoundly affected by monthly energy and water bills.

It is critical to document green practices on the front end and measure outcomes after completion. Our company hasn't been as disciplined as we should be in this regard, and other real estate companies should learn from our lapses. We all gain from learning what works and what does not.

This point about documentation and metrics leads almost intuitively to our last three points: letting green projects speak for themselves, being transparent as possible about lessons learned, and committing to an ongoing process of learning with others throughout the real estate industry. Arie de Geus, one-time head of Royal Dutch/Shell's strategic planning group and originator of the concept of the "learning company," notes that the growing complexity of work has

> . . . created a need for people to be a source of inventiveness, and to become distributors and evaluators of inventions and knowledge, through the whole work community. Judgment, on behalf of the company, [can] no longer be the exclusive prerogative of a few people at the top.[1]

It is well worth adding that, in the face of considerable concerns over global carbon emissions and climate change, it is not just the stand-alone business that can't afford to be a learning company; all of society cannot afford such insularity.

Those are the key take-aways from the narrower aspects of constructing a green building. But what, as my Uncle Scribe would say, about the broader context, the bigger picture? As I hope is evident from the structure and overall argument of this book, our company did not start from the idea of building a green building. In fact, it took many years of painstaking values-shaping and culture-building before we delivered our first LEED project. The key take-aways from this broader story, listed largely in chronological order, are:

- A green bottom line company begins with a focus on values.
- A green bottom line company uses those values to shape a culture built around a sense of shared meaning and purpose, largely comprised of a deep and abiding connection to both land and community.
- A green bottom line company assesses its own environmental footprint and develops a multi-year strategy to continually reduce its impacts.

- A green bottom line company invests in an ongoing learning program to develop expertise around green building practices.
- A green bottom line company draws a line in the sand by developing a different way.
- A green bottom line company broadens its outreach to include others in its paradigm-shifting practices, brokering innovative practices to a much broader segment of the community.
- A green bottom line company broadens that circle even farther by creating a marketing strategy largely focused on advocacy, education, and outreach, sharing both its successes and its failures with the community at large.

The story told in this book is more than a story about the bottom line(s) of green building. It is the story of constructing a green ethos. It is about the processes, described in the list above, that lay deep, long-lasting foundations that facilitate building green. Those processes involve a focus on being stewards of both the environment and the social context in which we work. Those processes entail a layered approach to investing in our own staff, our various stakeholders, and in the general community—a persistent raising of the bar of making a difference through shared learning and growth. And those processes also entail ongoing financial investment, as can be seen in Table 11.1, which provides an overview of the numbers crunching detailed throughout this book.

Over the past ten years, it is estimated that our company has spent—either in direct expenditures or by lost opportunity—somewhere in the neighborhood of $5 million on this journey into sustainability. And that does not include the specific investments we have made in stand-alone green projects, which have had to stand the test of meeting typical market threshold returns. It is not a trivial sum for even a fairly good-sized company. It has been a major investment for us. It has also paid off, as we have seen in the prior chapters, with revenues over that same ten-year period of $11.8 million, a net present value of $2.4 million, and a return on investment conservatively assessed to be around 28 percent, without factoring in numerous intangibles that would be incorporated into a sale and result in an overall return on all sustainable investments above 44 percent. A graphical representation of this layered investment in green can be seen in Figure 11.1.

A telling bit of information to be gleaned from Figure 11.1 is that while our company invested early on in learning about green concepts and practices, its investment has tapered off in latter years while the revenue generation from this focus has continued to increase. This trend line bears out our contention that investments in a green bottom line can (and should) be amortized over the long term, comprising part of a company's overall institutional expertise—an expertise that simply becomes more refined (and profitable) over time.

It is worthwhile looking a bit more in-depth at the value creation that a green bottom line company realizes over an extended time frame. In the case of our fictionalized company, Green, Inc., a bit over 50 percent of the $11.8 million in revenue creation comes from third-party fee generation (development, consulting, property management). An additional 15 percent of revenue creation comes from the reinvestment of such third-party fees. Savings (retention, operational, finance) account for just over

TABLE 11.1 **OVERVIEW OF FINANCIAL INVESTMENT IN BUILDING A GREEN COMPANY**

Chapter	Activity	Investment	Cumulative Investment	Cumulative Revenue	Cumulative Net Present Value	Total Return on Investment
1	Shaping values	$1,386,037	$1,386,037	$2,766,640	$441,189	22.50%
2	Shaping culture	$1,058,987	$2,445,024	$3,556,640	$335,955	20.85%
3	Environmental footprint	$355,862	$2,800,886	$3,589,640	$132,127	13.98%
4	Green expertise	$1,563,000	$4,363,886	$7,080,834	$2,300,994	28.38%
5–7	Green projects	per project	per project	per project	per project	per project
8	Sustainable brokerage	$412,500	$4,776,386	$11,795,474	$2,621,324	30.02%
9	Green legal expertise	per project	per project	per project	per project	per project
10	Green marketing	$358,000	$5,134,386	$11,795,474	$2,401,349	27.94%
1–10 with sale			$5,134,386	$31,795,474	$10,112,215	44.11%

20 percent, while the remaining 11 percent occurs as a result of value enhancement associated with the sustainable nature of Green, Inc.'s portfolio. An overview of these various categories of value creation can be seen in Table 11.2.

An overview of the $5 million expended on our sustainable program is equally revealing. The vast majority of expenditures concerns staff time: getting educated about the LEED program, becoming proficient at delivering a LEED certified project, hiring staff specifically devoted to sustainable initiatives, the lost opportunity cost of devoting staff time to managing this learning curve, etc. In all, only about 10 percent of the overall $5 million spent on sustainable initiatives concerned non-staff line items (such as building upgrades), programs (such as carbon sequestration), and marketing, as shown in Table 11.3.

I suppose I have to list as one of my failings the fact that it was not until the writing of this book that I had any clue as to precisely how much we have invested in all of our

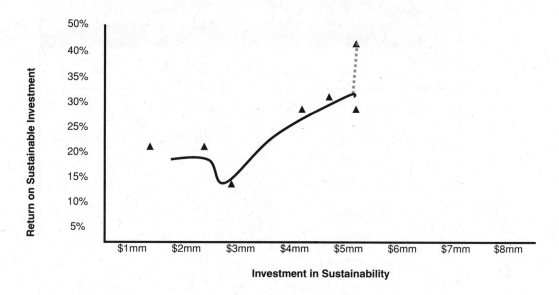

Figure 11.1   Overview of layered investment in Green.

TABLE 11.2 ANALYSIS OF VALUE CREATION		
Revenues	**$**	**%**
3rd-party development income	$4,500,000	38%
3rd-party consulting income	$1,125,000	9.5%
3rd-party property management	$750,000	6%
Reinvestment income	$1,801,974	15%
Staff retention	$1,003,000	8.5%
Operational savings	$428,000	4%
Finance savings	$937,500	8%
Enhanced value	$1,250,000	11%
Totals	**$11,795,474**	**100%**

TABLE 11.3   ANALYSIS OF EXPENSES FOR SUSTAINABLE INITIATIVES

Expenses	$	%
Lost opportunity	(1,773,774)	35%
Consultants/facilitators	(123,250)	2%
Staffing	(490,347)	10%
Continuing education	(449,000)	9%
HR benefits	(148,000)	3%
Staff commission fees	(112,500)	2%
LEED learning curve	(1,514,000)	29%
Sustainable programs	(105,000)	2%
Sustainable technologies & building improvements	(60,515)	1%
Marketing & public relations	(358,000)	7%
TOTAL	**(5,134,386)**	**100%**

various processes or what our return on that investment has been. It was simply the direction we needed to go; a certain moral imperative, if you will. There simply did not seem to be—nor is there today, in my opinion—any real alternative. And this leads me back to my Uncle Scribe.

The management of a sustainable business—tending to financial performance, environmental impacts, and social stewardship—tending to what we are calling throughout this book the green bottom line, is in many respects like my conversations with Uncle Scribe over the years. There's a restless, critical energy involved in the undertaking. There's a sense that no matter what you are doing, there is still much more you could be doing: making sure that your own environmental footprint as a company is being reduced; providing a bundle of services for your tenants that enable them to capture greater value through sustainable construction practices; building buildings that use less energy and water, emit fewer greenhouse gases, and reduce our impacts upon the natural environment; renovating existing buildings to be more sustainable; revising leases and lease structures so that there are financial incentives for both landlords and tenants to be more conservative of our resources; overhauling typical marketing practices so that a key part of a sustainable company's communication strategy becomes one of educating the market about building with greater sensitivities toward nature and society; developing processes in-house that foster a culture promoting a "land-community ethic," to use the naturalist Aldo Leopold's term.

Like my extended conversation with Uncle Scribe, it seems that no matter what positive strides you take to become more sustainable in your thoughts and actions, there is still so much more to do. You focus on the narrow challenges directly in front of you. And these narrow challenges lead to broader issues affecting our land and community. And these broader issues of land and community lead eventually to a question of consumption and ethics. *Never weigh over 150 pounds,* my Uncle Scribe admonished me, *otherwise there's no turning back.* He was right.

Beyond the narrow focus of green buildings and the broader focus of sustainable practices is the challenge of keeping our consumption within limits we can live with. A crash diet of constructing a few green buildings here and there won't do it. Like most crash diets, such an approach is largely cosmetic and does not get at the more foundational issues that call for a change in our entire regimen. We don't need a diet. We need a wellness program. The green bottom line company needs to be thinking seriously in the years to come about an entire change in what we consume and how we consume. And why. And how best to restore us all to a sense of land and community that is more in harmony with one another.

My maternal grandmother was fond of saying, "Honey, all you have is your health, make sure you take care of yourself." I think she would intuitively understand the wellness program that sits at the root of a sustainable orientation. It is all about making one's livelihood consonant with life itself.

NOTE

[1] Arie de Geus, *The Living Company: Habits for Survival in a Turbulent Business Environment* (Boston: Harvard Business School Press, 1997), p. 18.

INDEX